how to know the mosses and liverworts

The **Pictured Key Nature Series** has been published since 1944 by the Wm. C. Brown Company. The series was initiated in 1937 by the late Dr. H. E. Jaques, Professor Emeritus of Biology at Iowa Wesleyan University. Dr. Jaques' dedication to the interest of nature lovers in every walk of life has resulted in the prominent place this series fills for all who wonder **"How to Know."**

John F. Bamrick and Edward T. Cawley
Consulting Editors

The Pictured Key Nature Series

How to Know the
  AQUATIC INSECTS, Lehmkuhl
  AQUATIC PLANTS, Prescott
  BEETLES, Arnett-Downie-Jaques, Second Edition
  BUTTERFLIES, Ehrlich
  ECONOMIC PLANTS, Jaques, Second Edition
  FALL FLOWERS, Cuthbert
  FERNS AND FERN ALLIES, Mickel
  FRESHWATER ALGAE, Prescott, Third Edition
  FRESHWATER FISHES, Eddy-Underhill, Third Edition
  GILLED MUSHROOMS, Smith-Smith-Weber
  GRASSES, Pohl, Third Edition
  IMMATURE INSECTS, Chu
  INSECTS, Bland-Jaques, Third Edition
  LAND BIRDS, Jaques
  LICHENS, Hale, Second Edition
  LIVING THINGS, Jaques, Second Edition
  MAMMALS, Booth, Third Edition
  MARINE ISOPOD CRUSTACEANS, Schultz
  MITES AND TICKS, McDaniel
  MOSSES AND LIVERWORTS, Conard-Redfearn, Third Edition
  NON-GILLED FLESHY FUNGI, Smith-Smith
  PLANT FAMILIES, Jaques
  POLLEN AND SPORES, Kapp
  PROTOZOA, Jahn, Bovee, Jahn, Third Edition
  SEAWEEDS, Abbott-Dawson, Second Edition
  SEED PLANTS, Cronquist
  SPIDERS, Kaston, Third Edition
  SPRING FLOWERS, Cuthbert, Second Edition
  TREMATODES, Schell
  TREES, Miller-Jaques, Third Edition
  TRUE BUGS, Slater-Baranowski
  WATER BIRDS, Jaques-Ollivier
  WEEDS, Wilkinson-Jaques, Third Edition
  WESTERN TREES, Baerg, Second Edition

how to know the
mosses and liverworts

Second Edition

Henry S. Conard
formerly of Grinnell College and State University of Iowa

Revised by

Paul L. Redfearn, Jr.
Southwest Missouri State University

Boston, Massachusetts   Burr Ridge, Illinois   Dubuque, Iowa
Madison, Wisconsin   New York, New York   San Francisco, California   St. Louis, Missouri

WCB/McGraw-Hill
A Division of The McGraw-Hill Companies

Copyright © 1956 by H. E. Jaques

Copyright © 1979 by Wm. C. Brown Company Publishers

ISBN 0-697-04768-7 (Paper)
ISBN 0-697-04769-5 (Cloth)

Library of Congress Catalog Card Number: 78-52712

All rights reserved. No part of this publication may be reproduced, stored in a retrieval system, or transmitted in any form or by any means, electronic, mechanical, photocopying, recording, or otherwise, without the prior written permission of the publisher.

Printed in the United States of America

15 16 17 18 19 20 QPD 0 9 8 7 6 5 4 3 2

Contents

Preface   vii

About Mosses and Liverworts   1
   What They Are Not   1
   What Are They   2
   Life Cycle   2
   Where Bryophytes Grow   4
   Uses   4
Classification   6
How to Study Bryophytes   8
   What to Look For in Mosses   11
   What to Look For in Liverworts   12
General References   15
How to Use the Keys   18
Pictured-Key to the Bryophytes of North America   19
   Introduction Key   19
Musci—The Mosses   24
   Subclass 1. Sphagnidae   24
   Subclass 2. Andreaeidae   27
   Subclass 3. Bryidae—Key to the Genera   28
Class I. Anthocerotae—The Hornworts   230
Class II. Hepaticae—The Liverworts   232
   Order 1. Takakiales   232
   Order 2. Calobryales   232
   Order 3. Jungermanniales—Key to Genera   232
   Order 4. Metzgeriales—Key to Genera   238
   Order 5. Sphaerocarpales—Key to Genera   239
   Order 6. Marchantiales—Key to Genera   239

List of Synonyms   289
Index and Pictured-Glossary   293

Preface

Mosses and liverworts, as in the case of other small plants and animals, seem to look alike until they are examined closely. Then, remarkable differences are noted between different species. Who would say that the moss plants illustrated in Figure 1, or moss leaves in Figure 2, or moss capsules in Figure 3 look alike? There are at least 1,170 species of mosses and 524 species of liverworts reported from North America and all have been found to be different in some significant way. This book is about these different species and how they may be separated from each other. It has been designed for the beginning student of mosses and liverworts as well as the serious amateur. Though all of the species known from North America are not included, the vast majority of common ones, and even some rare but unusual ones, are included.

This edition, besides updating the taxonomic treatment of earlier editions, includes major changes in organization and style, and the inclusion of many addition species. Unlike previous editions, only couplets are used in the keys and major keys lead only to the genera. Species are keyed out separately under each genus, a feature which should aid the student in getting a "feel" for taxonomy of the bryophytes. One hundred twelve additional species that includes 42 additional genera have been added. Most are illustrated in the 117 new figures prepared by me. A majority of these additional species come from the southeastern United States and western North America. Also, the results of many studies since the appearance of the revised edition in 1956 have necessitated over 135 name changes. A "List of Synonyms" is included in order to aid one in associating the names used in this book with the names used in past editions or in other works on mosses and liverworts. Many new reference works and reports dealing with the mosses and liverworts of local areas have been added to the section on Books and Specimens as Means of Identification. Finally, the information for each species has been expanded to include a brief description, a discussion of habitat, and a statement about its general geographic range as reported in the literature.

Although this book represents an extensive revision, it still owes its existence to Dr. Henry S. Conard. It was his book that made it possible for so many of us to discover the fascination of collecting and studying mosses and liverworts.

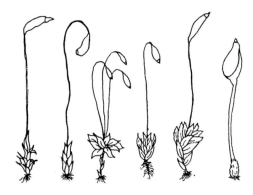

Figure 1 Moss Plants. From left to right: *Ceratodon purpureus, Funaria hygrometrica, Rhodobryum roseum, Bryum caespiticium, Aulacomnium heterostichum, Buxbaumia aphylla.*

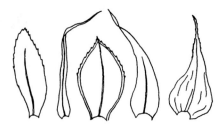

Figure 2 Moss Leaves. From left to right: *Aulacomnium heretostichum, Leptobryum pyriforme, Mnium cuspidatum, Drepanocladus aduncus, Rhytidiadelphus triquetrus.*

Figure 3 Moss Capsules. From left to right: *Physcomitrium pyriforme, Bryum argenteum, Dicranella heteromalla, Funaria hygrometrica.*

An enlarged photograph of the branches of *Mnium affine* var. *ciliare* bearing terminal clusters of antheridia.

Plate 45 from Species muscorum frondosorum by J. Hedwig, Leipzig 1801. This plate gives the name to our commonest moss (northeast), *Mnium cuspidatum* Hedw. Although Linnaeus, 1753, used the name cuspidatum for it, the Botanical Congress of 1930 ruled that names of true mosses should begin with Hedwig 1801. The figure with 4 capsules is our *Mnium medium*.

Thick mats of *Polystrichium juniperinum* are common.

About Mosses and Liverworts

WHAT THEY ARE NOT

The gray-green festoons that dangle from trees in the South, the "Spanish Moss", is not a moss. It is a near relative of the pineapple, with the same kind of hairs on the leaves, with flowers, and with silky seeds. No Moss has either flowers or seeds.

Similar festoons in the Northeast, where ". . . the murmuring pines and the hemlocks" are "bearded with moss", are made of lichens. They bear flat discs containing the spores. Many other leafless gray-green lichens are called mosses, for example "reindeer moss." Of course the moss roses, moss pink, flowering moss and any other "moss" with flowers, is not a moss.

Figure 5 *Usnea*.

Figure 6 Reindeer lichen, *Cladonia*.

Figure 4 Spanish moss.

Figure 7 *Portulaca*.

Figure 8 Seaweed.

Nor are there any Mosses in sea water. These are algae. Several Mosses grow in *fresh* water; they have stems with regularly arranged leaves.

WHAT THEY ARE

1. Bryophytes are small plants 1 mm–60 centimeters tall, mostly a few centimeters. Some are flat, scale-like growths (thallus plants) on earth or rocks or trees. Most of them have stem and leaves, the latter variously but regularly attached to the stem. Run through the pictures in this book to get a general idea.

2. All of the members of this great phylum are photosynthetic. They manufacture their own food out of constituents of earth and air, by means of chlorophyll, with the aid of sunlight. They are green, at least in part.

$$6CO_2 + 12H_2O + sunlight \rightarrow$$
$$C_6H_{12}O_6 + 6H_2O + 6O_2 + stored\ energy$$

3. All bryophytes are propagated and disseminated by spores, one-celled particles of living matter, by fragmentation of plants into small parts capable of growing into new plants, or by special structures called propagula (or gemmae) that range in size from a few cells to many cells.

4. At another period in their lives all are propagated by male and female germ cells, which fuse into a single cell (zygote) as in nearly all other plants and in animals. From this single-celled zygote an embryo develops and grows to a distinctive structure called the sporophyte.

LIFE CYCLE

The whole life of a moss runs this way: a spore, in a favorable spot, swells with water, bursts its shell, and puts forth a slender, branching, many-celled green thread, called protonema. This growth may cover several inches or feet of ground; it looks like a green alga. It is distinguished by having some of the partitions in the thread oblique, and by having branches going down into the ground, colorless or brown in color. In due time buds (or a bud) appear on this protonema, and each bud grows out as a leafy stem or flattened scale. In either case it is anchored and fed by numerous threads that grow into the soil (rhizoids). Thus we get new moss plants where previously there was none.

When such a plant reaches maturity it produces male and/or female germ cells. The male germ cells are minute colorless coiled bodies, driven by two cilia. They can swim

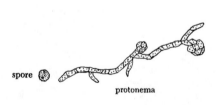

Figure 9

about in a drop of dew or rain for an hour or so. They are produced in oval sacs called antheridia (singular, antheridium) (Fig. 10). The antheridia are borne in a cluster of leaves, or in a pocket of the scale-like thallus. The egg cell is borne in the bottom of a long-necked vase called an archegonium. When the egg is ready for fertilization, the neck of the archegonium becomes a tube of mucilage, the tip opens, and the mucilage exudes, disseminating cane sugar (or some protein in liverworts). This exudate is overwhelmingly attractive to the spiral sperm. Sperms coming within the influence of the exudate swim toward and into the neck of the archegonium and eventually reach the egg. One sperm fuses with the egg, and a new phase in the life history of a moss, the sporophyte, is initiated.

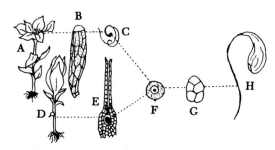

Figure 10 A, male plant; B, antheridium; C, sperm; D, female plant; E, archegonium containing the egg; F, zygote, the union of sperm and egg; G, embryo; H, mature sporophyte of *Funaria*.

This new cell, called the zygote, contains hereditary information from both the sperm and the egg. It remains in the archegonium while it divides into two, then into 4, 8, 16, many cells, and gradually shapes itself into the beginnings of a stalk (seta) and spore-case (capsule). At its maturity, the seta shoots forth, bursting the archegonium and pushing up the capsule into the air, whence the spores are liberated and float away. When the seta of a true moss (Musci) bursts the archegonium, the tip of the latter remains as a cover or cap over the tip of the capsule. This cap, the product of the upper end of the archegonium, is called a calyptra. The shapes of calyptras are often very characteristic.

The foot, seta and capsule compose the *sporophyte*. Being derived from a fertilized egg, the sporophyte is a diploid organism, in which heredity operates as in animals and flowers by pairs of genes. The thallus or leafy moss has only one set of genes; it is haploid. But its hereditary characters are just as precise and dependable as if it had its genes in pairs. This is a curious situation, applicable to all the mosses. Besides this, there are hybrid mosses, and triploids, tetraploids and octoploids. The genetics of mosses is a rich field, now being explored.

The sporophyte (seta and capsule) gets nearly all of its food from the mother plant. Most young sporophytes have a little chlorophyll (leaf-green) by means of which they manufacture a modicum of sugar for food. The capsules of true mosses often have a well developed system of chlorophyll cells and air spaces, served by true stomata, Figure 11.

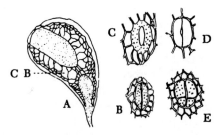

Figure 11 A, section of capsule of *Funaria* showing the netlike chlorophyll tissues; B, young stoma and C, mature stoma of *Funaria*; D, stoma of *Orthotrichum*; E, stoma of *Polytrichum commune*.

The life cycles of the Hepaticae (liverworts) and the Anthocerotae (hornworts) are similar to that of the Musci (mosses) except that in the liverworts and hornworts the haploid spore produces an ovoid, sphaerical, or plate-like protonema. There are also important structural differences between the sporophytes of mosses, liverworts, and hornworts. Specifically, the capsules of mosses are structurally more complex and usually open by means of lid or operculum and a peristome usually surrounds the resulting opening. Capsules of liverworts and hornworts are simpler, often opening by slits in the capsule wall and lacking both an operculum and peristome.

WHERE BRYOPHYTES GROW

Every state in the Union and Canada has its bryophytes. Iceland and Greenland have many more species. Mexico and tropical America are very rich in species. The evergreen forests of our west coast, from Santa Cruz to Dutch Harbor, have more and bigger mosses than any other part of the Continent.

Most bryophytes grow among trees. There are few mosses among the grasses of the Tallgrass Prairies or the Shortgrass Plains. Most spots or bare spots or wooded spots among these Plant Associations have their mosses. Some species grow best on exposed rocks in full sunshine. They are found at 14,000 feet in Colorado, and on boulders or sand beside the seashore. For some we wade waist-deep in ponds, or reach out from boats.

The kinds in each habitat are characteristic, and the assembly of species in each region is characteristic. Give me a list of the bryophytes of your region, and I will tell you where you live and what other vegetation is native there. Bryophytes, like other plants, have their own associations and associates. They *indicate* the natural conditions of their environment, and give clues to what naturally grows (or grew) there, and what crops can be grown. They even indicate acid or alkaline soils, and what, if any, treatment a soil needs to keep it normal. Unfortunately the indicator value of bryophytes has been very little studied.

USES

Perhaps no great group of plants has so few commercial or economic uses, as the bryophytes. The peat mosses (*Sphagnum*) are used for packing nursery stock. The moss holds moisture in quantity and keeps roots fresh for a journey across the continent. It is gathered from acres of bogs in northern regions, and sold in 50 lb. bales. Clean sprays of Sphagnum are sometimes wrapped in cheesecloth, sterilized and used as packing for seeping wounds. Sphagnum makes much of the peat, which is the fuel of Ireland and northwestern Europe. But the moss peat of Iowa is mostly *Drepanocladus*. Chopped sphagnum is an excellent cover for a seedbed, or an addition to soil to keep it moist and porous.

The big mosses of the west coast are good for packing crockery. They are soft and springy, and sufficiently long-stemmed and abundant.

In ancient and medieval medicine some bryophytes had a place. Because the thallus of several ribbon-like species is marked in polygonal areas, like a cross section of an animal's liver these bryophytes were believed to be good medicine for ailments attributed to malfunction of the liver. The plants were therefore called liverworts, a name that is still used for these plants and their kin. No bryophyte is credited with any medicinal virtues according to modern standards.

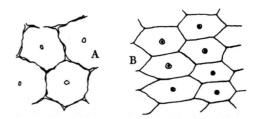

Figure 12 A, section of liver, after Encyclopedia Brittanica; B, surface of *Marchantia*, after Kny.

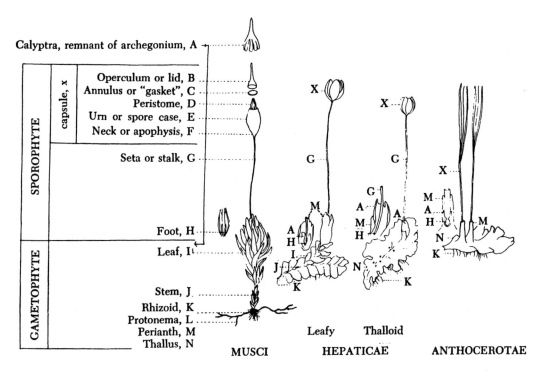

Figure 13 The whole Moss. Left to right: *Ptychomitrium incurvum, Lophocolea heterophylla, Pellia epiphylla, Phaeoceros laevis.*

Classification

The plants included under the general terms "mosses" and "liverworts" are of three distinct types, as shown in Fig. 13. They are all so much alike in structures and life-history that the same terms (with few exceptions) apply to all. The whole group is assigned to the Division Bryophyta. They belong to the large group of green land plants that are best characterized by their lack of specialized water conducting cells (tracheids and vessels) which make possible the life of larger land-plants.

The Division Bryophyta may be divided into three classes:

    Class I. Anthocerotae (Hornworts)
    Class II. Hepaticae (Liverworts)
    Class III. Musci (True mosses)

The Anthocerotae includes only one order and one family. The gametophytes are thallose with no distinct stem and leaf. The sporophyte is slender and it grows by division of cells at its base or foot region. *Phaeoceros laevis*, Fig. 13, 19.

The Hepaticae are assigned to six orders:

Order 1. Takakiales. A primitive leafy liverwort with leaves divided into two to three segments. Rhizoids are absent and sporophytes are unknown. Fig. 35A-C.

Order 2. Calobryales. A primitive leafy liverwort with undivided leaves. Rhizoids absent. Capsule wall only one cell thick. Fig. 26D, 35D-E.

Order 3. Jungermanniales. Leafy liverworts with rhizoids and a capsule 2-10 cells thick. *Lophocolea heterophylla,* Fig. 13, 30B.

Order 4. Metzgeriales. Mostly thalloid liverworts with thin thalli and stalked capsules. *Pellia epiphylla*, Fig. 13, 30A.

Order 5. Sphaerocarpales. Tiny thallose forms with Antheridia and archegonia in sacs on the upper *surfaces*, Fig. 33.

Order 6. Marchantiales. Thalloid liverworts with spongy (air-filled) thalli, and small sporophytes on umbrella-shaped receptacles, or imbedded in the thallus, Fig. 26C, 31.

The Musci are more diverse and require a more extensive classification, the outline of which is shown below:

Subclass 1. Sphagnidae, Order Sphagnales, Fig. 27
Subclass 2. Andreaeidae, Order Andreaeales, Fig. 28
Subclass 3. Bryidae

> Order 1. Fissidentales
> Order 2. Archidiales
> Order 3. Dicranales
> Order 4. Pottiales
> Order 5. Grimmiales
> Order 6. Funariales
> Order 7. Schistostegales
> Order 8. Eubryales
> Order 9. Isobryales
> Order 10. Hookeriales
> Order 11. Hypnobryales
> Order 12. Buxbaumiales
> Order 13. Tetraphidales
> Order 14. Polytrichales

The subclass Bryidae has been frequently classified into categories based on features of the peristome of the capsule and the location, in many cases, of the archegonia and later the sporophytes. This classification is outlined below:

1. Nematodonteae. Orders 12-14. Sporophytes with a single peristome not transversely barred or only faintly barred in *Buxbaumia*.
2. Arthodonteae. Order 1-11. Sporophytes with peristomes (if present) composed of teeth that are transversely barred.
   (1) Haplolepideae. Orders 1-5. Peristome composed of a single circle of teeth, or various reductions of this.
   (2) Diplolepideae. Orders 6-11. Peristome composed of an outer circle of hard teeth and an inner membraneous set of segments and/or cilia, or various reductions of this. This group is frequently, but imperfectly, divided into two sections based on the location of the archegonia.

   A. Acrocarpi. Orders 6-8. The archegonium and therefore the seta is borne at the tip of an ordinary, usually erect leafy stem with little or no modification of the adjacent leaves. A branch growing out below the sporophyte may give the seta the appearance of being lateral.
   B. Pleurocarpi. Order 9-11. Stems always abundantly branching, mostly creeping, with the archegonia and therefore the seta borne in a special lateral bud, with leaves very different from the vegetative leaves (perichaetium).

The rest of the classification, so far as it relates to the plants of this book, will appear in the following Pictured-keys. It has never been possible to make a useful key to the mosses and liverworts by tracing to the orders or families. It is a universal practice to key mosses and liverworts directly to the genus. Only the Introductory Key leads to Subclasses, Classes, and Orders. Subsequent keys for each of the Subclasses of Mosses and for the Orders of Liverworts lead directly to a genus. These genera are then arranged under their proper order and family and will contain, when necessary, a key to the individual species.

How to Study Bryophytes

To know the mosses go out and hunt for them. Collect all the kinds you can find. If they have no capsules, watch them until they do. But you will soon learn to recognize and identify them by the leaves.

Your equipment will be a carrying sack—any convenient receptacle—with small paper sacks (4 to 8 oz.) or flat pieces of paper. Many collectors use newspapers torn to about 5 × 8 or 8 × 10 inches and folded like packets (see below). A hand lens, 10× to 20×, on a shoestring to hang around your neck. An old knife, 3-inch blade. Bring in everything, and as often as you like. Record the date and place of collecting for every specimen.

The working equipment will be a table, dissecting microscope (preferably binocular), compound microscope (with 10× and 43× obj.), fine forceps, 2 fine dissecting needles (one of these may be ground down to a knife blade shape), micro-slides and covers (5/8 in. # 2), bottle of water, bottle of Hoyer's solution, labels, stiff backed razor blades, scalpel. Microscopes and dissecting equipment may be purchased from stores with camera departments or from biological supply houses. The Hoyer's solution is prepared by the following formula:

Distilled water	50 milliliters
Gum arabic (U. S. P. Flake)	30 grams
Chloral hydrate	200 grams
Glycerin	20 milliliters

The ingredients should be mixed in the above order at room temperature. Gum arabic goes into solution slowly during the mixing operation and for that reason U. S. P. Flake is specified. Upon standing bubbles formed during mixing will disappear. The solution should be stored in airtight bottles.

Specimens collected in the field should be dried until you have the opportunity to study and identify them. Many collectors simply open the individual sacks or packets and allow the plants to dry slowly in the air. A day or two is sufficient in a warm room. After the plants are dry, they are placed in folders or packets (Fig. 14).

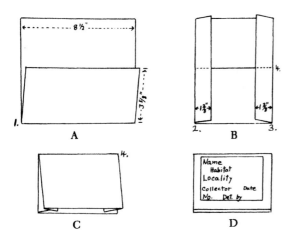

Figure 14 Folding of packets from 8½" × 11" paper. A, fold at 1; B, next fold at 2 and 3; C, final fold made at 4; D, folded packet with label.

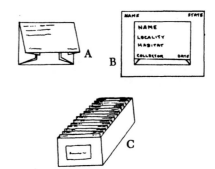

Figure 15 Filing system. A, packet; B, packet pasted on card; C, cards filed in drawer or box.

The folders, packets and the labels should be made of 8 1/2" × 11" white paper that is 50% rag or better and 16 lb. or heavier. Many collectors adopt a standard size packet that can be filed upright in large shoe boxes or in boxes that can be filed side by side into one standard herbarium shelf. Others use packets of several sizes to suit the specimens, and then paste each on a card 5 9/16 × 4 1/4 inches. These cards may be of different colors indicating different regions of the world, white for your own local area (state or province), manila for the rest of North America, green for Europe, red for Asia, etc. The cards are then stored, catalog fashion, in shoe boxes specially made to fit a herbarium case. Many collectors mount the packets by pasting them on standard herbarium sheets, one species to a sheet; finally the sheet is completely covered with packets. These sheets can be placed in genus covers and stored in the usual way. Mosses do not readily mould, and they are seldom eaten by the insects that so often destroy specimens of flowering plants.

Labeling of each collection is important. The label may be written on the outer flap of the packet or prepared on a separate slip of paper. It should be typed or printed with permanent ink. It is best to establish a uniform procedure for preparing your labels. The basic information that should be included is:

1. Name of species, with authority. Name or names of accompanying species.
2. Habitat: substratum, moisture, exposure, plant community, and any other significant or helpful details.
3. Elevation, Latitude, Longitude when known.
4. Locality: Country, State, or Province; State, County or District; nearest town, place, river, lake, mountain, canyon, etc. Do not use abbreviations.
5. Collector's name, date and collection number.
6. Determiner's name and date, particularly if different from collector. If an authority has verified the collection it should be noted. If an authority has identified or verified the collection by examining a duplicate of the collection, the statement "Duplicate identified (or verified) by..." should be made. An example of a properly prepared label is shown in Fig. 16.

```
         Hypnum curvifolium Hedw.
   Abundant on north-facing, shaded sandstone
ILLINOIS: Jackson County. 20 August 1965.
          Beech-maple forest dissected by
          sandstone gorges. Ca. 4 miles west
          of Etherton. Sec. 1, T. 10 S.,
          R. 3 W. Alt. ca. 500 feet.
Collected by Paul L. Redfearn, Jr.
No.   18251    Det.    PLR
```

Figure 16 Example of a properly prepared label.

When we are ready to identify a moss, we select a good shoot with a capsule, pluck it out with the forceps and soak it in water until it is "as good as new"—5 minutes to 15 hours as required. A moment in boiling water will do the trick.

Lay a leafy shoot of the soaked moss on a glass slide under the dissecting microscope. Hold it firmly near the apex with a needle or forceps, and scrape the stem rather forcibly from apex toward base to remove a lot of leaves. Remove the stem, spread out the leaves, cover and examine with the compound microscope. If papillae are to be sought, a soaked twig may be examined, mounted in water under a coverglass. By looking with low power at the profile of a fold of a leaf, any papillae that are present will appear as tiny projections from the surface of the fold. Very thin sections of the leaf are desirable. And the experienced eye can detect papillae simply by focusing on the flat surface of a leaf. Fig. 17.

The capsule may be cut off from the seta and laid on a slide. Carefully pry off the lid, if it has not already fallen. Watch for the annulus, or "gasket," as the lid comes off. Cut off the upper end of the capsule, bearing the peristome. Split this ring-shaped end into 2 or more pieces; lay at least one piece with outer side up, and one with inner side up. Cover and examine. Fig. 18. Or, cut the capsule lengthways in half and mount one piece with the outside up, the other with the inside up.

When the slide prepared as directed above is examined, you are likely to find that the peristome is obscured by a blur of air bubbles and spores. This condition can be avoided by boiling the capsule *under water* before dissecting. My practice is to use water at room temperature as directed. Then if the mount is obscured, I hold the slide one or two inches above a lighted match (or alcohol lamp or candle) until the water under the coverglass boils. Then examine and the air bubbles are gone and the spores dispersed. The boiling must be just momentary, and very gentle. Even so the desired pieces of peristome may float out from under the coverglass. If they do they must be coralled and covered again. Only glass covers will do for this; plastic covers crumple.

As the water dries away from under the coverglass, replace it with Hoyer's solution—a droplet placed at the edge of the cover. Such a mount will last forever! It may be labeled and stored.

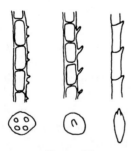

Figure 17

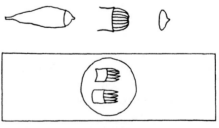

Figure 18

In some cases a cross section of a leaf is absolutely necessary for identification; in many cases a section is very helpful. Lay a wet leaf on a glass slide and hold it with the end of another slide laid so as to cover a part of the leaf. A sharp razor blade simply brought down upon the leaf will cut off the free end. Another cut closer to the upper glass slide will give a cross section. Try several times and drag the best sections to one side (or remove the worse), cover and examine. If the leaf is laid on a piece of stiff transparent plastic (plastic coverglass) the razor blade will last longer.

WHAT TO LOOK FOR IN MOSSES

The Leaf. Is it broad or narrow or filiform? Is the margin entire or toothed plane or rolled upward or rolled backward (revolute) or reflexed? Has it a single midrib (costa) or double or none? Does the midrib stop at the middle of the leaf, or near the tip, or does it extend clear to the tip (percurrent) or does it extend beyond the leaf as an awn or bristle? Are the cells isodiametric or elongate or long-hexagonal or spindle-shaped? Are they smooth or papillose? Are the cells at the basal angles of the leaves (alar cells) just like the rest, or are they small and rectangular (quadrate) or greatly swollen (inflated) and transparent, or swollen and colored golden brown? Does the leaf stop abruptly at the stem, or do the edges of the leaf continue down the stem like wings (decurrent)? Does the margin of the leaf consist of long, thick-walled cells ?

The Capsule. Is it straight, erect, inclined, nodding, symmetrical, smooth or ribbed, strumose(s) or contracted under the mouth when dry ?

The Seta. Is it long or short, smooth or rough, and of what color?

The Calyptra. Is it cucullate or mitrate, hairy or smooth?

The Peristome. Is it single or double? Are the teeth entire or split or irregular or absent? Are cilia present or absent? Fig. 19.

How to Study Bryophytes 11

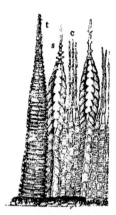

Figure 19 Peristome of *Bryum* with appendiculate cilia, c; segment, s; tooth, t.

Spores. Sometimes the size of the spores should be measured, or their surface noted: smooth, granulated, prickly, sculptured.

The Antheridia. Wherever there is a sporophyte, or seta, there has been an egg in an archegonium. Dissect away the leaves from the base of the seta and we find the dead, unsuccessful archegonia. The antheridia may be found among the archegonia; the plant is *synoicous*. The antheridia may be just below the archegonia, all around; *paroicous*. The antheridia may be in a special cluster or bud somewhere along the stem; *autoicous*. If you find no antheridia on the plant that has archegonia and/or sporophytes, the species is *dioicous*; they will be found on another plant.

The key should do the rest. Each number in the key offers two sets of conditions a and b. The plant cannot be like both a and b. We must decide which statement best describes it. At the end of this statement is a number to which we go next; there we again find two choices, a and b. And so we proceed until we find the genus and species of our moss, and a picture of it. We record the name on the packet—*Polytrichum commune* Hedw. And that packet is ready for the herbarium.

WHAT TO LOOK FOR IN LIVERWORTS

First, some liverworts are mere green scales or ribbons lying on the ground, or floating in water. Others have stems with two rows of leaves. (See Fig. 13.) Of these, some have a third row of small leaflike growths on the under side of the stem (underleaves).

The leaf of a liverwort may be almost perfectly round in outline, or oval or variously lobed or divided. Many species have a notch at the tip of the leaf. The shape of this notch and the shape of the two lobes are important. Sometimes there are two notches (three lobes), or four, or many. In fact the leaf may be completely divided into threadlike rows of cells. (See Figs. 389 to 478.)

The most critical details are (1) how the leaves are attached to the stem, (2) whether the leaf is simple or "complicate bilobed" and (3) whether the walls of the leaf cells are thin or thickened.

1. The leaf is nearly always attached obliquely to the stem. As the stem lies horizontally this means that one edge of the leaf is attached along the upper side of the stem, the other along the lower. If that edge of the leaf that is attached to the upper side of the stem is also nearer to the tip of the stem the arrangement is called *incubous,* and each leaf seems to ride up over the edge of the next leaf toward the tip of the stem. If the reverse is true and the front edge of each leaf runs under the rear edge of the leaf next nearer the tip of the stem the condition is called *succubous*. This is by far the commoner condition. Figs. 20, 21.

Figure 20   Incubous leaves of *Calypogeja*.

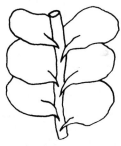

Figure 21   Succubous leaves of *Lophocolea heterophylla*.

Figure 22   Complicate bilobed leaf with the smaller lobe above the larger; *Scapania*.

Figure 23   Complicate bilobed leaves seen from below, the smaller lobe being under the larger: From left to right, *Porella, Radula, Frullania*.

These two terms may be kept separate in one's mind if these leaf arrangements were looked upon as shingles on a roof. If the arrangement is incubous, the rain runs in, if the arrangement is succubous, the rain runs down.

2. A "complicate bilobed" leaf is made in two lobes and these are folded close against each other, like a creased piece of paper. If the smaller lobe is upper the condition is easily seen: a small leaf seems to lie upon a larger one. If the smaller lobe is underneath, it can be seen by looking at the stem from beneath. This smaller lobe may be flat and leaflike, or it may be formed into a hollow sac, or it may lie against the larger lobe so as to enclose a space. This sac or space is a device that holds water. Figs. 22, 23.

3. The cells of the leaf may be equally thin-walled all around (Fig. 24a), or equally thick-walled all around (b), or with the walls thickened in the corners of the cells so as to form small (c), medium (d) or large (e) triangular masses of wall substance. Such triangular masses are called "trigones"; they are sometimes so large as to bulge out into the cells (e).

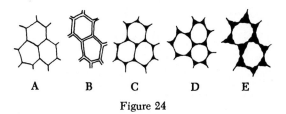

Figure 24

All liverworts have round or oval or rod-shaped capsules. The round and oval ones are short-lived: they may break to pieces at maturity letting out the spores, or more often they split into four spreading quarters discharging both spores and spirally-banded threads called *elaters*. The seta may be an inch long, but it is weak and watery and quickly withers away. Thus, a collection with open capsules is a rare catch, for you must find it on just the right day.

How to Study Bryophytes   13

The capsule originates in an archegonium (See Fig. 10) as in the case of mosses proper (Musci). One or more archegonia are surrounded by a leaflike sac, the perianth, (Fig. 25). This is a long-lived object, and is more useful for identification than the sporophyte itself. For there are many kinds of perianths: smooth, triangular, ridged or plaited, with the mouth plane, flat, or tubular, and perhaps fringed in various ways.

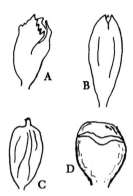

Figure 25 A, Perianths of *Lophocolea;* B, *Jungermannia;* C, *Frullania;* D, *Porella.*

The leaves adjacent to the perianth are usually quite different from those of the vegetative shoot. These adjacent leaves are called bracts. They are often divided, lobed or toothed, even when the stem-leaves are entire. Frequently we find underleaves as bracts, even when no underleaves occur elsewhere.

Liverworts with perianths can be named fairly readily. Without perianths the problem is much more difficult.

Antheridia, oval or globular bodies containing the male germ cells (sperms), are borne in pits of the surface of the thalloid liverworts. In leafy liverworts they occur, 1 to 3 or 4, in the axil of a leaf. Such leaves usually bulge out over the antheridia, and occur in a group of 6-12 along the stem.

General References

The achievement of Bryologists is preserved for us in books and collections. This key is a possibility only because many workers have made their discoveries available. And this Key does not attempt to name every species known from North America. Besides, it is likely that species will be found that have not previously been known from this continent. Probably species new to science are still waiting to be discovered. It is certain that much collecting and identifying will have to be done before we know accurately the ranges of the different species.

The serious student or hobbyist will want to consult additional books. The following titles are particularly valuable and most may be purchased though some may now only be available in college or university libraries.

Anderson, L. E. & H. Crum. Mosses of Eastern North America. Vol. I & II. (This work is expected to be published by Columbia University Press in 1980. It will contain keys, complete descriptions and illustrations.)

Andrews, A. LeRoy. 1957. The bryophyte flora of the Upper Cayuga Lake Basin, New York. 87 pp. Memoir 352. Cornell Univ. Agriculture Station. (Excellent keys and distributional data.)

Bartram, Edwin B. 1933. Manual of Hawaiian mosses. 275 pp. Bull. 101, Bernice P. Bishop Museum, Honolulu, Hawaii. (Keys and descriptions of Hawaiian mosses.)

Bird, C. D. & Won Shick Hong. 1969. Hepaticae of southwestern Alberta. Can. Jour. Bot. 47:1727-1746. (Annotated list with phygeographic discussions.)

Breen, Ruth Schornherst. 1963. Mosses of Florida, an illustrated manual. 273 pp. University of Florida Press. Gainesville, Florida. (Excellent keys, descriptions, and photographs.)

Crosby, M. R. & R. E. McGill, 1977. A dictionary of mosses. 43 p. Missouri Botanical Garden. (Alphabetical listing of genera indicating familial disposition.)

Crum, Howard. 1976. Mosses of the Great Lakes Forest. Revised Edition. 404 pp. Contributions Univ. of Michigan Herbarium 10. Ann Arbor, Michigan. (Fine keys, descriptions, and illustrations as well as useful morphological, distributional, ecological, and systematic information.)

Crum, H. A., W. C. Steere, & L. E. Anderson. 1973. A new list of mosses of North America north of Mexico. Bryologist 76:85-130. (A list of currently acceptable scientific names for the mosses of North America. Includes a catalog of synonyms and excluded names.)

Darlington, H. T. 1964. The mosses of Michigan. 212 pp. Cranbrook Institute of Science Bull. No. 47. Bloomfield Hills, Michigan. (Keys, descriptions and illustrations.)

Flowers, Seville. 1973. Mosses of Utah and the West, edited by Arthur Holmgren. 567 pp. Brigham Young Univ. Press, Provo, Utah. (Keys, descriptions, and line drawings of mosses of the Rocky Mountain region.)

Frye, T. C. and Lois Clark. 1937-47. The Hepaticae of North America. Vols. 1-6. 1019 pp. Univ. of Washington Press, Seattle. (Keys, illustrations, and descriptions of limited usefulness because of frequent errors and poor quality illustrations.)

Grout, A. J. 1903-10. Mosses with hand-lens and microscope. 416 pp. Published by the author, Brooklyn, New York. (Abundant illustrations and excellent text for mosses east of the 100th meridian and north of North Carolina.)

———. Editor. 1928-1941. Moss flora of North America north of Mexico. Vols. 1-3. Published by the editor. Newfane, Vermont. (Describes all species then known to occur in the continental United States, Canada, Alaska, Newfoundland, and Greenland and illustrating nearly every species not illustrated in *Mosses with Hand-lens and Microscope*.)

Hong, Won Shic. 1977. An annotated checklist of the Hepaticae of Wyoming. Bryologist 80: 480-491.

Ireland, R. R. & R. F. Cain. 1975. Checklist of the mosses of Ontario. National Museums of Canad Pub. in Botany No. 5 (Comprehensive list of mosses, synonyms, and County and District records. Good bibliography.)

Jennings, O. E. 1951. A manual of the mosses of Western Pennsylvania and adjacent regions. 2nd. Ed. 323 pp. Univ. of Notre Dame Press, Notre Dame, Ind. (Keys, descriptions, and excellent line drawings, useful for western Pennsylvania, extreme southwestern New York, eastern Ohio, and northern West Virginia.)

Lawton, Elva. 1971. Moss flora of the Pacific Northwest. 262 pp. Pl. 1-195. The Hattori Botanical Laboratory, Nichinan, Japan. (Keys, descriptions, and line drawings, covering Washington, Oregon, Idaho, western Montana, Wyoming and the Canadian provinces of British Columbia through the Rocky Mountains and north to about the fifty-second parallel.)

Magill, Robert E. 1976. Mosses of Big Bend National Park, Texas. Bryologist 79:269-313. (Keys and ecological information, useful in western Texas, adjacent Arizona and Mexico.)

Moul, E. T. 1952. Taxonomic and distributional studies of mosses of central and eastern Pennsylvania. Farlowia 4:139-233. (Keys and habitat information.)

Redfearn, Jr. P. L. 1972. Mosses of the Interior Highlands of North America. Ann. Missouri Bot. Garden 59:1-103. (Keys, generic descriptions, and habitat information, useful for southern Missouri and Illinois, Arkansas, and eastern Oklahoma.)

Reese, W. D. 1972. List of mosses of Louisiana. Bryologist 75:290-298. (Contains a species list and pertinent literature.)

Schofield, W. B. 1969. Some common mosses of British Columbia. 262 pp. Handbook No. 28. British Columbia Provincial Museum, Dept. of Recreation and Conservation. Victoria, B.C. (Keys, descriptions, and illustrations of the more common mosses.)

Schuster, R. M. 1949. The ecology and distribution of Hepaticae in central and western New York. Amer. Midl. Nat. 42:513-712. (Keys, descriptions, and excellent illustrations.)

———. 1953. Boreal Hepaticae, a manual of the liverworts of Minnesota and adjacent regions. Amer. Midl. Nat. 49:257-684. (A fine manual on the liverworts with excellent discussions on collection, morphology, and good keys, descriptions, and illustrations. Indispensable for the serious student, available from the University of Notre Dame Press, Notre Dame, Ind.)

———. 1963. An annotated synopsis of the genera and subgenera of Lejeuneaceae. 203 pp. Nova Hedwigia Vol. 9. J. Cramer, Weinheim, Germany. (Keys and descriptions to the genera and subgenera of a difficult group of leafy liverworts.)

———. 1966-74. The Hepaticae and Anthocerotae of North America. Vols. 1-3. Columbia Univ. Press, New York. (An extensive monographic treatment of the hepatics and hornworts east of the 100th meridian.)

Sharp, A. J. 1939. Taxonomic and ecological studies of eastern Tennessee bryophytes. Amer. Midl. Nat. 21:267-354. (Keys, county distributional records, and ecological and geographic discussions.)

Steere, W. C. 1940. Liverworts of southern Michigan. 97 pp. Bull. 17. Cranbrook Institute of Science, Bloomfield Hills, Michigan. (Keys, descriptions, and illustrations.)

———. 1978. The mosses of Arctic Alaska. Bryophytorum Bibliotheca 14:i-x, 1-508. J. Cramer, Lehre, Germany.

——— and Hiroshi Inove. 1978. The Hepaticae of Arctic Alaska. Journ. Hattori Bot. Lab. 44: 251-345. (Taxonomic list with detailed discussions on distribution, many illustrations.)

Stotler, R. and Barbara Crandall-Stotler. 1977. A checklist of the liverworts and hornworts of North America. Bryologist 80:405-428. (A list of currently acceptable scientific names for the liverworts and hornworts reported from North America and Greenland. Includes a catalog of synonyms and excluded names.)

Watson, E. V. 1967. The structure and life of bryophytes. 2nd. ed. Hutchinson & Co. LTD, New York. (An introductory treatment of the anatomy, morphology, cytology, ecology and systematics of bryophytes.)

Welch, Winona H. 1957. Mosses of Indiana. 478 pp. Dept. of Conservation, State of Indiana. Bookwalter Co., Indianapolis, Ind. (Keys, descriptions, and illustrations, useful for Indiana and surrounding states in the midwest.)

Weber, W. A. 1973. Guide to the mosses of Colorado. 48 pp. Occasional Paper No. 6. Institute of Arctic and Alpine Research. Univ. of Colorado, Boulder. (Keys and habitat information.)

Whitehouse, Eula. 1954. The mosses of Texas. Bryologist 57:53-146. (Checklist and country distributional records.)

The Checklist by Crum, Steere, and Anderson (1973) is available from the Secretary-Treasurer of the American Bryological and Lichenoligical Society who will also accept applications for membership in the society. Membership in the society entitles one to a subscription to its journal, *The Bryologist* as well as eligibility to join the Moss and Hepatic Exchanges. Membership in these exchanges is a good way to accumulate a reference collection of named specimens. The secretary-treasurer can also provide you with the names of bryologists in your area that can help you with difficult identifications. If you wish to learn more about the activity and location of bryologists, you should consult:

The Directory of Bryologists and Bryological Research by S. R. Gradestein. Regnum Vegetabile Vol. 88. International Association for Plant Taxonomy, Utrecht, Netherlands.

How to Use the Keys

The keys in this book are intended to enable a person to determine the names of most mosses and liverworts that are usually collected. Complete specimens with sporophytes are desirable, but if sporophytes are not present, most plants may be identified to genus and often to the proper species. Each step in the key involves two statements labeled a and b. You should read each statement carefully and choose the one which best fits the specimen. This statement will direct you to another pair of statements or to the name of a large group of bryophytes sharing some common characteristics, to the genus, or to the scientific name of a species.

The arrangement of the keys involves three levels of decision. The first, the Introductory Key (page 19) will aid you in placing your specimen in the correct Class of Bryophytes, Subclass of Mosses, Order of Liverworts, or in one case to a specific genus. A page number will direct you to a second level of decision where additional keys, when necessary, will direct you to the page number of the correct genus. Here further keys, when needed, aid you in making your final decision, the determination of the species. At this point you will find a brief description of the species and its general distribution pattern in North America and, in most cases, a reference to an illustration. As you become more and more familiar with mosses and liverworts, you will find that it is not always necessary to begin your process of determination with the Introductory Key, but that you will begin directly with the keys leading to genera or even species.

The specific name consists of two names written in italics, the genus and the species and followed by an abbreviation of the author or authors who first described this plant or who assigned it to its present genus and species. At this point a comparison of the unknown specimen with the illustration should be carefully made as an important step in confirming the identification. If further confirmation is desirable, you should compare your collection with a specimen named by an expert if such is available in your own herbarium. If you cannot name a collection from this manual it may be that you have discovered a species new to science or most probably, one not included in this book. In such cases you should send two packets of your unknown to an expert for determination with a request that he return one of the duplicates to you.

Pictured-Key to the Bryophytes of North America

INTRODUCTION KEY

1a Plants with stems and leaves; leaves equally spaced around the stem or, if in two opposite rows, leaf cells elongate or isodiametric and rhizoids multicellular; sporophyte persisting for weeks or months with no elaters present in the capsule. Class III MUSCI. Fig. 26A. .. 2

1b Plants thalloid or, if with stems and leaves, the larger leaves in two rows on the stem and a third row of leaves often present on the underside of the stem (amphigastra), never on the upper side of the stem, leaf cells isodiametric and rhizoids unicellular; sporophyte short-lived with elaters present or absent in the capsule. Class II. HEPATICAE. Fig. 26B, C, D. 4

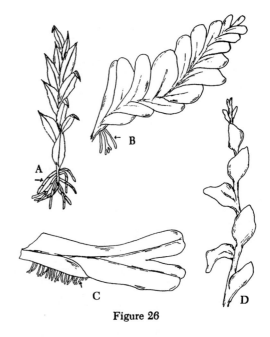

Figure 26

Figure 26 A, *Mnium*, a moss with multicellular rhizoids (indicated by arrow); B, *Chiloscyphus*, a leafy liverwort with unicellular rhizoids; C, *Marchantia*, a thalloid liverwort with unicellular rhizoids; D, *Haplomitrium*, a leafy liverwort without rhizoids.

2a Plants with many spreading, recurved branches along the stem and clustered

at the tip; leaf cells in one layer, of two kinds of cells, narrow linear chlorophyllose-cells forming the meshes of a network enclosing large rhomboidal hyaline cells. Fig. 27. ....
.... (p. 24) Subclass 1. SPHAGNIDAE

Figure 27 A, Habit sketch of *Sphagnum palustre*; B, ventral surface of the leaf cells of *S. imbricatum*; C, cross section, branch leaf of *S. imbricatum*; D, cross section of branch leaf of *S. palustre*; E, cross section of branch leaf of *S. magellanicum*.

2b Plants not as above. 3

3a Plants in cushions or tufts, 0.5-10 cm high, brownish or reddish green to nearly black; cells of leaves thick-walled, often sinuose; capsule raised on pseudopodium and opening by four vertical slits in the sides; growing on non-calcareous rocks. Fig. 28. ....
.... (p. 27) Subclass 2. ANDREAEIDAE

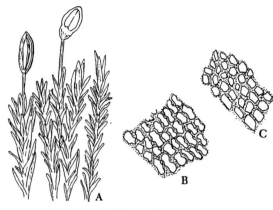

Figure 28

Figure 28 *Andreaea rothii*. A, Habit sketch of plant; B, thick walled leaf cells of *A. rupestris*; C, thick, sinuose leaf cells of *A. rothii*.

3b Plants not as above. ....
.... (p. 28) Subclass 3. BRYIDAE

4a Plants thalloid, with only one chloroplast per cell; sporophyte long, rod-like, splitting into two parts above as it grows up from the base. Fig. 29. ....
.... (p. 230) Class I. ANTHOCEROTAE

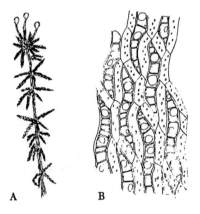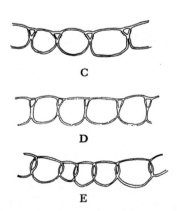

Figure 27

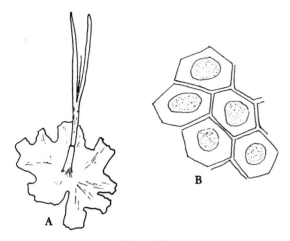

Figure 29

Figure 29 *Anthoceros punctatus.* A, fertile thallus; B, thallus cells with large chloroplasts.

4b Plants thalloid or leafy, with more than one chloroplast per cell; sporophyte with a sphaerical or ellipsoidal capsule. Fig. 30. 5

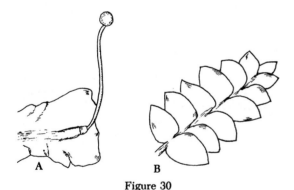

Figure 30

Figure 30 A, *Pellia nessiana*, a thalloid liverwort with a sphaerical capsule; B. *Calypojeia trichomanes*, a leafy liverwort.

5a Plants strongly flattened, thalloid, without distinction between stem and leaf. Fig. 30A. 6

5b Plants slightly flattened, distinctly divided into stem and leaf. Fig. 30B. 8

6a Plants with opaque thallus that is divided into an epidermis, loose tissue beneath the epidermis containing air spaces, and a lower solid, parenchyma-like tissue; rhizoids of two kinds, one with smooth wall and one with peg-like thickenings on inner wall; sporophytes with capsules not dividing regularly into four valves. Fig. 31. ....
 .. (p. 239) Order 6. MARCHANTIALES

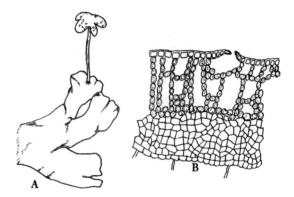

Figure 31

Figure 31 *Reboulia hemisphaerica.* A, plant bearing archegoniophore; B, cross section of opaque thallus.

6b Plants with a translucent or transparent thallus, internal tissue homogenous; (Fig. 32) rhizoids all smooth on internal wall; sporophytes with capsule dividing regularly into four valves or dehiscing irregularly and retained in a sac-like structure. Fig. 33. 7

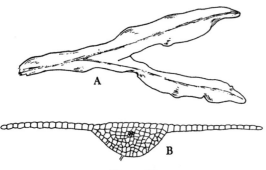

Figure 32

Figure 32 *Pallavicinia lyellii*. A, thallus; B, cross section of thin, transparent thallus.

7a Thalli appearing as small rosettes or scales, with surface covered with pear-shaped sacs (involucres) each containing a capsule that dehisces irregularly; no elaters in capsule. Fig. 33. .... (p. 239) Order 5. SPHAEROCARPALES

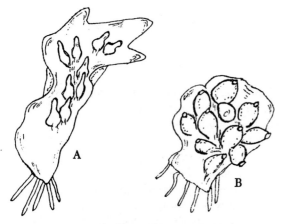

Figure 33

Figure 33 *Sphaerocarpus*. A, male plant with involucres; B, female plant with involucres.

7b Thalli ribbon-shaped, not covered on the upper surface with pear-shaped sacs; capsules regularly splitting into four lobes; elaters present. Fig. 32. .... (p. 238) Order 4. METZGERIALES

8a Archegonia on dorsal surface, not limiting the growth of the thallus, several developing, one behind the other on the same plant; surrounded at the base by a campanulate pseudoperianth; sporophytes with capsules not rupturing into four regular valves. Fig. 34A. .... (p. 275) *Fossombronia*

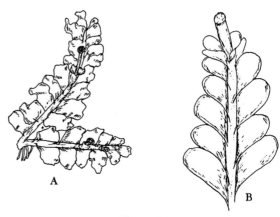

Figure 34

Figure 34 A, *Fossombronia wondraczekii* with subterminal sporophytes; B, *Jungermannia leiantha* with terminal perianth.

8b Archegonia at the end of the thallus, terminating its further growth, with only one developing into a sporophyte, usually surrounded at the base by a perianth; sporophyte with capsule rupturing regularly into four valves. Fig. 34B. 9

9a Rhizoids present; wall of capsule 2-10 cells thick. .... (p. 232) Order 3. JUNGERMANNIALES

9b Rhizoids absent; when present, capsule with wall 1-cell thick. 10

10a Leaves typically divided into 2-3 segments to base. Fig. 35A, B, C. ....
.... (p. 232) Order 1. TAKAKIALES

10b Leaves unlobed. Fig. 35D, E. ....
.... (p. 232) Order 2. CALOBRYALES

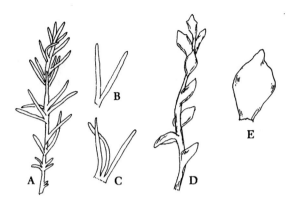

Figure 35

Figure 35 *Takakia*. A, plant; B-C, leaves; *Haplomitrium;* D, plant; E, leaf.

Musci
The Mosses

SUBCLASS 1. SPHAGNIDAE
Sphagnum—Peat or Bog Mosses

About fifty species of *Sphagnum* are reported from North America.

1a   Cortical (outer) cells of the stem and branches with fibril-bands. Fig. 36A. 2

1b   Cortical cells of the stem and branches without fibril-bands. Fig. 37A. 5

2a   Green cells of branch leaves entirely enclosed by the colorless (emtpy) cells or exposed equally on both surfaces. Fig. 36D, F. 3

2b   Green cells of branch leaves exposed more broadly on the upper (ventral) surface of the leaf. Fig. 36C, E. 4

3a   Green cells entirely enclosed. Fig. 36D. .... *Sphagnum magellanicum* Brid.

Typically of open bogs, Labrador to Alaska, south to the Gulf of Mexico and California.

3b   Green cells exposed equally on both surfaces. Fig. 36F. .... *Sphagnum papillosum* Brid.

Open, wet bogs, often brownish tinged, Labrador to British Columbia, Alaska, south to North Carolina and Indiana.

4a   Green cells of the branch-leaves isosceles-triangular in section. Fig. 36 A-C. .... *Sphagnum palustre* L.

Shaded to semi-open bog forests, throughout the United States and Hawaii.

4b   Green cells of branch leaves equalateral-triangular in section. Fig. 36E. .... *Sphagnum imbricatum* Hornsch *ex* Russ.

Shaded to open wet acid soil and stream banks, Newfoundland southward to the Gulf Coast, west to Illinois and Alaska.

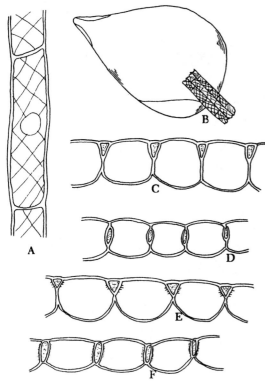

Figure 36

Figure 36 *Sphagnum palustre*. A, cortical cell of branch showing fibril bands; B, branch leaf; C, cross section of a branch leaf. *Sphagnum magellanicum*. D, cross section of branch leaf. *Sphagnum imbricatum*. E, cross section of branch leaf. *Sphagnum papillosum*. F, cross section of branch leaf.

5a Green cells completely enclosed by the colorless empty cells. Fig. 37A, B. .... *Sphagnum compactum* DC. *ex* Lam. & DC.

Relatively dry habitats, edges of bogs and ditches, Labrador to the Great Lakes south to the Gulf of Mexico and in British Columbia and Alaska.

5b Green cells not completely enclosed by the colorless empty cells. Fig. 37 C, D, E. 6

6a Green cells of branch leaves exposed more broad on the lower (dorsal) surface or more or less exposed equally on both surfaces. Fig. 37C, D, E. 7

6b Green cells of branch leaves exposed more broadly on the upper (ventral) surface. Fig. 38E. 10

7a Green cells triangular in section. Fig. 37C. .... *Sphagnum recurvum* P.-Beauv.

Common in neutral moist habitats, Newfoundland to Alaska south to the Gulf of Mexico and California.

7b Green cells truncately-elliptic to trapeziodal in section. Fig. 37D, E. 8

8a Green cells exposed more or less equally on both surfaces. Fig. 37E. 9

8b Green cells exposed more broadly on the lower surface. Fig. 37D. .... *Sphagnum cuspidatum* Ehrh. *ex* Hoffm.

Relatively acid bogs, ditches, and drainage channels, Newfoundland to the Great Lakes south to eastern Gulf States.

9a Colorless cells with a single row of pores in the middle of the outer wall, without fibrils. Fig. 37G. .... *Sphagnum macrophyllum* Bernh. *ex* Brid.

Open water of ponds and swamps, in drainage ditches, Maine to Florida, Arkansas, and Louisiana.

9b Colorless cells with two rows of pores in the outer wall, with fibrils. Fig. 37E, F. .. *Sphagnum subsecundum* Nees ex Strum

Bogs, sedge weadows, and other various moist habitats, cosmopolitan.

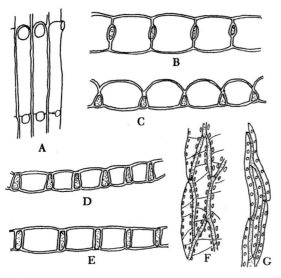

Figure 37

Figure 37 *Sphagnum compactum*. A, cortical cells of a branch without fibril bands; B, cross section of a branch leaf. *Sphagnum recurvum*. C, cross section of a branch leaf. *Sphagnum cuspidatum*. D, cross section of a branch leaf. *Sphagnum subsecundum*. E, cross section of a branch leaf; F, outer surface of a branch leaf. *Sphagnum macrophyllum*. G, outer surface of a branch leaf.

10a Stem leaves somewhat to conspicuously fringed. Fig. 38A, B. 11

10b Stem leaves not fringed. 12

11a Stem leaves spoon-shaped, fimbriate-lacerate on upper margins and apex. Fig. 38B. ....
Sphagnum fimbriatum Wils. ex J. Hook.

Along streams, in low lying places and bogs, Labrador to Alaska south to Maryland, Missouri, and California.

11b Stem leaves tongue-shaped, fimbriate-lacerate only across the apex. Fig. 38A. .... *Sphagnum girgensohnii* Russ.

Swamps and bogs, Labrador to the Great Lakes south to North Carolina, Tennessee, and Illinois; Alaska south to Oregon, Idaho, and Montana.

12a Outer surface of colorless cells of branch leaves near apex with very small, round, strongly ringed pores on the surface, Fig. 38C. .... *Sphagnum warnstorfii* Russ.

Open, rich bogs and fens, Labrador to Alaska south to the Great Lakes, Illinois, Florida and Colorado.

12b Outer surface of colorless cells near apex with larger, elliptic pores. Fig. 38D. 13

13a Plants brown. Fig. 38D. ....
.. *Sphagnum fuscum* (Schimp.) Klinggr.

Acid bogs, Labrador to Alaska south to Virginia, the Great Lakes, Colorado, and Washington.

13b Plants green. Fig. 38E. ....
.... *Sphagnum capillaceum* (Weiss.) Schrank.

Open to shady moist acid habitats, Labrador to Alaska, south to North Carolina, the Great Lakes, Arkansas, Colorado, and Washington.

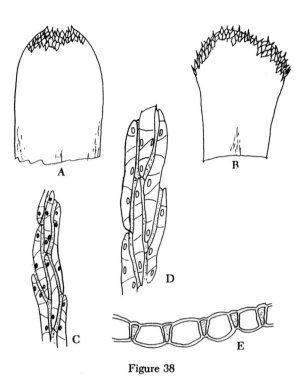

Figure 38 *Sphagnum girgensohnii.* A, leaf of stem. *Sphagnum fimbriatum.* B, leaf of stem. *Sphagnum warnstorfii.* C, cells of outer surface of stem leaf near apex. *Sphagnum fuscum.* D, cells of outer surface of stem leaf near apex. *Sphagnum capillaceum.* E, cross section of stem leaf.

Figure 38

SUBCLASS 2. ANDREAEIDAE
Andreaea

Six species of Andreaea are reported from North America.

1a Leaves without a midrib (costa) of elongated cells. Fig. 38B, 39A-C. .... *Andreaea rupestris* Hedw.

On moist, non-calcareous rocks, Ontario south to Georgia, Great Lakes area, throughout the Pacific Northwest.

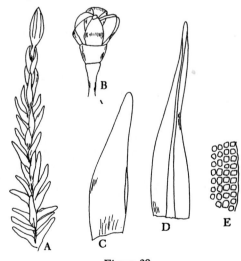

Figure 39

Figure 39 *Andreaea rupestris.* A, plant with unopen capsule; B, open capsule; C, leaf. *Andreaea rothii.* D, leaf; E, basal leaf cells.

1b Leaves with a midrib (costa). 2

2a Plants not brittle when dry; upper leaf margins usually distinctly serrate and papillose. Fig. 40A-C. ....
.... *Andreaea nivalis* Hook.

On wet rocks of subalpine to alpine regions, Alaska, British Columbia, and Washington.

2b Plant brittle when dry; upper leaf margins not normally serrate or papillose. .. 3

3a Midrib weak at base; basal leaf cells elongate. Fig. 40D, E. ....
.... *Andreaea blyttii* B.S.G.

On moist rocks of arctic alpine regions, Labrador, Alaska south to California.

3b Midrib strong at base; basal leaf cells short except near midrib. Fig. 28A, C. 39D, E. .... *Andreaea rothii* Web. & Mohr

On moist rocks, lowland to arctic alpine, Labrador, Great Lakes region south to Tennessee and Georgia, Colorado, British Columbia south to Oregon and California.

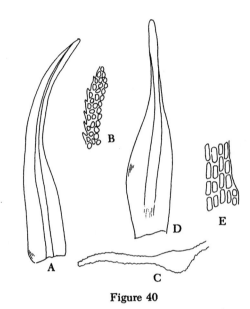

Figure 40

Figure 40 *Andreaea nivalis.* A, leaf; B, marginal cells of upper leaf; C, cross section of leaf near base. *Andreaea blyttii.* D, leaf; E, basal leaf cells.

SUBCLASS 3. BRYIDAE

KEY TO THE GENERA

1a Whitish plants with cells of two kinds throughout the leaf; large empty cells in 2 or 3 layers, with small green cells in between. Fig. 41. 2

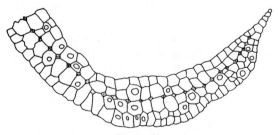

Figure 41

Musci—The Mosses

Figure 41 *Leucobryum glaucum*: cross section of leaf.

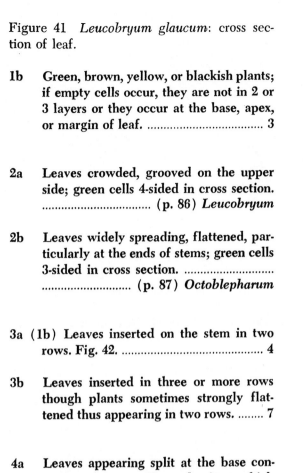

Figure 42 Plants of *Fissidens*, a genus with leaves inserted in two rows along the stem.

1b   Green, brown, yellow, or blackish plants; if empty cells occur, they are not in 2 or 3 layers or they occur at the base, apex, or margin of leaf. 3

2a   Leaves crowded, grooved on the upper side; green cells 4-sided in cross section. .... (p. 86) *Leucobryum*

2b   Leaves widely spreading, flattened, particularly at the ends of stems; green cells 3-sided in cross section. .... (p. 87) *Octoblepharum*

3a (1b) Leaves inserted on the stem in two rows. Fig. 42. 4

3b   Leaves inserted in three or more rows though plants sometimes strongly flattened thus appearing in two rows. 7

4a   Leaves appearing split at the base consisting of two vaginant laminae which clasp the stem as well as the base of the leaf above; wing-like laminae extending dorsally and apically. Fig. 42. .... (p. 62) *Fissidens*

4b   Leaves not divided into vaginant, dorsal, and apical laminae. 5

5a   Sterile plants with leaves pinnately lobed and connected by broad decurrent bases, not sheathing or subulate; growing in damp caves and basements. .... (p. 119) *Schistostega*

5b   Leaves filiform from a sheathing base and narrowed into a linear awn. 6

6a   Linear awn smooth; peristome absent. .... (p. 72) *Bryoxiphium*

6b   Linear awn rough; peristome present. .... (p. 72) *Distichium*

7a (3b) Plants with a peristome of 4 teeth; sterile plants with stems commonly ending in a gemma cup; leaves costate and with isodiametric cells. Fig. 43. .... (p. 223) *Tetraphis*

Figure 42

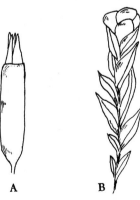

Figure 43

Musci—The Mosses

Figure 43 *Tetraphis*. A, capsule with four teeth; B, stems with gemmae cups at apex.

7b Plants with a peristome of more than 4 teeth or the peristome absent; gemmae when present at the tips of sterile stems not arranged in a cup. 8

8a Leaves small, rarely seen; plants bearing a flat, red-purple capsule obliquely or horizontally attached to a rough seta 1-2 cm long. .... (p. 222) *Buxbaumia*

8b Leaves well developed and persistent. 9

9a Capsule sessile (without a seta), inclined, asymmetric, immersed in bristle-tipped perichaetial leaves; leaves of sterile stems limgulate, rounded at apex, with bulging cells in more than 1 layer, crisped when dry. .... (p. 222) *Diphyscium*

9b Capsule and leaves not as above. 10

10a Plants stemless or with a very short stem; capsules without an operculum; ephemeral. 11

10b Plants with or without a distinct stem; capsule when present with an operculum or regularly dehiscing near middle; may or may not be ephemeral. 17

11a Leaf margins involute. Fig. 44b. .... (p. 90) *Astomum*

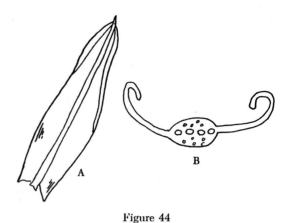

Figure 44

Figure 44 *Weissia controversia*. A, Leaf, dorsal view showing involute margins near apex; B, cross section of leaf near apex.

11b Leaf margins not involute. 12

12a Capsules short-pyriform, emergent to shortly exserted. Fig. 45. .... (p. 74) *Bruchia*

Figure 45

Figure 45 *Bruchia flexuosa*.

12b Capsules more or less spherical, without a neck. 13

13a Plants bulbiform, often appearing 3-angled, reddish; leaves broadly ovate, deeply concaved or keeled; spores large, 40-50 μ in diameter. .... (p. 98) *Acaulon*

13b Plants not bulbiform or 3-angled; leaves linear to lanceolate, flat to slightly concave. 14

14a Leaves setaceous. ..... (p. 68) *Pleuridium*

14b Leaves narrow or broad but not setaceous. 15

15a Plants surrounded by branching, velvety protonema; leaf cells smooth or coarsely unipapillose. Fig. 46. 16

Figure 46

Figure 46 *Ephemerum serratum*.

15b Plants not surrounded by conspicuous protonema; leaf cells pleuri-papillose; leaves broadly lanceolate to oblong-lanceolate, ending in a smooth, yellow awn. .... (p. 98) *Phascum*

16a Calyptra covering 1/4 to 1/2 of the capsule, more than 0.2 mm in overall length; capsule cleistocarpous; leaves costate or ecostate. .... (p. 113) *Ephemerum*

16b Calyptra minute, less than 0.2 mm in length, covering only the tip of the capsule; capsule dehiscing near the middle; leaves ecostate. .. (p. 114) *Micromitrium*

17a (10b) Plants minute, almost stemless; ephemeral; capsules immersed; growing on soil. 18

17b Plants minute to large; capsules immersed to exserted; if plants minute with immersed capsules, growing on rocks, trees or soil and not ephemeral. 20

18a Outer cells of capsule wall (exothecial) very lax and thin walled. Fig. 47A. .... (p. 114) *Physcomitrella*

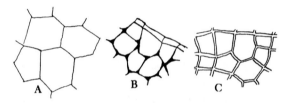

Figure 47

Figure 47 Exothecial cells. A, *Physcomitriella*; B, *Aphanorhegma*; C, *Physcomitrium*.

18b Exothecial cells of capsule wall evenly thickened or thickened at the corners. Fig. 47A, B. 19

19a Capsules dehiscing at the middle; exothecial cells thickened at the corners. .... (p. 114) *Aphanorrhegma*

19b Capsules with a distinct operculum; exothecial cells not thickened at the corners. .... (p. 115) *Physcomitrium*

20a (17b) Leaves with filiform green photosynthetic outgrowths on the upper surface or with lamina-like vertical lamellae on the upper surface, at least on the costa. Fig. 48. 21

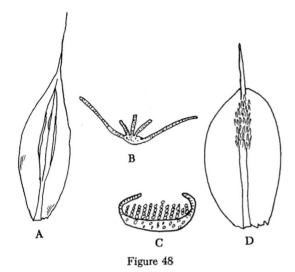

Figure 48 Leaves with green photosynthetic outgrowths. A, *Pterogoneurum*, dorsal view of leaf; B, cross section of leaf; C, *Alonia*, cross section of leaf; D, *Crossidium*, dorsal view of leaf.

20b Leaves without such outgrowths; sometimes with deciduous gemmae. 28

21a Leaves with filiform green photosynthetic outgrowths on the upper surface. 22

21b Leaves with vertical lamellae on the upper surface. 23

22a Filaments from the midrib only. .... (p. 99) *Crossidium*

22b Filaments from the upper surface of the leaf. .... (p. 101) *Aloina*

23a Leaves entire, oval, thin, with lamellae on distal half of costa; costa prolonged into a hair (awn). .... (p. 99) *Pterigoneurum*

23b Leaves toothed on the margin or with translucent sides folded up over the upper surface; peristome teeth 32-64, with tips attached to a transverse membrane. 24

24a Lamellae on the upper surface of the leaf 1-20, calyptra smooth or with only a few hairs. 25

24b Lamellae on the upper surface of the leaf more than 20; calyptra hairy. 27

25a Cilia borne on the upper part of the leaf sheath; lamina of leaf bistratose. Fig. 379A, B. .... (p. 225) *Bartramiopsis*

25b Cilia absent; lamina unistratose or bristratose along the margin. 26

26a Margin of leaf bordered with long narrow cells. .... (p. 223) *Atrichum*

26b Leaves not bordered; often with lamellae on the back. .. (p. 225) *Oligotrichum*

27a (24b) Capsule cylindrical; calyptra densely hairy; leaves serrate at least in the upper half; marginal cells of lamellae papillose, or if smooth, then the leaves thin and strongly crisped when dry. .... (p. 226) *Pogonatum*

27b Capsule 4-6 angled; calyptra slightly to densely hairy; leaves entire to serrate, never thin and crispate when dry; marginal cells of lamellae never papillose. .... (p. 227) *Polytrichum*

28a (20b) Plants growing on rich organic soil, dung, bones, or other organic matter of animal origin; peristome teeth 2 or 3 cells thick, the outer cells quadrate; teeth forming 32 curley threads, or 16 teeth 2 cells wide, or 8 teeth 4 cells wide, often strongly reflexed when dry, or capsules cleistocarpous and peristome absent; hypothesis (lower part of capsule) well developed in operculate species and often larger than the urn. Fig. 49. 29

28b Plants not having the above combination of characters; peristome when present single or double, composed of teeth formed by strips of cell-wall split apart through the cell cavities; growing on various substrates but not restricted to rich organic soil, bones, or other organic matter. 33

29a Capsule cleistocarpous, without peristome or hypothesis; growing on peaty soil of arctic-alpine regions of Alberta, Alaska, and Colorado. .... (p. 118) *Voitia*,

29b Capsule with operculum and hypothesis. 30

30a Leaves ovate to ovate-spatulate, the apex broad and rounded; basal cells with long cilia on the margins; peristome wanting; on soil in rock crevices in the mountains of Washington and Alaska. .... (p. 117) *Oedipodium*

30b Leaves without the above combination of characters. 31

31a Hypothesis of capsule not wider than the urn and lighter in color. .... (p. 118) *Tayloria*

31b Hypothesis of capsule wider than urn when moist. 32

32a Hypothesis distinctly wider than the urn, fresh or dry, usually darker in color; columella exserted; teeth of peristome 3-layered. Fig. 49A. .. (p. 118) *Splachnum*

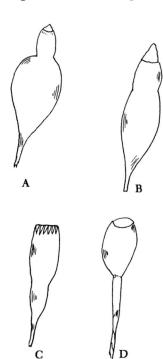

Figure 49

Figure 49 Capsules. A, *Splachnum*; B, *Tetraplodon*; C, *Tayloria*; D, *Oedipodium*.

32b Hypothesis when fresh and moist only slightly wider than urn, not conspicuously colored; columella not exserted. Fig. 49B. .... (p. 118) *Tetraplodon*

33a (28b) Stems erect, unbranched, or with branches erect, generally in tufts; archegonia and sporophytes terminal (may appear lateral because of innovative branches arising below the sporo-. phytes). Fig. 50. 34

Figure 50 Examples of acrocarpous mosses. A, *Tortella*; B, *Funaria*; C, *Orthotrichum*; D. *Bryum*.

33b Stems creeping or ascending, usually extensive branched, in interwoven mats: archegonia, antheridia, and sporophytes lateral or at the ends of branches. Fig. 51. 139

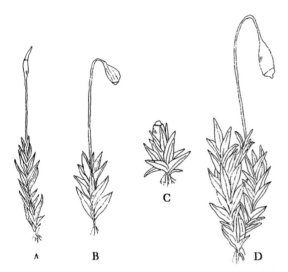

Figure 50

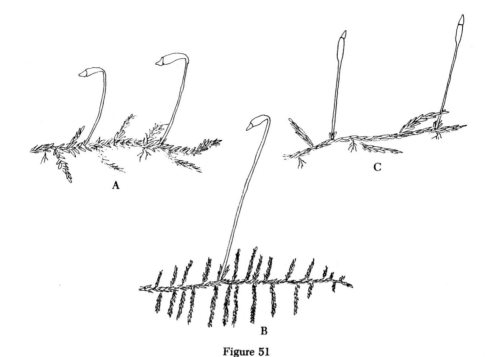

Figure 51

Figure 51 Examples of pleuricarpous mosses. A, *Amblystegium;* B, *Thuidium;* C, *Entodon.*

34a Leaves lanceolate to spatulate with sheathing bases that have a conspicuous area of large, inflated, hyaline cells next to the costa and with marginal or intramarginal border of elongated hyaline cells or with a thickened border. Fig. 52. 35

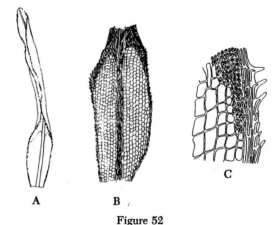

Figure 52

Figure 52 *Syrrhopodon texanus.* A, leaf; B, sheathing leaf base with hyaline cells; C, leaf base showing elongated marginal cells.

34b Leaves, especially the bases, not as above. 36

Musci—The Mosses 35

35a Leaves with intramarginal border of hyaline or yellowish elongated cells. .... (p. 88) *Calymperes*

35b Leaves without intramarginal border of elongated cells; border marginal, composed of elongated hyaline cells or of thickened, double wing-like green cells. .... (p. 87) *Syrrhopodon*

36a (34b) Leaf cells conspicuously bulging or bulging mammillose, at least on the upper (ventral) side, or papillose with papillae not, or scarcely, including the cell cavity, or roughened by fine longitudinal ridges (striolate). Fig. 53. 37

36b Leaf cells smooth. 83

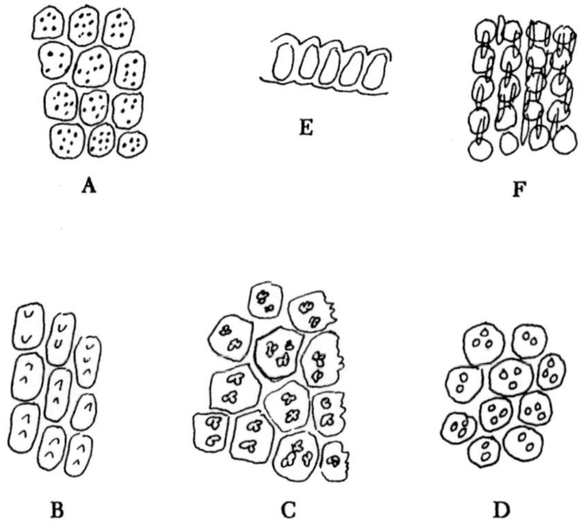

Figure 53

Figure 53 A-D, papillose cells; E, cross-section of mammillose cells; F, longitudinally striate cells.

37a Leaf cells conspicuously bulging mammillose, at least on upper surface. 38

37b Leaf cells papillose or striolate. 43

38a Leaf cells mammillose on upper side; smooth or slightly papillose on lower side. 39

38b Leaves mammillose on both sides. 41

39a Leaves 2 cells thick, the upper (ventral) cells mammillose, the lower (dorsal) cells smooth. .... (p. 93) *Timmiella*

39b Leaf cells mainly in one layer. 40

40a Leaves broadly lanceolate from a broader base; calyptra often adhering to the tip of seta after falling from the mature capsule. .... (p. 139) *Timmia*

40b Leaves broadly oblong-ovate, apex obtuse, slightly apiculate; brood bodies (gemmae) often borne on axillary filaments. .... (p. 95) *Hyophila*

41a (38b) Plants very small; leaves narrow, slenderly lingulate (tongue shaped), apex rounded, costa near the apex obscured by bulging cells. .... (p. 101) *Luisierella*

41b Plants not as above. 42

42a Leafy shoots 3-5 mm tall, from creeping stems; leaves entire; growing on trees in the southeastern United States. .... (p. 148) *Groutiella*

42b Leafy shoots 1-10 cm tall, not growing from creeping stems; leaves nearly entire to finely serrate or coarsely serrate; growing mainly on rocks, not on trees. .... (p. 79) *Dichodontium*

43a (37b) Cells at basal angles (alar cells) of leaves enlarged and clearly differentiated, cell walls colored brown or hyaline. Fig. 54. 44

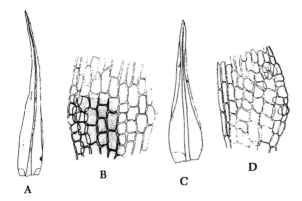

Figure 54

Figure 54 Leaves with enlarged alar cells. *Dicranum condensatum*. A, leaf; B, cells of alar region. *Dicranum flagellare*. C, leaf; D, cells of alar region.

43b Alar cells not enlarged and clearly differentiated. 45

44a Leaf cells longitudinally striate (Fig. 53F); perichaetial leaves truncate or short pointed; capsules erect and symmetric. .... (p. 80) *Dicranoweisia*

44b Leaf cells papillose; perichaetial leaves subulate; capsules inclined to suberect, usually distinctly asymmetric. .... (p. 81) *Dicranum*

45a (43b) Stems repeatedly branching, appearing more or less pleuricarpous; sporophytes borne at intervals. Fig. 55. 46

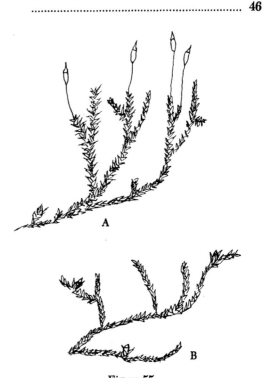

Figure 55

Figure 55 A, *Rhacomitrium*; B, *Hedwigia*.

45b Stems strictly upright, sparsely branched; sporophytes terminal. 54

46a Cells of leaf narrow with extremely wavy walls; leaves lanceolate, ending in a serrate, white hair, or hair obsolete. .... (p. 110) *Rhacomitrium*

46b Cells wall not wavy. 47

47a Leaves without a costa, colorless at apex, often awned. 48

Musci—The Mosses   37

47b   Leaves with costa, lanceolate to ovate lanceolate; capsules with broad peristome teeth; tufted blackish mosses growing on trees or rocks. 50

48a   Papillae forked; seta shorter than leaves. .... (p. 153) *Hedwigia*

48b   Papillae simple; seta longer than leaves. 49

49a   Leaves acuminate, ending in an elongate, hyaline, flexuous apex; capsule sulcate when dry. .... (p. 154) *Pseudobraunia*

49b   Leaves muticous (without a hyaline point); capsule smooth when dry. .... (p. 154) *Braunia*

50a   (47b) Leaves not crisped and contorted when dry. .... (p. 141) *Orthotrichum*

50b   Leaves crisped and contorted when dry. 51

51a   Calyptra cucullate, never hairy; peristome absent. Fig. 56B. 52

51b   Calyptra campanulate, glabrous or hairy. Fig. 56A. 53

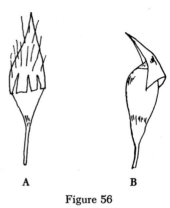

Figure 56

Figure 56   A, Campanulate calyptra of *Ulota*; B, cucullate calyptra of *Amphidium*.

52a   Costa with central guide cells, or the leaves toothed; gemmae absent; growing on rocks and trees. Fig. 57A. .... (p. 141) *Amphidium*

52b   Costa without guide cells or the guide cells ventral; gemmae common; growing on trees and rocks. Fig. 57B. .... (p. 140) *Zygodon*

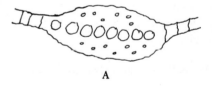

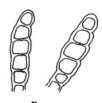

Figure 57

Figure 57   A, Cross section of costa showing central row of large guide cells; B, gemmae of *Zygodon*.

53a (51b) Calyptra smooth or with few hairs; stomata immersed in capsule wall; basal marginal cells not particularly differentiated; gemmae absent. .... (p. 141) *Orthotrichum*

53b Calyptra hairy; stomata superficial; basal cells differentiated, very thick walled, yellow, arranged in rows fanning out from the costa; sterile plants usually with septate gemmae on the leaf tips. .... (p. 147) *Ulota*

54a (45b) Plants in small, cushion-like tufts; setae often very short; capsules immersed to short exserted; growing on rocks and trees. .... Return to 50

54b Plants usually in more extensive tufts; usually setae longer and capsule long exserted; **growing on rocks or soil.** 55

55a Leaves contracting from an ovate, clasping base into a narrow subulate awn roughened by papillae formed by projecting cell ends. .... (p. 69) *Trichodon*

55b Leaves not as above. 56

56a Leaves broad, widest at middle or above; costa with dorsal stereids or none, often ending in an awn; cells of lower 1/4 of leaf colorless. 57

56b Leaves lanceolate to linear-lanceolate, widest below the middle, long tapering to the apex; cells of lower 1/4 of leaf various. 64

57a Leaves opaque in upper half with densely crowded coarse rounded papillae; lower leaf cells transparent, large, with thick transverse walls (usually brown); calyptra cylindrical, beaked, persistent, covering the capsule. .... (p. 88) *Encalypta*

57b Papillae dot-like or columnar or c-shaped, o-shaped, or forked; clear cells at base thin walled; calyptra smaller, split on one side and deciduous. 58

58a Margins of leaf thicker than median part due to deeper cells or more than one layer, or margins composed of greatly elongated cells. 59

58b Margins of leaf not differentiated as above. 60

59a Border of leaf composed of deep cells or more than one layer of cells. .... (p. 90) *Scopelophila*

59b Border of leaf composed of greatly elongated cells. .... (p. 101) *Tortula*

60a (58b) Plants with leaves very much broken, the fragments serving as gemmae. .... (p. 101) *Tortula*

60b Plants with leaves not broken into fragments. 61

61a Plants with gemmae. 62

61b Plants without gemmae. 63

62a   Gemmae club-shaped, in axils of leaves. .... (p. 96) *Barbula*

62b   Gemmae resembling small serrate leaves or occurring on borne or the surface of the leaf. .... (p. 101) *Tortula*

63a (61b) Peristome absent or composed of 16 teeth, slender and acute, split and broken off. .... (p. 99) *Desmatodon*

63b   Peristome of 32 spirally twisted thread from a basal membrane, colorless cells at base of leaf extending up along the costa. .... (p. 101) *Tortula*

64a (56b) Margins of leaves distinctly involute. 65

64b   Margins of leaves revolute or plane. .. 66

65a   Peristome present, inserted at the mouth of the capsule; costa 60 µ or more wide at the base; leaves linear-lanceolate from a broadly oblong to oblong-lanceolate base. .... (p. 92) *Trichostomum*

65b   Peristome absent; if present, inserted below the mouth of the capsule; costa 50 µ or less wide at base, or if wider, then leaves narrowly oblong-lanceolate. .... (p. 91) *Weissia*

66a (64b) Stems matted together with brown, branching rhizoids. 67

66b   Stems free and separate. 72

67a   Many stems ending in a naked stalk bearing a cluster of gemmae; cells isodiametric with a single large papillae near the middle of the cell; capsule cylindrical, curved and ribbed. .... (p. 133) *Aulacomnium*

67b   Without gemmae; papillae much smaller. 68

68a   With one low papillae over the cell cavity; capsule inclined, curved, cylindric; peristome of 16 red, forked teeth. .. .. (p. 79) *Cynodontium*

68b   With papillae at one or both ends of the cell; capsule globose to ovoid, inclined to erect, wrinkled to often longitudinally ribbed when dry. 69

69a   Plants of wet, seepy placed; leaves lanceolate to ovate-lanceolate; capsule ribbed and furrowed. .... (p. 137) *Philonotis*

69b   Plants of drier habitats as shaded rocks or banks; leaves linear, narrowly-lanceolate to acicular, at least in the upper half. 70

70a   Leaves 4-7 mm long, curly or straight; capsule ribbed and furrowed when dry. .... (p. 136) *Bartramia*

70b   Leaves 0.8 to 3.5 mm long, erect, appressed; capsule smooth to furrowed when dry. 71

71a Stems with conspicuous clusters of reddish brown rhizoids; leaves 2.5-5.0 mm long; capsule erect, smooth to somewhat wrinkled but never strongly ribbed when dry, operculum conic. .... (p. 136) *Anacolia*

71b Stems with less conspicuous clusters of rhizoids; leaves small, to 1.5 mm long, stiff; capsule inclined, operculum beaked. .... (p. 137) *Conostomum*

72a (66b) Margins of leaf revolute. 73

72b Margins plane, at least in upper half. 78

73a Costa with a single stereid band. Fig. 58. 74

73b Costa with two stereid bands. 75

74a Leaves linear-lingulate; costa ending below the apex, spurred above. .... (p. 92) *Husnotiella*

74b Leaves ovate to oblong-lanceolate, costa excurrent, not spurred above. .... (p. 98) *Phascum*

75a (73b) Leaf margins 2-3 cells thick. .... (p. 95) *Didymodon*

75b Leaf margins 1-cell thick. 76

76a Peristome of 32 spirally twisted, papillose threads, or entirely absent. .... (p. 96) *Barbula*

76b Peristome of 16 erect, slenderly triangular teeth, cleft or perforated. 77

77a Leaves entire throughout. .... (p. 95) *Didymodon*

77b Leaves denticulate to strongly denticulate at apex; plants usually brick-red below. .... (p. 95) *Bryoerythrophyllum*

78a (72b) Colorless cells at base of leaf extending up along the margin. 79

78b Cells of leaf base not as above. 80

79a Hyaline cells of leaf base not reaching the costa; leaves serrate above. .... (p. 94) *Pleurochaete*

79b Hyaline cells of leaf base reaching the costa; leaves entire. .... (p. 93) *Tortella*

80a (78b) Leaves toothed on the margin where the colorless cells meet the green ones. .... (p. 92) *Eucladium*

80b Leaves with entire or wavy margins. .. 81

81a Leaves slenderly tapering to apex, to 4 mm long; peristome teeth short, irregular. .... (p. 92) *Trichostomum*

81b Leaves acute to obtuse, smaller, to 1.5 mm. 82

82a Leaves bistratose in patches; archegonial inflorescences lateral. .... (p. 89) *Anoectangium*

82b Leaves not bistratose; archegonial inflorescences terminal. .... (p. 91) *Gymnostomum*

83a (36b) Plants growing on bark of trees in long creeping stems with crowded, erect (2-8 mm) densely leafy branches on which sporophytes are borne. 84

83b Plant growing on soil, rocks, or trees but not forming long creeping stems with erect, branches densely leafy. 85

84a Cells near base of leaf small, rounded; capsule ovoid-globose; widespread in the eastern United States. .... (p. 148) *Drummondia*

84b Cells near base of leaf linear; capsule oblong-cylindric; southeastern coastal plain. .... (p. 148) *Scholtheimia*

85a (83b) Very small mosses, less than 4 mm tall; extremely difficult to identify without capsules. 86

85b Larger mosses, stems usually 4 mm or more tall, if minute, with persistent protonema. 91

86a Capsules immersed in or barely emergent from the upper leaves; seta short; peristome absent. 87

86b Seta longer; capsule not immersed in or barely emergent from the upper leaves. 89

87a Spores very large, 100-300 μ in diameter, 4-60 in a spherical capsule; calyptra very small. .... (p. 67) *Archidium*

87b Spores smaller, 20 to 70 μ in diameter. 88

88a Spores 45-70 μ in diameter; calyptra 4-angled, completely covering the capsule. .... (p. 115) *Pyramidula*

88b Spores smaller, 20-40 μ in diameter; calyptra smaller, not 4-angled or covering the capsule at maturity. .... Return to 10

89a (86b) Plants dark green in color, minute (1-2 mm tall), growing as velvety films on limestone rocks; peristome teeth not split (or lacking). .... (p. 72) *Seligeria*

89b Plants green to yellow-green in color, larger, usually growing on soil or on rocks; peristome teeth split half way down or divided into threads. 90

90a Teeth divided into threads. .... (p. 69) *Ditrichum*

90b Teeth split half way down. .... (p. 75) *Dicranella*

91a (85b) Leaves distinctly in three ranks; stems triangular with outer cells thin-walled and colorless; capsules globular; peristome double. .... (p. 135) *Plagiopus*

91b Leaves in 5 ranks; stems not as above. 92

92a Leaves 6 to 20 times as long as broad, mostly long-tapering to a slender upper half. 93

92b Leaves 1-5 times as long as broad. 111

93a Cells at basal angles of leaf (alar) enlarged, often with brown walls, or if not clearly differentiated, leaves with a clasping base or with multicellular gemmae on underside of leaf near base. 94

93b Alar cells not notably enlarged. 103

94a Peristome teeth undivided; capsule cylindrical, erect, 1.5-2 mm long; leaves strong crisped, (curley) when dry. .... (p. 80) *Dicranoweisia*

94b Peristome teeth split at the tip, split half way down, or undivided; if undivided capsule pear-shaped, erect, about 1 mm long and leaves erect or bent to one side when dry. 95

95a Peristome teeth undivided or split just at the tip. .... (p. 73) *Blindia*

95b Peristome teeth split half way down. .. 96

96a Costa narrow, 1/3 to 1/10 of leaf at base. 97

96b Costa wide, 1/3 to 7/8 of leaf at base. 100

97a Leaves very brittle, mostly broken off; costa without stereids. .... (p. 81) *Dicranum*

97b Leaves not easily broken; costa with or without stereids. Fig. 58. 98

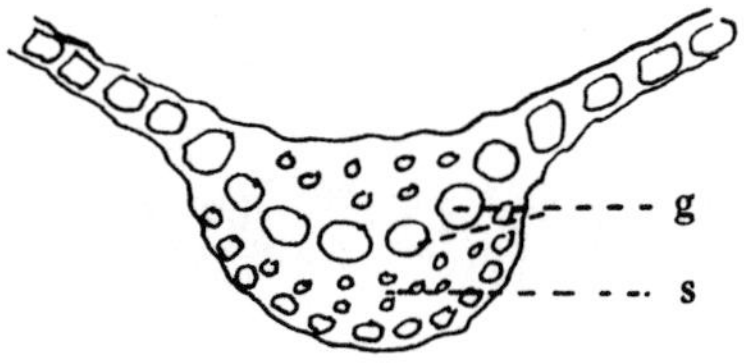

Figure 58

Figure 58 Cross section of a costa showing guide cells (g) and stereid cells (s).

98a Costa without stereids; leaves crisped when dry. .... (p. 81) *Kiaeria*

98b Costa with stereids. 99

99a Base of leaf broad, clasping, abruptly narrowed to the slender upper portion; capsule strumose. .... (p. 80) *Oncophorus*

99b Leaves evenly tapering from base upward; capsules not strumose. .... (p. 81) *Dicranum*

100a (96b) Costa with stereid bands. 101

100b Costa without stereid bands or guide cells, very wide. 102

101a Costa with dorsal and ventral stereid bands, with guide cells in between; surface cells large and similar on both sides. Fig. 59B. .... (p. 79) *Dicranodontium*

101b Costa with large clear cells ventrally; dorsally with strands of stereids alternating with large clear cells. Fig. 59A .... (p. 77) *Campylopus*

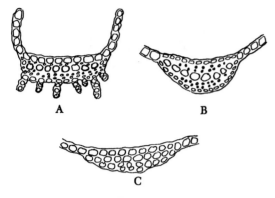

Figure 59

Figure 59 Cross sections of costae. A, *Campylopus;* B, *Dicranodontium;* C, *Paraleucobryum.*

102a (100b) Plants with clusters of fusiform gemmae whose surface cells are clear and empty. Fig. 60 .... (p. 79) *Brothera*

102b Plants without gemmae; costa of 3 or 4 layers of uniform cells. Fig. 59C .... (p. 78) *Paraleucobryum*

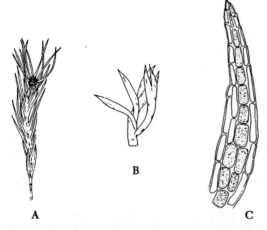

Figure 60

Figure 60 *Brothera leana.* A, plant with cluster of gemmae at apex; B, gemmae cluster; C, gemma showing clear surface cells.

103a (93b) Leaf cells long-rhomboidal to linear, 6-12 times as long as wide; capsules inclined to nodding, pyriform or oblong-pyriform. 104

103b Leaf cells shorter; capsules erect or inclined, not pyriform. 105

104a Leaves linear-subulate; costa filling about 1/2 of the base and most of the slender tip. .... (p. 123) *Leptobryum*

104b Leaves oblong-lanceolate; costa narrow. .... (p. 120) *Pohlia*

105a (103b) Plants restricted to wet seepage areas in Florida, Georgia, and Louisiana; leaves narrowly linear-lanceolate, widely spreading when moist, giving plants a feathery appearance, costa narrow, percurrent; margins of leaves without elongated cells but with paired teeth from base to apex; peristome double. .... (p. 132) *Rhizogonium*

105b Plants not possessing the above combination of characters. 106

106a Neck of capsule about as long as urn. .... (p. 75) *Trematodon*

106b Without a noticeable neck. 107

107a Teeth 16, split half way down or to base into threads. 108

107b Teeth 16, not split at all; leaves 2-3 mm long, erect, not filiform; margins plane... .... (p. 79) *Rhabdoweisia*

108a Teeth split to base or nearly so. 109

108b Teeth split to the middle. .... (p. 75) *Dicranella*

109a Leaves narrowly lanceolate to subulate; margins neither revolute throughout nor toothed near the apex. .... (p. 69) *Ditrichum*

109b Leaves lanceolate or ovate-lanceolate; margins nearly revolute throughout and irregularly toothed near apex. 110

110a Plants bluish or glaucous. .... (p. 71) *Saelania*

110b Plants dark-green to red-brown. .... (p. 71) *Ceratodon*

111a (92b) Leaves dense and blackish, never broadly spatulate to lingulate; upper cells mostly isodiametric, small and thick-walled; peristome single or lacking. 112

111b Leaves bright or dark green. 115

112a Walls of leaf cells closely and evenly wavy; peristome teeth 16, split nearly to base into 2 or 3 slender prongs. .... (p. 110) *Rhacomitrium*

112b Walls smooth or irregularly wavy at the middle of leaf. 113

113a Leaves bearing rhizoids at base; seta shorter than leaves; on rocks in streams; western United States and Alaska. .... (p. 104) *Scouleria*

113b Leaves without rhizoids. 114

114a Leaves crisped when dry, linear from a lanceolate base; margins plane, entire, or serrate above; apices obtuse to almost blunt; calyptra nearly covering capsule, split into several linear strips. Fig. 61A .... (p. 140) *Ptychomitrium*

114b Leaves usually imbricate when dry, ovate to linear-lanceolate; margins plane to narrowly or strongly recurved; apices muticous to ending in a colorless hair point: calyptra various. Fig. 61B .... (p. 105) *Grimmia*

Figure 61 A, Crisped leaves of *Ptychomitrium*; B, imbricate leaves of *Grimmia*.

115a (111b) Upper leaf cells elongate rectangular, rhombic-hexagonal, to linear, 6-12:1. 116

115b Upper leaf cells quadrate, hexagonal, rhombic, or short-rectangular, 6 times or less as long as wide. 118

116a Leaves bordered by elongate, often reddish cells; leaves of sterile stems more or less dimorphous, smaller in the dorsal row and the larger more or less distichous. .... (p. 122) *Epipterygium*

116b Leaves not as above. 117

117a Sporophytes arising from a short basal shoot over-topped by a long innovation; peristome delicate and single. .... (p. 120) *Mielichhoferia*

117b Sporophytes terminal; peristome double and well developed. .... (p. 120) *Pohlia*

118a (115b) Upper leaf cells quadrate, small (to 15 μ in diameter). 119

118b Upper leaf cells hexagonal, rhombic or short-retangular, larger (to 800 μ long). 125

119a Margins of the leaves revolute. 120

119b Margins of leaf plane (occasionally slightly revolute near base). 121

120a Leaves entire. .... (p. 95) *Didymodon*

120b Leaves toothed at the apex. .... (p. 71) *Ceratodon*

121a (119b) Leaves entire. 122

121b Leaves toothed at apex. .... (p. 133) *Aulacomnium*

122a Leaves half sheathing at base, sharply bent back (squarrose) at middle. .... (p. 75) *Dicranella*

122b Leaves erect, not sheathing or squarrose. 123

123a Leaves oblong-spatulate, rounded at apex; marginal cells somewhat thick-walled forming a distinct border; usually growing on copper bearing rocks. .... (p. 90) *Scopelophila*

123b Leaves not as above. 124

124a Leaves narrowly lance-acuminate, slenderly acute; capsules minute (0.5 mm), horizontal, and black. .... (p. 135) *Catoscopium*

124b Leaves lanceolate, ovate-lanceolate, oblong-lanceolate to abovate, ending in an apiculus or awn; capsules minute to small (0.5-1.5 mm), orange to red-brown. .... (p. 98) *Pottia*

125a (118b) Marginal cells similar to those of lamina. 132

125b Marginal cells slender, fusiform, not similar to cells of lamina. 126

126a Capsules erect; peristome single, composed of 32 teeth. 127

126b Capsules nodding to erect; peristome double, composed of 16 teeth made of cell walls only split apart through the cavity; inner peristome sometimes absent. 128

127a Leaves oval-oblong or oblong-lanceolate; margins serrate; costa bearing a few short obsolete lamellae; peristome composed of 32 stout teeth attached to a transverse membrane, each tooth made up of many cells. .... (p. 223) *Atrichum*

127b Leaves oblong or oblong-obovate; margins entire; costa without lamellae; peristome composed of 32 spirally twisted teeth. .... (p. 101) *Tortula*

128a (126b) Leaves to 6 mm long, in dense terminal rosettes; stems connected by a tough underground "rhizomes." .... (p. 128) *Rhodobryum*

128b Leaves not both large and in dense terminal rosettes; stems not connected by underground "rhizomes." 129

129a Upper leaf cells prosenchymatous, rhomboidal, rhomboid-hexagonal to linear; capsules nodding to erect, pear-shaped (broadest near mouth). 130

129b Upper leaf cells parenchymatous, mostly hexagonal to rounded; capsules nodding, barrel-shaped, narrowed equally at both ends. 131

130a Upper leaf cells rhomboidal-hexagonal, 2-5 times as long as wide; capsule inclined to pendant. Fig. 62A, B, D, E. .... (p. 123) *Bryum*

130b Upper leaf cells short-rhomboidal, 1-2 times as long as broad; capsule erect. Fig. 62C, F. ...... (p. 123) *Brachymenium*

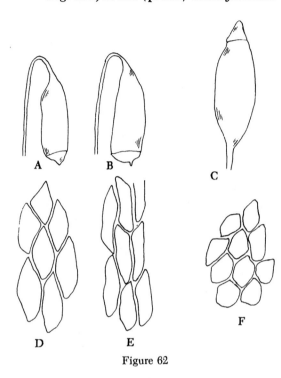

Figure 62

Figure 62 A-B, Capsules of *Bryum*; C, capsule of *Brachymenium*; D-E, leaf cells of *Bryum*; F, leaf cells of *Brachymenium*.

131a (129b) Margins of leaf entire, abruptly cuspidate; outer peristome shorter than the inner. .... (p. 132) *Cinclidium*

131b Margins serrate or if entire leaves not abruptly ending in a long, cuspidate point; outer peristome as long as the inner. .... (p. 129) *Mnium*

132a (125a) Upper leaf cells with pointed ends. 133

Musci—The Mosses   47

132b Upper leaf cells with flat, transverse ends. 134

133a Plants dendroid, large, to 8 cm tall. .... (p. 128) *Leucolepis*

133b Plants not dendroid, smaller, to 5 cm tall though usually smaller. .... (p. 129) *Mnium*

134a (132b) Capsules with neck nearly as long as urn, nearly erect; peristome double, the inner peristome taller. 135

134b Neck of capsule not noticeable; outer peristome stronger than inner, or peristome single or absent. 136

135a Leaf cells large and thin-walled, 20-40 μ wide; plants soft and delicate. .... (p. 134) *Amblyodon*

135b Leaf cells small, 15 μ or less wide; plants coarse, more rigid. .... (p. 134) *Meesia*

136a (134b) Capsule unsymmetric, horizontal; peristome of 16 teeth, each one split from base nearly or quite to apex; costa wanting or weak, not extending to apex; plants very small with persistent green protonema. .... (p. 112) *Discelium*

136b Capsules erect or nodding, symmetric or asymmetric; peristome absent, single, or double; leaves with a distinct costa from base to apex or occasionally excurrent. 137

137a Capsule unsymmetrical (arcuate); peristome double. .... (p. 116) *Funaria*

137b Capsule erect, symmetrical. 138

138a Peristome absent; calyptra mitrate. .... (p. 115) *Physcomitrium*

138b Peristome single, often rudimentary; calyptra cucullate. .... (p. 117) *Entosthodon*

139a (33b) Leaves papillose, at least on the dorsal (lower) side and on the upper half of the leaf. 140

139b Leaves smooth; more or less glossy. .. 174

140a Papillae formed by projecting angles of cell wall. 141

140b Papillae located over the cell-cavity; if located at the end of the cell not formed by a projecting angle. 150

141a Costa of leaves single, extending to the middle of the leaf or beyond. 142

141b Costa of the leaves absent, short and double or if extending beyond the middle of the leaf, double. 144

142a Leaves large, to 3-5 mm long, rugose and plicate, secund (all bent to one side). .... (p. 219) *Rhytidium*

142b Leaves small, to 1.2 mm long, not rugose or plicate. 143

143a Leaves sharply serrate, translucent, decurrent; costa not prominent; cells fusiform, 1:4-6, thin-walled; on soil. .... (p. 197) *Bryhnia*

143b Leaves entire or nearly so, mostly revolute on margin, opaque, not decurrent; costa prominent on back of leaf; cells small, angular, rather thick walled; paraphyllia numerous; on rocks. .... (p. 169) *Pseudoleskea*

144a (141b) Paraphyllia lacking. 145

144b Paraphyllia few and small to abundant and large. 148

145a Plants small, creeping, forming mats; leaves 1.5-2.5 mm long. 146

145b Plants large, erect and spreading, to 15 cm tall; leaves long, cordate-triangular, widely spreading, wet or dry; costa double reaching the middle of the leaf or beyond; upper cells spinose-papillose on back. .... (p. 220) *Rhytidiadelphus*

146a Leaves broadly cordate at base, abruptly long-acuminate, strongly falcate-secund. .... (p. 218) *Ctenidium*

146b Leaves ovate to ovate-lanceolate; acuminate to abruptly acuminate, not broadly cordate at base or strongly falcate-secund. 147

147a Stem leaves distant, spreading, ovate-lanceolate, abruptly acuminate; upper leaf cells about 8.1, fusiform; southern Georgia and Florida. .... (p. 219) *Mittenothamnium*

147b Stem leaves close, julaceously imbricate when dry, ovate, acuminate; upper leaf cells 3-4:1, irregularly rhomboidal to oblong; British Columbia to California. .... (p. 158) *Pterogonium*

148a (144b) Plants small and slender, sparsely and irregularly branched; leaves small, to 0.6 mm long; paraphyllia few, small, filiform or branched. .... (p. 170) *Pterigynadrum*

148b Plants larger, not slender, stems to 20 cm long; paraphyllia large and abundant. 149

149a Leaves rugose (transversely wrinkled), plicate, falcate-secund, irregularly branched; Pacific northwest. .... (p. 220) *Rhytidiopsis*

149b Leaves not rugose or falcate; shoots 2-3 times pinnately branched in one plane, each year's growth arising from the middle of the preceding. .... (p. 221) *Hylocomium*

150a (140b) Plants with creeping, leafless stems and secondary stems erect, fern-like or dendroid. 151

150b Plants not conspicuously fern-like or dendroid. 152

151a On hardwood trees in Florida and Louisiana. .... (p. 159) *Pireella*

151b On trees and rocks, western United States. .... (p. 155) *Dendroalsia*

152a (150b) Leaves oval, rounded at apex, flattened with two rows of dorsal leaves and much smaller ventral leaves. Fig. 63 .... (p. 140) *Erpodium*

152b Leaves not as above. 153

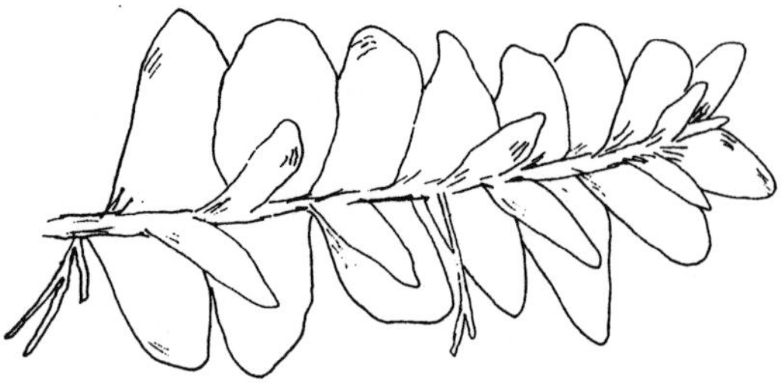

Figure 63

Figure 63 *Erpodium biserratum,* habit sketch showing flattened stem with dorsal (larger) leaves and ventral (smaller) leaves.

153a Leaves without a costa or with a short or long double costa. 154

153b Leaves with a single costa reaching the middle of the leaf or beyond in at least some leaves. 159

154a Costa long and double. .... (p. 163) *Callicostella*

154b Costa short and double to absent. 155

155a Median leaf cells linear-flexuose, 10-12:1, with a single row of minute papillae down the middle of the dorsal side; Florida. .... (p. 209) *Taxithelium*

155b Median leaf cells oval to short-rhomboidal, 1-8:1, papillae not in a single row or dorsal side. 156

156a Paraphyllia present, linear or lanceolate, not branched, few and small. .... (p. 171) *Heterocladium*

156b Paraphyllia absent. 157

157a Leaves ovate or narrower, acuminate to ovate-lanceolate; margins entire to serrulate; plants light olive- or yellow-green. 158

157b Leaves round-ovate, deeply concave, margins spinose-dentate from base to apex; plants glaucous-green. .... (p. 164) *Myurella*

158a Leaf margins serrulate from base to apex; median leaf cells 1-2:1, unipapillose; costa absent; widespread in the eastern United States. .... (p. 166) *Schwetschkeopsis*

158b Leaf margins slightly toothed at apex; median leaf cells 8:1, with 1 or 2 low papillae; costa absent or slender and single, ending about mid-leaf; Florida. .... (p. 157) *Leucodontopsis*

159a (153b) Plants complanate (stems flattened parallel to the substratum). .... (p. 205) *Stereophyllum*

159b Plants not complanate. 160

160a Plants with numerous capsules immersed in perichaetial leaves; costa protruding on dorsal side of leaf; leaf cells narrowly oval or elliptical with very thick walls; quadrate alar cells numerous; on trees. .... (p. 154) *Cryphaea*

160b Capsule when present exserted beyond the perichaetial leaves; costa and leaf cells not as above. 161

161a Median leaf cells with papillae forming a single row over the center of the cells; Florida and Louisiana. 162

161b Papillae not forming a single row over the center of the cells. 163

162a Leaves auriculate, decurrent; median leaf cells in rows oblique to the costa. Fig. 64. .... (p. 159) *Papillaria*

163a (161b) Leaf cells with a single very large papilla, as tall as the diameter of the cell, often forked into 2, 3 or 4; shoots julaceus; leaves nearly circular, deeply concave; United States east of the Plains. Fig. 65 .... (p. 164) *Thelia*

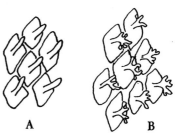

Figure 65

Figure 65 Papillae. A, *Thelia hirtella*; B, *T. asprella*.

163b Papillae smaller, often several on one cell. 164

164a Paraphyllia absent or few, linear, scale-like or lanceolate. 165

164b Paraphyllia numerous, filamentous, mostly branched. 172

165a Leaves ending in a hairpoint. 166

165b Leaves not ending in a hairpoint. 168

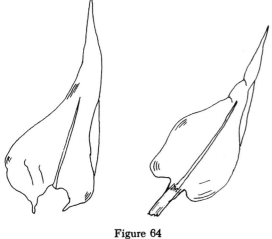

Figure 64

Figure 64 Auriculate, decurrent leaves of *Papillaria nigrescens*.

162b Leaves not auriculate or decurrent; median leaf cells paralleling the costa. .... (p. 160) *Barbella*

166a Plants usually growing on trunks of trees as scattered strands; leaves squarrose when moist; cells unipapillose. .... (p. 167) *Lindbergia*

166b Plants in dense mats on rocks and bases of trees, or on soil; unipapillose or pluripapillose. 167

Musci—The Mosses 51

167a Leaves entire. .... (p. 171) *Anomodon*

167b Leaves dentate-serrate; western United States. .... (p. 167) *Claopodium*

168a (165b) Leaves clasping the stem at base. 169

168b Leaves not clasping, mostly ovate, acute or obtuse, not at all complanate, closely appressed when dry, spreading when wet; peristome segments narrow, keeled. 170

169a Medium to large matted or tangled mosses. .... (p. 171) *Anomodon*

169b Mosses in small thin patches on bark; leaves much broken and appressed tightly against the stem when dry. .... (p. 171) *Haplohymenium*

170a (168b) Leaves with costa variable, ending below the middle, short or double or both or absent; cells with low papillae or even smooth. .... (p. 169) *Pseudoleskeella*

170b Leaves with a single costa usually extending beyond the middle of the leaf; cells papillose or smooth. 171

171a Median cells of leaves isodiametric, papillose; peristome teeth sharply bent inward at the base when dry. .... (p. 167) *Leskea*

171b Median cells longer than wide, oval-hexagonal, smooth or slightly papillose; teeth straight when dry. .... (p. 168) *Leskeella*

172a (164b) Paraphyllia matted together on stem and present on base of leaf. .... (p. 176) *Helodium*

172b Paraphyllia not attached to leaves. .. 173

173a Apical cell of branch leaves crowned with 2-4 papillae. .... (p. 174) *Thuidium*

173b Apical cell of branch leaves with one terminal papillae. .... (p. 174) *Haplocladium*

174a (139b) Paraphyllia or multicellular propagula present in leaf axils numerous and conspicuous. 175

174b Paraphyllia or multicellular propagula few or none, usually not seen, if numerous, then plants complanate. 182

175a Plants with creeping stems, erect scaly stems, and cluster of spreading leafy branches; dendroid in habit except in semi-aquatic situations. .... (p. 153) *Climacium*

175b Plants without a dendroid habit, if somewhat dendroid, growing on trees. 176

176a Leaves strongly wrinkled (rugose), all bent toward one side of stem (secund); tips of branches hooked. .... (p. 220) *Rhytidiopsis*

176b Leaves not both strongly rugose and secund. 177

177a Plants with complanate-foliate branches ending in a slender flagella; paraphyllia lanceolate, serrulate; capsule longer than seta. .... (p. 161) *Metaneckera*

177b Plants not complanate-foliate. 178

178a Costa single and strong, extending beyond the middle of the leaf. 179

178b Costa absent, single or double, not extending much beyond the middle of the leaf when present. 180

179a Costa very strong, ending in the tip (percurrent), or excurrent; alar cells large and clear. .... (p. 177) *Cratoneuron*

179b Costa not quite percurrent; alar cells small, quadrate. .... (p. 169) *Pseudoleskea*

180a (178b) Leaves revolute and strongly toothed in upper half. .... (p. 221) *Hylocomium*

180b Leaves plane and entire to serrulate in upper half. 181

181a Plants with slender creeping stems, naked or with a few scale-like leaves; secondary stems few, uncrowded, erect and simple; Florida. .... (p. 158) *Jaegerina*

181b Plants in mats, creeping stems not as above; secondary stems more numerous, ascending, sometimes flagelliform; British Columbia to California. .... (p. 155) *Alsia*

182a (174b) Costa double, ending well beyond the middle of the leaf; leaf cells thin-walled, pellucid, hexagonal; Florida. .... (p. 163) *Cyclodictyon*

182b Costa single and strong to short and/or double to absent. 183

183a Costa strong with one or two short accessory ribs on each side at base of leaf. .... (p. 157) *Antitrichia*

183b Costa single, reaching the middle of the leaf or beyond to lacking, or short and/or double. 184

184a Costa single, reaching the middle of the leaf or beyond. 185

184b Costa lacking, or short and/or double. 230

185a Plants aquatic, normally completely submerged, at least sometime of the year. 186

185b Terrestrial, on dry or wet substrates, not normally submerged. 192

186a Leaves complanate (lying in two opposite rows) or more or less falcate-secund. 187

186b Leaves not complanate or falcate-secund but appressed or erect or spreading. 189

187a Leaves complanate. .... (p. 179) *Leptodictyum*

187b Leaves more or less falcate-secund; shoots hooked at tip. 188

188a Leaves 3-ranked, sharply keeled (folded along the costa). ..... (p. 152) *Dichelyma*

188b Leaves not keeled, flat or tubulose, smooth or plicate. .... (p. 184) *Drepanocladus*

189a (186b) Leaves 3-ranked, sharply keeled (folded along the costa), lanceolate. .... (p. 151) *Brachelyma*

189b Leaves flat or concave, not keeled. .. 190

190a Leaves finely serrate all round, broadly ovate to orbicular. .... (p. 200) *Eurhynchium*

190b Leaves entire, flat or concave. 191

191a Leaves thin, cordate-ovate to rounded orbicular, costa strong, usually ending below the apex; alar cells inflated and usually forming a well-marked area. .... (p. 188) *Calliergon*

191b Leaves thick, opaque, lanceolate to ovate; costa very stout, excurrent. .... (p. 181) *Hygroamblystegium*

192a (185b) Plants with horizontal "rhizome", erect wiry stems and many leafy branches (dendroid). .... (p. 162) *Thamnobryum*

192b Plants not dendroid. 193

193a Leaves distinctly bordered with elongated, thick-walled cells in 2 layers; median cells rhomboid. .... (p. 182) *Sciaromium*

193b Leaves not distinctly bordered. 194

194a Leaves nearly circular, appressed, deeply concave, with abrupt slender tips; shoots fat and cylindric (julaceus). .. 195

194b Leaves narrowly lanceolate to ovate. 196

195a Apex ending abruptly in a short twisted point. .... (p. 198) *Bryoandersonia*

195b Apex ending in a long, narrow acumination, not twisted. .... (p. 198) *Cirriphyllum*

196a (194b) Leaves narrowly lanceolate, long-tapering, falcate-secund to circinate. .... (p. 184) *Drepanocladus*

196b Leaves ovate to lanceolate, rarely falcate, never circinate. 197

197a Median leaf cells short, 1-5:1. 198

197b Median leaf cells elongated, 5-20:1. .. 214

198a Cell walls very thick, often irregularly so being thicker at the upper and lower ends of the cells. Fig. 66A-E. 199

198b Cell walls thin, of equal thickness all round. Fig. 66F-I. 205

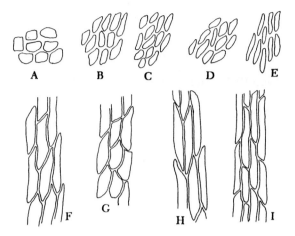

Figure 66

Figure 66 Cell wall types. A, *Neckeropsis*; B, *Stereophyllum*; C, *Clasmatodon*; D, *Forsstroemia*; E, *Isothecium*; F, *Brachythecium*; G, *Amblystegium*; H. *Rhynchostegium*; T, *Calliergon*.

199a Branches complanate. 200

199b Branches not complanate. 201

200a Leaves lingulate, truncate or notched at apex, or spathulate, asymmetrical, obtuse to rounded-obtuse. 213

200b Leaves oblong-ovate, acuminate; leaves dimorphous. ...... (p. 205) *Stereophyllum*

201a (199b) Leaves entire near base, very shallowly denticulate above. 202

201b Leaves sharply serrate above. 203

202a Leaves 1.5-2 mm long; seta shorter than perichaetial leaves. ....
.... (p. 155) *Forsstroemia*

202b Leaves 0.5-0.7 mm long; seta 3-4 mm long, much longer than perichaetial leaves. .... (p. 167) *Clasmatodon*

203a (201b) Leaves ovate, abruptly short-acuminate, margins plane and sharply serrate from base to apex; alar cells numerous and very small and thick-walled; Pacific slope only. ....
.... (p. 191) *Isothecium*

203b Leaves with tapering acumen, sharply serrate above. 204

204a Teeth of leaf margin broad, ascending; southeastern United States. ....
.... (p. 173) *Herpetineuron*

204b Leaves revolute below, sharply serrate above with long, slender, reflexed teeth; Pacific coast only. ....
.... (p. 157) *Antitrichia*

205a (198b) Leaves distinctly serrate above the middle; seta rough. 206

205b Leaves entire or slightly denticulate on lower half. 207

206a Branch leaves 0.9 mm or less long, scarcely plicate; capsule horizontal. ....
.... (p. 191) *Brachythecium*

206b Branch leaves 1.3-2 mm long, strongly plicate; capsule nearly erect; western United States. ....
.... (p. 190) *Homalothecium*

207a (205b) Leaves squarrose-recurved, especially when wet. .... (p. 177) *Campylium*

207b Leaves erect, spreading or appressed, not squarrose spreading. 208

208a Leaves nearly at right angles to stem. .... (p. 179) *Leptodictyum*

208b Leaves erect or appressed (imbricate). 209

209a Capsules erect; slender mosses on trees; leaves 0.8-1.5 mm long. 210

209b Capsules curved, strongly contracted under the mouth when dry. 211

210a Costa reaching the tip of leaf; cells small, 8-12 μ in diameter, dense. .... (p. 168) *Leskeella*

210b Costa weak, ending near the middle of leaf; cells 12 μ wide, thin-walled, clear. .... (p. 166) *Anacamptodon*

211a (209b) Costa very strong, extending to apex or beyond. .... (p. 181) *Hygroamblystegium*

211b Costa strong or weak, ending below the apex. 212

212a Margins entire. .... (p. 182) *Amblystegium*

212b Margins serrulate. .... (p. 200) *Rhynchostegiella*

213a (200a) Leaves truncate or notched at apex; Florida. .... (p. 161) *Neckeropsis*

213b Leaves short-obtuse to rounded-obtuse. .. .. (p. 162) *Homalia*

214a (197b) Shoots complanate-foliate. 215

214b Shoots not complanate-foliate. 216

215a Leaves entire, the base often obliquely attached to stem. .... (p. 179) *Leptodictyum*

215b Leaves sharply serrate, the apex twisted; seta smooth. .. (p. 200) *Rhynchostegium*

216a (214b) Alar cells thin-walled, clear, inflated. 217

216b Alar cells little or not at all enlarged; leaves acute or acuminate, straight or nearly so, not recurved. 221

217a Leaves entire, rounded at apex. .... (p. 188) *Calliergon*

217b Leaves distinctly pointed at apex. 218

218a Inflated alar cells broadly decurrent. Fig. 67B. .... (p. 191) *Brachythecium*

218b Inflated alar cells clustered, not decurrent. Fig. 67A 219

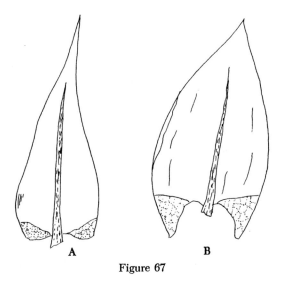

Figure 67

Figure 67   Leaves with inflated alar cells. A, *Drepanocladus;* B, *Brachythecium rivulare.*

219a Leaves acuminate. 220

219b Leaves blunt or with short tips. ....
.... (p. 186) *Hygrophypnum*

220a Leaves slenderly acute, falcate-secund, at least at the hooked tips of the stem. ....
.... (p. 184) *Drepanocladus*

220b Leaves widely spreading with a very long acumination, entire; in very wet places. .... (p. 177) *Campylium*

221a (216b) Leaves evenly tapering to a slender point, plicate. 222

221b Leaves ovate to lanceolate, with curved outlines, not plicate or with only two longitudinal folds. 223

222a Capsules oblong-cylindric, more or less curved; peristome perfect; large matted mosses, often yellowish. ....
.... (p. 190) *Tomenthypnum*

222b Capsules erect and symmetric; inner peristome imperfect, even the segments reduced. .... (p. 190) *Homalothecium*

223a (221b) Alar cells little or not at all differentiated; beak of operculum as long as urn; plants generally glossy in appearance. 224

223b Alar cells quadrate; beak conic or long-conic; plants not glossy. 225

224a Apical cells of leaf short-rhomboidal to circular. .... (p. 200) *Eurhynchium*

224b Apical cells not differentiated; seta smooth; robust pinnately branched plants; leaves clasping the stem. ....
.... (p. 201) *Stokesiella*

225a (223b) Leaves with a large area of thick- or thin-walled alar cells. 226

225b Alar cells fewer, not occupying a large area at the base of the leaf; sometimes inflated. 227

226a Alar cells thick-walled, rounded; cilia of peristome well-developed. ....
.... (p. 191) *Isothecium*

226b Alar cells thin-walled, quadrate, extending from margin to midrib; cilia of peristome lacking; capsules erect. ....
.... (p. 191) *Brachythecium*

227a (225b) Small glossy mats on trees; leaves about 1 mm long; serrate; leaves ovate to elliptical-oblong, abruptly acuminate; calyptra hairy; inner peristome adherent; eastern United States. .... (p. 191) *Homalotheciella*

227b Plants larger, not possessing the above characteristics. 228

228a Branches julaceus (cylindrical, densely and closely leafed); leaves smooth, concave; seta rough; western North America. .... (p. 198) *Scleropodium*

228b Branches not julaceus; if sub-julaceus the leaves plicate. 229

229a Leaves lanceolate, small quadrate alar numerous, not plicate, sharply serrate from base to apex. .... (p. 190) *Homalothecium*

229b Leaves ovate or ovate-lanceolate, small alar cells not numerous; if leaves lanceolate or quadrate alar cells numerous, not sharply serrate from base to apex, smooth or plicate. .... (p. 191) *Brachythecium*

230a (184b) Plants aquatic, long, and dangling, often in dense tufts; alar cells inflated. .... (p. 149) *Fontinalis*

230b Plants terrestrial; xeric, mesic, or hydric, not in areas where subject to long periods of submergence in flowing water or if growing in water, costa short and double. 231

231a Median leaf cells short, 2-5:1. 232

231b Median leaf cells long to very long, 5-20:1. 239

232a Cell walls very thick, the lumen elliptic to linear; alar cells very numerous. Fig. 66 A-E. 233

232b Cell walls thin, equally thick all around. Fig. 66 F-G. 234

233a Secondary stems little branched; calyptra smooth. .... (p. 156) *Leucodon*

233b Secondary stems freely and often pinnately branched; calyptra hairy. .... (p. 155) *Forsstroemia*

234a (232b) Leaves entire. 237

234b Leaves finely denticulate to coarsely toothed. 235

235a Leaves denticulate to very finely denticulate. 236

235b Leaves coarsely toothed with whole projecting cells to serrate-dentate. .... (p. 165) *Fabronia*

236a Branches not flattened. .... (p. 177) *Campylium*

236b Branches flattened; Florida. .... (p. 218) *Vesicularia*

237a (234a) Cells very large and clear, 50-60 μ; alar cells not differentiated; plants of moist, wet, rock ledges. .... (p. 163) *Hookeria*

237b Cells smaller, 7-8 μ wide; alar cells small and numerous. 238

238a Leaves minute, to 0.6 mm long. .... (p. 183) *Platydictya*

238b Leaves larger, 0.8-1.2 mm long, concave. .... (p. 211) *Homomallium*

239a (231b) Leaves with one inflated cell at extreme basal angle, and many quadrate cells; tips of shoots crowded with gemmae. .... (p. 209) *Platygyrium*

239b Leaves not as above; gemmae if present, not forming clusters at the tips of shoots. 240

240a Leaves with several (2 or more) distinctly inflated alar cells. 241

240b Leaves with small quadrate alar cells or with little or no differentiation of alar cells or alar cells slightly inflated and hyaline. 252

241a Alar cells in a transverse row of 3 or 4 inflated cells; operculum with a beak as long as urn. Fig. 68A .... (p. 208) *Sematophyllum*

241b Alar cells in a cluster, inflated, 3 or 4 transversely and 3 or 4 up margin of leaf, or less inflated, thicker walled, often colored. Fig. 68B. 242

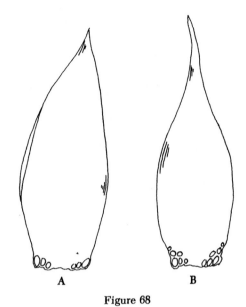

Figure 68   Alar cells. A, *Sematophyllum*; B, *Brotherella*.

242a Alar cells inflated, thin-walled, in a cluster, 3 or 4 transversely and 3 or 4 up margin. 243

242b Alar cells more numerous, less inflated, thicker walled, often colored. 244

243a Alar cells scarcely decurrent; leaves complanate and falcate-secund, sharply serrate on the slender acumination; capsules inclined smooth. .... (p. 209) *Brotherella*

243b Alar cells decurrent; leaves squarrose, finely serrulate above; capsule longitudinally furrowed when dry. .... (p. 218) *Herzogiella*

244a (242b) Leaves squarrose-recurved, long acuminate, entire. .... (p. 177) *Campylium*

244b Leaves not as above. 245

245a Leaves erect, straight, broad, blunt, sometimes more or less falcate. 246

245b Leaves complanate. 249

246a Leaves curved so that shoots are hooked at tips, rugose (wrinkled) when dry. .... (p. 189) *Scorpidium*

246b Leaves not as above. 247

247a Leaves ovate to orbicular, acute or abruptly short-acuminate, spreading, more or less falcate-secund. .... (p. 186) *Hygrohypnum*

247b Leaves deeply concave, spoon-shaped, obtuse and rounded at apex, not falcate-secund. 248

248a Tips of stems ending in a firm acute bud; alar cells much inflated, clear. .... (p. 189) *Calliergonella*

248b Tips loose; stems red; alar cells somewhat inflated, usually colored and opaque. .... (p. 204) *Pleurozium*

249a (245b) Leaves shiny, not falcate-secund; alar cells in a cluster, enlarged and clear. 250

249b Leaves falcate-secund. 251

250a Leaves entire. ...... (p. 216) *Callicladium*

250b Leaves serrate, especially on the slender acumen. .... (p. 207) *Heterophyllium*

251a (249b) Plants rather large, handsome, distinctly pinnately branched to form triangular fronds; leaves strongly plicate. .... (p. 219) *Ptilium*

251b Plants small to medium-sized, not particularly handsome in appearance, variously branched, but not forming triangular fronds; leaves smooth to only slightly plicate. .... (p. 211) *Hypnum*

252a (240b) Leaves with small quadrate alar cells. 253

252b Leaves with little or no differentiation of alar cells, or alar cells slightly inflated and hyaline. 260

253a Leaves appressed, broad and entire; shoots julaceous to slightly flattened; peristome teeth with few, long, smooth joints. .... (p. 203) *Entodon*

253b Leaves spreading to loosely imbricate, complanate or falcate secund. 254

254a Leaves spreading. 255

254b Leaves complanate, falcate-secund, or loosely imbricate. 256

255a Leaves slenderly acute, finely denticulate; stem leaves much larger than branch leaves. .... (p. 220) *Rhytidiadelphus*

255b Leaves nearly orbicular, rounded and obtuse and denticulate at apex; cilia present in peristome. .... (p. 186) *Hygrophypnum*

256a (254b) Leaves loosely imbricate and crowded; plants rather stout and robust, sparcely branched. .... (p. 189) *Scorpidium*

256b Leaves complanate to falcate secund. ... 257

257a Branches bent upward by reason of upwardly pointed leaves; cilia of peristome lacking. .... (p. 210) *Pylaisiella*

257b Leaves curved downward or backward, or not curved and shoots very flat (complanate-foliate). 258

258a Shoots complanate-foliate. .... (p. 203) *Entodon*

258b Shoots curved downward or backward. ... 259

259a Shoots essentially cylindric, but the large leaves falcate. .... (p. 220) *Rhytidiadelphus*

259b Shoots flat, the leaves both complanate and falcate. .... (p. 211) *Hypnum*

260a (252b) Leaves imbricate to erect open. ... 261

260b Leaves complanate-foliate or falcate-secund with branches hooked at tip. 263

261a Stems and branches long, slender and stoloniferous at the ends and brearing small distant leaves. .... (p. 186) *Hygrohypnum*

261b Stems and branches not as above. 262

262a Leaves deeply plicate. .... (p. 202) *Orthothecium*

262b Leaves not deeply plicate. .... (p. 205) *Plagiothecium*

263a (260b) Leaves falcate-secund, the branches hooked at tips. .... (p. 220) *Rhytidiadelphus*

263b Leaves complanate-foliate, not falcate. ... 264

264a Leaves ovate to ovate-lanceolate; cells thin-walled, spindle-shaped. 265

264b Leaves ovate to oblong; cells thick-walled, linear-flexuose. .... (p. 160) *Neckera*

265a Leaves decurrent. .... (p. 205) *Plagiothecium*

265b Leaves not decurrent. 266

266a Pseudoparaphyllia present, leaf-like. Fig. 69A. .... (p. 217) *Taxiphyllum*

266b Pseudoparaphyllia absent or if present, filamentous. Fig. 69B 267

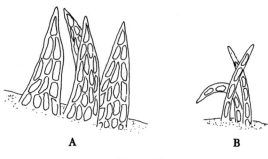

Figure 69

Figure 69 Pseudoparaphyllia. A, *Taxiphyllum*; B, *Isopterygium*.

267a Outer layer of stem cells large and thin-walled in cross section; leaves entire or minutely serrulate. Fig. 70A .... (p. 216) *Isopterygiopsis*

267b Outer layer of stem cells small and thick-walled in cross section; leaves serrulate to strongly serrate in upper half. Fig. 70B .... (p. 216) *Isopterygium*

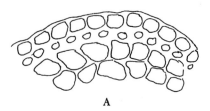

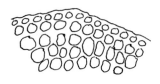

Figure 70

Figure 70 Stem cells in cross section. A, *Isopterygiopsis*; B, *Isopterygium*.

SUBCLASS 3. BRYIDAE
Order Fissidantales

FISSIDENTACEAE
Fissidens

1a Plants submerged in running water. 2

1b Plants on earth, trees, or rocks, not normally submerged. 4

2a Leaves large (about 2 mm), stiff, 2-layered in spots, bordered with elongated cells; border and costa reddish. Fig. 71. .... *Fissidens ventricosus* Lesq.

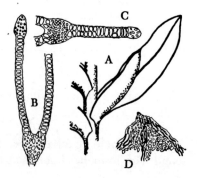

Figure 71

Figure 71. *Fissidens ventricosus*. A, leaf; B, section of leaf below middle; C, section of dorsal lamina; D, apex of leaf.

A large (1-3 cm) coarse aquatic plant, on rocks in streams, California, Oregon, Washington, Idaho, British Columbia. Rare.

2b Leaves not bordered. 3

3a Main portion of leaf cartilaginous, of 4 or more layers of cells; stems in dense tufts. Fig. 72A-B. ....
.... *Fissidens grandifrons* Brid.

Figure 72

Figure 72 *Fissidens grandifrons*. A, plant; B, cross section of leaf; C, plant of *F. fontanus*; D, cross section of leaf.

A stiff, dark green moss, 2-12 cm long, in swiftly flowing calcareous water, widespread in the northern States and Canada south to Tennessee, Alabama, Arkansas, Utah and California.

3b Leaves thin, slender and soft; stems much branched, 5 to 10 cm long. Fig. 72C-D. ....
.... *Fissidens fontanus* (B.-Pyl.) Steud.

A limp and soft moss commonly on rocks and woods in streams, Washington, Oregon, Idaho, California east to Ontario, south to Texas and Florida.

4a Leaves bordered at least in part with long, narrow cells. 5

4b Leaves not bordered with long cells. .. 11

5a Border confined to vaginant lamina (often only on perichaetial leaves). 6

5b Border extending to the upper part of the leaf. 8

6a Leaf cells papillose. 7

6b Leaf cells smooth. Fig. 73 A-B. ....
.... *Fissidens obtusifolius* Wils.

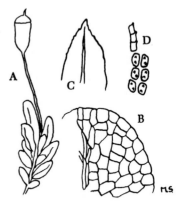

Figure 73

Figure 73 *Fissidens obtusifolius*. A, plant; B, apex of leaf; C, apex of leaf of *F. ravenelii*; D, cells of same.

Small to minute glaucous mosses growing on moist, shaded rocks, often along the edges of creeks and streams, Idaho, Michigan, Minnesota, Indiana south to Florida, Texas and Arizona.

7a Costa ending 2-several cells below apex. Fig. 74C-E. ....
.... *Fissidens garberi* Lesq. & James

Small plants, 2-4 mm high, on rocks, rotten wood, Missouri, Illinois south to Texas, Louisiana, and Florida.

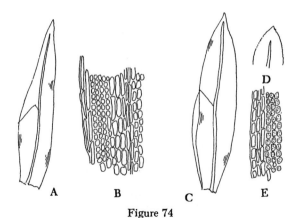

Figure 74

Figure 74 *Fissidens keglianus*. A, leaf; B, vaginant lamina showing enlarged cells near costa. *Fissidens garberi*. C, leaf; D, apex of leaf; E, border on lower vaginant lamina.

7b Costa percurrent (ending at apex) to excurrent (extending beyond apex). Fig. 73C-D. .... *Fissidens ravenelii* Sull.

Small plants, 2-4 mm tall, on shaded soil, rotten wood, and rocks, North Carolina to Missouri and Arkansas, south to Florida, Louisiana, and Texas.

8a Cells of vaginant lamina larger, especially along costa. Fig. 74A-B. ....
.... *Fissidens kegelianus* C. Muell.

Small plants, 2-5 mm tall, on limestone and calcareous soil, Florida.

8b Cells of vaginant lamina not larger. 9

9a Costa usually joining with borders of leaf; antheridia conspicuous in axils of leaves. Fig. 75E. ....
.... *Fissidens bryoides* Hedw.

Small plants, 2-7 mm tall, on moist soil and rock, often along and in small creeks, occasionally on rotten wood, widely distributed in North America.

9b Border not reaching the apex. 10

10a Lower inner margin of leaf bordered with small quadrate cells outside of the long narrow cells; western. Fig. 75A-D. .... *Fissidens limbatus* Sull.

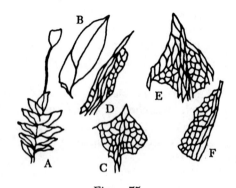

Figure 75

Figure 75 *Fissidens limbatus*. A, plant; B, leaf; C, apex of leaf; D, margin of vaginant lamina; E, apex of leaf of *F. bryoides*; F, margin of vaginant lamina of *F. viridulus*.

Plants small, 5-20 mm tall, on moist shaded soil, British Columbia south to California and New Mexico.

10b Leaves not so bordered, mainly eastern North America. Fig. 75F. ....
.... *Fissidens viridulus* (Sw.) Wahlenb.

Plants small, 3-6 mm tall, on moist shaded soil and rocks, widely distributed in eastern North America from southern Canada south to the Gulf States.

11a (4b) Margins of leaves entire to slightly crenulate-serrate. 12

11b Margins of leaves coarsely to finely and evenly serrate. 14

12a Margins entire; seta terminal; stems 2-3 mm tall. Fig. 73A-B. ....
.... (p. 63) *Fissidens obtusifolius* Wils.

12b Margins entire or slightly wavy; seta axillary or terminal; stems 1-5 cm long. ..
.... 13

13a Leaves changing only slightly upon drying with only a few leaves becoming curled and rolled; leaf cells do not bulge conspicuously on both surfaces. Fig. 76C-D. ....
.... *Fissidens polypodioides* Hedw.

Plants 2-5 cm high, a rich green in color, on shaded calcareous soil and rocks, Tennessee, North Carolina south to Florida and Texas.

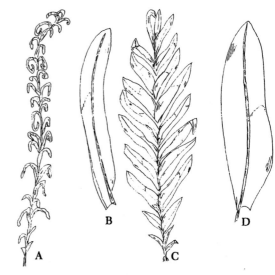

Figure 76

Figure 76 *Fissidens asplenioides*. A, habit of dry plant; B, leaf. *Fissidens polypodioides*. C, habit of dry plant; D. leaf.

13b Leaves conspicuously curled and rolled when dry; leaf cells conspicuously bulging on both surfaces. Fig. 76A-B. ....
.... *Fissidens asplenioides* Hedw.

Plants 1-2.5 cm tall, on moist soil and vertical rocks, North Carolina, South Carolina, Mississippi and Louisiana.

14a (11b) Margins of leaf coarsely and irregularly (doubly) toothed near apex. ....
.... 15

14b Margins finely and evenly serrate. 17

15a Costa ending several cells below apex and covered with low mammillose cells. Fig. 77E. .. *Fissidens subbasilaris* Hedw.

Plants 5-10 mm tall, leaves curved downward when dry, on bases of trees and rocks, usually shaded, Ontario south to Florida, Louisiana and Texas, west to Illinois, Missouri and Oklahoma.

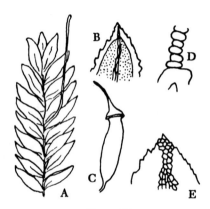

Figure 77

Figure 77 *Fissidens cristatus*. A, plant; B, apex of leaf; C, capsule; D, section of dorsal lamina; E, apex of leaf of *F. subbasilaris*.

15b Costa percurrent and not covered with mammillose cells. 16

16a Leaf cells obscure, enlarged and elevated tumid cells appearing singly or in groups in surface view near the apex of the leaf; leaf cells 7-9 μ in diameter; border of paler or thicker-walled cells often distinct; leaves crowded on stem, rolled inward from tips when dry. Fig. 77A-D, 78C. .... *Fissidens cristatus* Wils. ex Mitt.

Plants 1-3 cm long, on soil, bark, humus, bases of trees, and rocks, throughout eastern North America.

16b Leaf cells clear, tumid cells mostly absent; leaf cells 13-16 μ in diameter; border of leaves indistinct; leaves distant on stem, curving underneath stem from tips when dry. Fig. 78A-B. ....
.... *Fissidens adiantoides* Hedw.

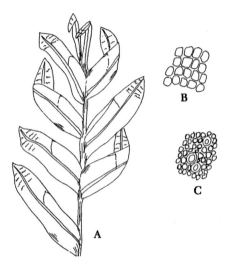

Figure 78

Figure 78 *Fissidens adiantoides*. A, plant; B, cells of upper leaf. *Fissidens cristatus*. C, upper leaf cells showing tumid cells.

On moist soil and rock, especially along streams, Alaska to Labrador south to California and Florida.

17a Costa ending several cells below apex; sporophyte terminal. Fig. 79D. ....
.... *Fissidens osmundoides* Hedw.

Plants 5-12 mm tall (rarely larger), on soil, humus or logs, in woods and along streams, Alaska to Nova Scotia south to Oregon and the Gulf of Mexico.

17b Costa percurrent to shortly excurrent; sporophyte lateral. 18

18a Costa stout, usually filling the apex, short-excurrent; cells of upper leaf conically papillose or mammillose. Fig. 79A-B. .... *Fissidens taxifolius* Hedw.

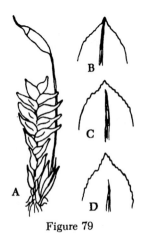

Figure 79

Figure 79 *Fissidens taxifolius*. A, plant; B, apex of leaf; C, apex of *F. bushii*; D, of *F. osmundoides*.

Plants small, 4-8 mm tall, on calcareous soil and rocks, bases of trees; occasionally on tree trunks, British Columbia to California and Arizona, widespread in eastern North America south to the Gulf of Mexico.

In the absence of sporophytes the apex and crenulation of the leaf, and length of midrib will help to distinguish the species. *F. taxifolius* has cells on the outside of the vaginant lamina prominently bulging, as seen in cross sections; the other species are relatively smooth. Typically *F. taxifolius* has the midrib excurrent as a cylindrical mucro. *F. bushii* has the upper margin (typically) slightly unequally toothed, suggesting *F. cristatus*, and the midrib ending 5 or 6 cells below the apex of the leaf. Many plants, especially young ones, cannot be identified with certainty.

18b Costa slender, not filling the apex, percurrent; cells of upper leaf pluripapillose. Fig. 79C. .... *Fissidens bushii* (Card. & Thér.) Card. & Thér.

On clayey soils of banks and forest openings, southern Manitoba and southern Ontario south to the Gulf of Mexico.

Order Archidiales

ARCHIDIACEAE
Archidium

Treatment of this genus is based upon a recent revision by Jerry A. Snider (Journ. Hattori Bot. Lab. 39:105-201. 1975).

1a Plants with archegonia and antheridia on separate short lateral branches (autoicous). Fig. 80. .... **Archidium ohioense Schimp. *ex* C. Muell.**

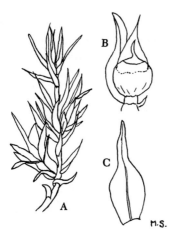

Figure 80

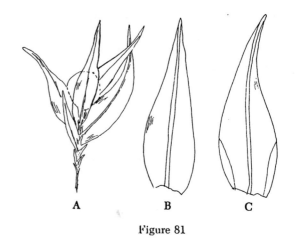

Figure 81

Figure 80 *Archidium ohioense.* A, plant; B, capsule; C, leaf.

Plants 2-20 mm high, on moist sandy or loamy soil, Minnesota to New England south to Florida and Texas.

1b Plants with antheridia in axils of perichaetial leaves below archegonia (paroicous). Fig. 81. ....
 Archidium alternifolium (Dicks. *ex* Hedw.) Schimp.

Figure 81 *Archidium alternifolium.* A, Plant with globose sporangium; B, stem leaf; C, perichaetial leaf.

Plants 3-20 mm tall, on moist sandy or loamy soil, especially around temporary pools and banks of roadside ditches, Missouri, Illinois east to Ohio and North Carolina, south to Florida, Louisiana, and Texas. Five other species are known in the eastern United States, mostly in the southeast and Texas.

Order Dicranales

DITRICHACEAE
Pleuridium

Pleuridium subulatum (Hedw.) Rabenh. (Fig. 82)

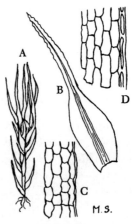

Figure 82

Figure 82 *Pleuridium subulatum.* A, plant; B, leaf enlarged; C, cells of leaf; D, ?

Small plants, 3-7 mm high, found usually in the spring on soil of old fields, pastures, and rocky ledges, Nova Scotia to Minnesota south to Gulf States and Arizona. There are five other species of *Pleuridium* in North America.

Trichodon

Trichodon cylindricus (Hedw.) Schimp. (Fig. 83)

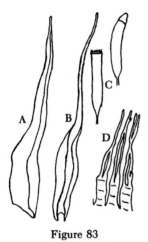

Figure 83

Figure 83 *Trichodon cylindricus.* A, median leaf; B, upper leaf; C, capsules; D, teeth.

Plants small, to 5 mm high, leaves with a very rough awn, on clay banks, Northwestern North America, south to Oregon and with disjunct populations in Michigan and Labrador.

Ditrichum

1a Leaves long, linear-subulate, more or less erect to loosely spreading, awn slender, as long as to much longer than broad base; apex acute. 2

1b Leaves not long, linear-subulate, appressed; awn shorter than broad base; apex obtuse. Fig. 84 and 86E. .... *Ditrichum lineare* (Sw.) Lindb.

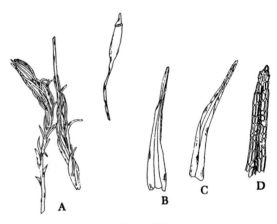

Figure 84

Figure 84 *Ditrichum lineare.* A, Gametophyte with attached sporophyte; B-C, leaves; D, apex of leaf.

Plants small, 5-10 mm high, on base mineral soils, frequently on road banks, Nova Scotia to Michigan south to Alabama and Missouri.

2a Seta bright yellow. Fig. 85. .... *Ditrichum pallidum* (Hedw.) Hampe

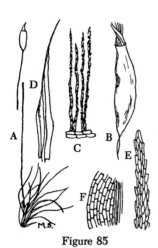

Figure 85

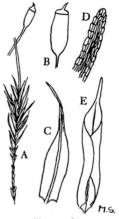

Figure 86

Figure 85 *Ditrichum pallidum*. A, plant; B, capsule; C, peristome; D, leaf; E, apex of leaf; F, basal cells of leaf.

Plants small with stem to 6 mm long, on dry bare soils in woods, open fields, and road sides, occasionally on thin soil over rocks, common throughout eastern North America from Labrador and southern Canada south to the Gulf States.

2b   Seta red. 3

3a   Plants 5-10 mm tall; leaves erect to somewhat spreading, margins more or less revolute. Fig. 86A-D. ....
 .... *Ditrichum pusillum* (Hedw.) Hampe.

Figure 86 *Ditrichum pusillum*. A, plant; B, capsule; C, leaf; D, apex of leaf; E, perichaetial leaf of *D. lineare*.

Small plants growing on bare mineral soil of shaded banks or on thin soil over rocks, Alaska, British Columbia and eastern North America from Labrador to the Gulf States.

3b   Plants 2-10 cm tall; leaves usually bent to one side (secund), margins plane or incurved. Fig. 87. .... *Ditrichum flexicaule* (Schwaegr.) Hampe

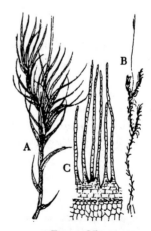

Figure 87

70   Musci—The Mosses

Figure 87 *Ditrichum flexicaule.* A, plant, enlarged; B, plant natural size; C, peristome.

Large plants in dense tomentose mats, on alkaline soils of moist shaded banks and ledges. Alaska, British Columbia, Ontario, Michigan, Iowa, Colorado and Nova Scotia to New England. In the Pacific Northwest a similar species occurs, *Ditrichum heteromallum* (Hedw.) Britt. It is distinguished from *D. flexicaule* by the presence of long marginal cells near the middle of the leaf. In *D. flexicaule* these cells are short, isodiametric or transversely elongate. In the eastern United States another species with red setae may be encountered, *Ditrichum rhynchostegium* Kindb. It otherwise looks very similar to *D. pallidum.*

Saelania

Saelania glaucescens (Hedw.) **Bomanss. & Broth.** (Fig. 88)

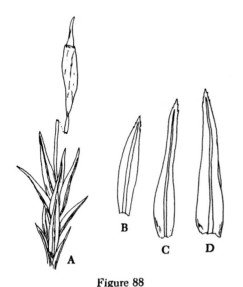

Figure 88

Figure 88 *Saelania glaucescens.* A, gametophyte with sporophyte; B-D, leaves.

Small plants, 4-15 mm high, bluish-green or glaucous because of a fine, whitish, mealy or filamentous covering on leaves, usually growing on calcareous soil of steep banks especially beneath overhanging turf, also on rocks or soil in rock crevices, Alaska, British Columbia to Ontario and Nova Scotia south to Arizona, Colorado, Iowa, Michigan, and New Jersey.

Ceratodon

Ceratodon purpureus (Hedw.) Brid. (Fig. 89)

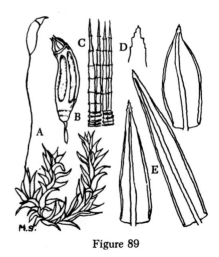

Figure 89

Figure 89 *Ceratodon purpureus.* A, *plant;* B, capsule; C, peristome; D, apex of leaf; E, extreme forms of leaf.

A cosmopolitan species charaterized by leaves with revolute margins below and plane, slightly toothed at the apex, with inclined, furrowed capsules, red-purple in color; the peristome is also red-purple in color, common on sterile soils, rock walls, sidewalks, and even in lawns, may even grow in swamps. Extremely variable in size and shape of leaves.

Distichum

1a Capsules cylindric, symmetric, and erect; spores 16-25 μ. Fig. 90A-C. ....
Distichum capillaceum (Hedw.) B.S.G.

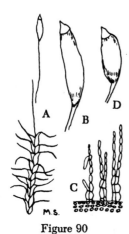

Figure 90

Figure 90 *Distichium capillaceum.* A, plant; B, capsule; C, peristome; D, capsule of *D. inclinatum.*

Plants 1-6 cm high, in silky tufts, leaves 2-ranked with a broad clasping base and filiform apex; in cool rocky places on rocks and in crevices, Alaska to Nova Scotia and New England south to California, Nevada, Arizona, Michigan, Iowa, and New York.

1b Capsules ovoid-cylindric, asysmetric, inclined to horizontal; spores 30-45 μ. Fig. 90D. ....
.. *Distichum inclinatum* (Hedw.) B.S.G.

Plants 0.5 to 1.5 cm high, leaves 2-ranked but not as clearly as in the preceeding species, on soil or soil over rocks, humus, of sheltered banks, usually in calcareous regions from British Columbia to Quebec south to California, Nevada, Minnesota, Michigan and New England.

BRYOXIPHIACEAE
Bryoxiphium

Bryoxiphium norvegicum (Brid.) Mitt. (Fig. 91)

Figure 91

Figure 91 *Bryoxiphium norvegicum.* A, plant; B, leaf; C. capsule.

On shaded sandstone cliffs and under overhanging ledges, usually in dense, closely adhering sheets. Washington to Minnesota, Wisconsin, Ohio south to Arkansas, Alabama, Tennessee, and North Carolina, and a disjunct population in Arizona.

SELIGERIACEAE
Seligeria

Based on a revision of this genus by Dale H. Vitt (Lindbergia 3:241-275. 1976).

1a Capsules without a peristome; margins of leaves denticulate in the lower 1/3.
.... *Seligeria donniana* (Sm.) C. Muell.

Plants minute to very small, often less than 1 mm high, on moist limestone and sandstone rocks, Washington, British Columbia, Yukon to Michigan and Ontario south to Pennsylvania with disjunct populations in Iowa and Missouri.

1b Capsules with a peristome; margins of leaves entire to serrulate. 2

2a Costa excurrent, filling the subulate apex. Fig. 92E. ....
   .... *Seligeria calcarea* (Hedw.) B.S.G.

Plants minute to very small, often less than 1 mm high, on moist limestone rocks, New York and southern Ontario west to southern Manitoba south to Minnesota, Arkansas and Tennessee.

3a Plants light-green; cells of leaf pellucid; capsule tending to be widest at mouth. Fig. 92D. ....
   .... *Seligeria pusilla* (Hedw.) B.S.G.

Minute, delicate, yellow-green plants growing on limestone, Wisconsin, Minnesota south to Iowa, Indiana, Missouri, Arkansas, and Tennessee.

3b Plants olive- to dark-green; cells of leaf opaque; capsule not usually widest at mouth. Fig. 92A-C. ....
   .... *Seligeria campylopoda* Kindb.

Minute plants with entire, lanceolate-ligulate leaves ending in an obtuse tip, growing on damp rocks, usually limestone, British Columbia, Washington, Alberta to Michigan, Ontario southward to Montana, Michigan, Ohio and Tennessee.

Blinda

Blinda acuta (Hedw.) B.S.G. (Fig. 93)

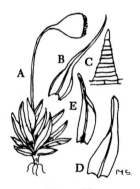

Figure 92

Figure 92 *Seligeria campylopoda*. A, plant; B, leaf; C, peristome tooth; D, leaf of *S. pusilla*; E, leaf of *S. calcarea*.

2b Costa (except in perichaetial leaves) ending just below the apex. 3

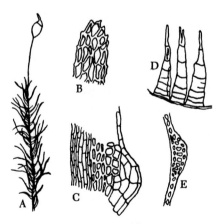

Figure 93

Figure 93 *Blinda acuta*. A, plant; B, apex of leaf; C, base of leaf; D, teeth; E, section of midrib.

Plants 1-10 cm high in glossy, loose tufts, dark green to blackish, growing on wet rocks, Alaska, British Columbia to Ontario, Labrador south to California, Colorado, Wisconsin, Michigan, Tennessee, and South Carolina.

DICRANACEAE
Bruchia

Twelve species of this small "pigmy moss" are recorded for North America. Most are small, not over 3 mm tall, gregarious, and fruiting the spring.

1a Calyptra papillose to spinose. Fig. 94D. .... ***Bruchia ravenelii*** Wils. ex Sull.

Usually on sandy soil, South Carolina, to Florida, Louisiana, and Texas.

1b Calyptra smooth. 2

2a Seta shorter than capsule. 3

2b Seta longer than capsule. 4

3a Neck equal to or longer than capsule. Fig. 94E; 95C. .... ***Bruchia brevifolia*** Sull.

On moist sandy soil, North Carolina to Texas.

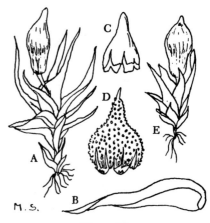

Figure 94

Figure 94 *Bruchia sullivanti*. A, plant; B, leaf; C, calyptra; D, calyptra of *B. ravenelii*; E, plant of *B. brevifolia*.

3b Neck shorter than the capsule. Fig. 94A-C; 95B. .... ***Bruchia sullivantii*** Aust.

Sandy soil of bank, old fields, and on thin soil over rock exposures, Minnesota to Maine south to Florida and Texas.

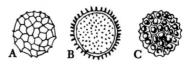

Figure 95

Figure 95 *Bruchia texana*. A, spore; *B. ravenelii* is similar; B, spore of *B. sullivanti*; *B. flexuosa* is similar; C, spore of *B. brevifolia*.

4a Spores spinose; neck 1/2 the length of the capsule. Fig. 45. .... ***Bruchia flexuosa*** (Sw. ex Schwaegr.) C. Muell.

Habitat and range the same as the preceding, with which it often grows.

4b Spores reticulate, neck reaching 1/2 the length of capsule. Fig. 95A. .... ***Bruchia texana*** Aust.

Sandy soil of banks, old field, and open areas, Illinois to Maryland south to Georgia and Texas.

Trematodon

1a Neck about twice as long as the urn. Fig. 96A-D. .... ***Trematodon longicollis*** Michx.

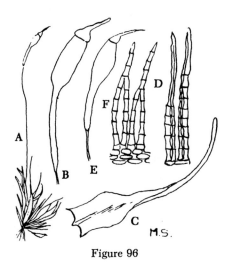

Figure 96

Figure 96 *Trematodon longicollis*. A, plant; B, capsule; C, leaf; D, teeth of peristome; E, capsule of *T. ambiguus;* F, peristome.

Plants gregarious, up to 1 cm tall, on moist sandy or clayey soil of banks, old fields, waste places, southern Ontario, Illinois to Massachusetts south to Florida and Texas.

1b Neck about 1/2 the length of the urn. Fig. 96E-F. .... ***Trematodon ambiguus*** (Hedw.) Hornsch.

Gregarious plants up to 1 cm tall, leaves 2-4 mm long growing on sanding or clayey soils of old fields and meadows, Alaska to Newfoundland south to British Columbia, Wisconsin, Michigan, Pennsylvania, New York and West Virginia. *T. boasii* Schof. occurs in British Columbia and may be separated from *T. ambiguus* by its undivided peristome teeth and shorter leaves.

Dicranella

There are twelve species of this genus reported from North America. Most are small, growing in loose, green, yellowish, or brownish tufts with leaves that tend to be erect-flexuose to secund, at least above.

1a Leaves squarrose from an enlarged, clasping base. 2

1b Leaves not squarrose, base not enlarged and clasping. 3

2a Leaves entire except for the slightly crenulate, broad, obtuse apex, bases slightly decurrent. .. ***Dicranella palustris*** (Dicks.) Crundw. *ex* Warb.

Plants 2-10 cm tall, growing in streams, bogs, and wet places, British Columbia, Washington and Oregon to Nova Scotia and New England.

2b Leaves serrulate at the acute to ± obtuse apex and slightly serrate at the top of the sheathing base, bases not decurrent. Fig. 97. .... ***Dicranella schreberiana*** (Hedw.) Schimp.

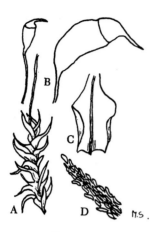

Figure 97

Figure 97 *Dicranella schreberiana.* A. plant; B, capsule with lid; C, base of leaf; D, apex of leaf.

Plants in silky tufts, 2-3 cm tall, on damp soil, Alaska, British Columbia to Labrador, Quebec and south to Oregon, South Dakota, Ohio, New York, Pennsylvania and New Jersey.

3a Apex of most leaves mostly blunt, rounded with costa ending below it. Fig. 98. .... *Dicranella hiliariana* (Mont.) Mitt.

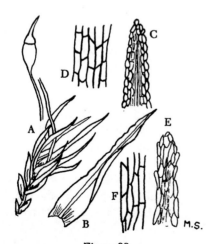

Figure 98

Figure 98 *Dicranella hilariana.* A, plant; B, leaf; C, apex of leaf; D, median cells of leaf; E, leaf apex, and F, median cells.

Plants in pale yellowish-green mats, 2-5 mm tall, on sandy soils, southern Coastal Plain from Florida to Texas.

3b Apex of leaves mostly acute, costa percurrent to excurrent. 4

4a Mature capsules erect, symmetric or nearly so, seta red-brown. Fig. 100E, G. .. .... *Dicranella rufescens* (With.) Schimp.

Plants small, 5-10 mm tall, yellow-green to reddish, on moist soil of banks and in open areas in woodlands, British Columbia south to Oregon to northeastern Canada south to Missouri and Tennessee.

4b Mature capsules inclined, unsymmetric; if erect then distinctly furrowed when dry. 5

5a Capsule plicate or furrowed when dry and empty; seta yellow. Fig. 99. .... *Dicranella heteromalla* (Hedw.) Schimp.

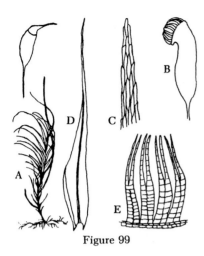

Figure 99

Figure 99 *Dicranella heteromalla.* A, plant; B, capsule; C, apex of leaf; D, leaf; E, peristome.

Plants small, yellow to dark green in shiny tufts, usually about 1 cm tall, on bare soil of roadside, waste places, edge of woods, widespread in North America.

5b Capsules smooth or only slightly furrowed when dry and empty; seta reddish. Fig. 100A-D, F. ....
.... *Dicranella varia* (Hedw.) Schimp.

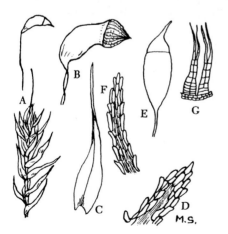

Figure 100

Figure 100 *Dicranella varia.* A, plant; B. capsule, old; C, leaf; D, F, apex of leaf; E, capsule of *D. rufescens*; G, peristome.

Plants small, 4-10 mm high, green to yellow-green, on moist soil of banks, over rocks, and in bare open places, Alaska, British Columbia to Nova Scotia, Quebec and Ontario south to California, Texas, Louisiana, and Florida.

Campylopus

Eleven species of *Campylopus* are reported from North America. The following two are likely to be encountered.

1a Leaves, particularly older ones, ending in a hyaline hair point; narrow rectangular hyaline cells extending up the margin from base to apex as in *Tortella*. Fig. 101.
.... *Campylopus pilifera* Brid.

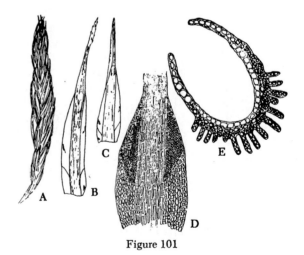

Figure 101

Figure 101 *Campylopus pilifera.* A, habit of gametophyte; B-C, leaves; D, lower leaf; E, cross section of leaf.

Plants dark to golden green above, in dense tufts, 0.5 to 4 cm tall, on open sandstone rocks and rock ledges, Arizona, Texas, Arkansas, Tennessee, and North Carolina south to the Gulf States.

1b Leaves without hyaline hair points; no hyaline cells extending up the margin. Fig. 102E-F. ....
.... *Campylopus flexuosus* (Hedw.) Brid.

Plants in dense green tufts, 1-4 cm tall, on bare, exposed rocks, usually sandstone, British Columbia, Arizona, Missouri, Arkansas, Louisiana, Tennessee and North Carolina south to Florida.

Dicranodontium

Dicranodontium denudatum (Brid.) Britt. *ex* Williams (Fig. 102A-D)

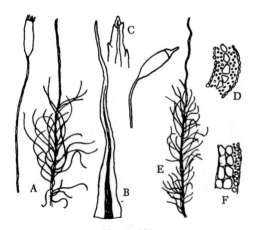

Figure 102

Figure 102 *Dicranodontium denudatum*. A, plant; B, leaf; C, apex of leaf; D, section of midrib; E, plant of *Campylopus flexuosus*; F, section of midrib.

Plants in compact pale green, silky tufts, 4-8 cm tall, on moist rocks, ledges, and thin soil over rocks, occasionally on decaying logs, Michigan, southern Ontario to New England south to North Carolina, Tennessee, and Alabama. Two other species are known from North America.

Paraleucobryum

Paraleucobryum longifolium (Hedw.) Loeske (Fig. 103)

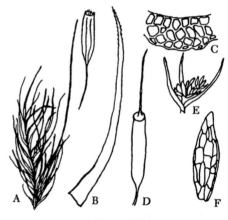

Figure 103

Figure 103 *Paraleucobryum longifolium*. A, plant; B, leaf (4-7 mm long); C, section of midrib; D, capsule; E, gemmae of *Brothera leana*; F, one gemma.

Whitish plants in dense silky, light to dark green tufts, 3-4 cm tall, on tree trunks, logs, rocks and humus, Alaska south to British Columbia, Montana, Arizona to Minnesota, Michigan, Ontario, Quebec, New England south to Tennessee, North Carolina and Georgia. Two other species of *Paraleucobryum* are known from North America.

Brothera

Brothera leana (Sull.) C. Muell. (Fig. 60; 103E-F)

Small whitish plants resembling *Leucobryum*, in yellow-green cushions less than 1 cm high, on soil, bark, bases of trees, decaying wood, and rocks, Minnesota to Pennsylvania south to Arkansas, Illinois, Ohio, Tennessee and North Carolina.

Rhabdoweisia

Rhabdoweisia crispata (With.) Lindb. (Fig. 104)

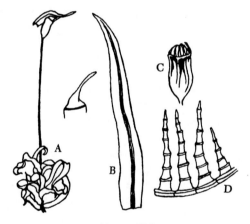

Figure 104

Figure 104 *Rhabdoweisia crispata*. A, plant; B, leaf; C, capsule, old; D peristome.

Plants in loose to dense, yellowish to dark green tufts, less than 1 cm tall, on dry to moist rock ledges, Alaska to British Columbia, Newfoundland south to Missouri, Arkansas, Tennessee, North Carolina and Georgia. A rare species, *R. crenulata* (Michx.) Jameson, is found in a single locality in North Carolina.

Cynodontium

Cynodontium strumiferum (Hedw.) Lindb. (Fig. 105)

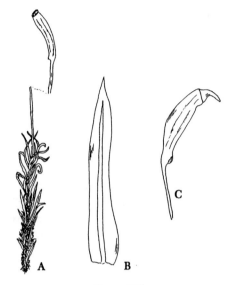

Figure 105

Figure 105 *Cynodontium strumiferum*. A, plant; B, leaf; C, capsule.

Plants small, 0.5 to 1 cm high, in green to yellow-green tufts, on soil over rocks and in rock crevices, British Columbia, Alberta, Montana, Alaska to Minnesota, Labrador, Quebec and New England. Three other species are reported from North America.

Dichodontium

Dichodontium pellucidum (Hedw.) Schimp. (Fig. 106)

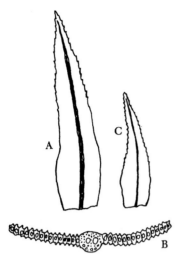

Figure 106

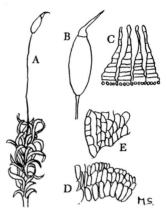

Figure 107

Figure 106 *Dichodontium pellucidum*. A, leaf; B, section of leaf; C, leaf of Oreoweisia serrulata.

Plants 1.5-5.5 cm tall, with spreading to squarrose leaves when moist, crisped and contorted when dry, on moist soil banks, wet cliffs and near streams, Alaska, British Columbia south to California and Arizona, Texas, Utah, Colorado, Alberta, Idaho, Montana, northeastern Canada south to Ohio, Pennsylvania, Michigan, North Carolina and Tennessee. *D. olympicum* Ren. & Card. occurs in the Pacific Northwest and has leaves with plane margins instead of margins that are recurved below.

Dicranoweisia

1a Leaf margins recurved; alar cells not inflated. Fig. 107A-D. .... *Dicranoweisia cirrata* (Hedw.) Lindb. *ex* Milde

Figure 107 *Dicranoweisia cirrata*. A, plant, dry; B, capsule; C, peristome; D, alar cells; E, alar cells of *D. crispula*.

Plants 1-2 cm high in dense green to yellow-green tufts, on decaying wood, tree trunks and rocks, Alaska, British Columbia, Washington, Idaho south to California and Arizona.

1b Leaf margins not recurved, alar cells inflated. Fig. 107E. .... *Dicranoweisia crispula* (Hedw.) Lindb. *ex* Milde.

Plants 1-2 cm high, in dense green to yellow-green tufts, on rocks and in crevices of cliffs, tree trunks, and logs, British Columbia south to California, Montana, Utah, and Colorado east to Michigan, Maine, and New York.

Oncophorus

Oncophorus wahlenbergii Brid. (Fig. 108)

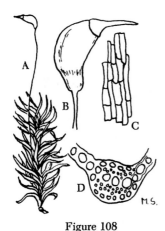

Figure 108

Figure 108 *Oncophorus wahlenbergii* Brid. A, plant; B, capsule; C, alar cells; D, cross section of leaf.

Plants in dense tufts, 5 cm or more in height, on rotten logs in moist cool forests, Alaska, British Columbia, Alberta south to Montana, Colorado, South Dakota, Minnesota, Michigan, Quebec and West Virginia. Two other species *Oncophorus* are reported from North America.

Kiaeria

Kiaeria starkei (Web. & Mohr.) Hag. (Fig. 109D)

Small, 1-3 cm high, dark green to brownish green tufted plants, on soil and soil over rocks, Alaska, British Columbia south to Oregon, Alberta, Montana, Idaho, to Quebec and New England. Three other species of *Kiaeria* are present in North America.

Dicranum

At least 24 species of this genus are present in North America. Many show intergrating forms which make separation difficult. The following are ones most likely to be encountered.

1a Leaf points not mostly broken off, or if broken off, then the plants fulvous to brownish-green in color. 3

1b Leaf points mostly broken off, plants green to yellow-green in color. 2

2a Leaves straight, plants restricted to western North America. Fig. 109A-C. .... *Dicranum tauricum* Sapeh.

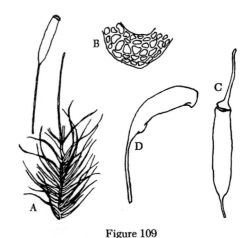

Figure 109

Figure 109 *Dicranum tauricum*. A, plant; B, section of midrib; C, capsule; D, capsule of *Kiaeria starkei*.

Plants in dense tufts, 0.5-3 cm high, on rotten wood, tree bases, and humus over rocks, throughout the Pacific Northwest south to California, Idaho, Wyoming, Colorado, and Utah.

2b Leaves falcate-secund to crispate when dry, green, restricted to eastern North America. Fig. 110D. .... *Dicranum viride* (Sull. & Lesq. *ex* Sull.) Lindb.

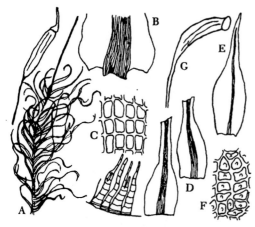

Figure 110

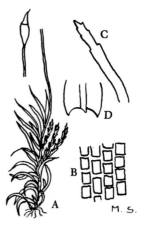

Figure 111

Figure 110 *Dicranum fulvum*. A, plant, dry; B, base of leaf; C, median cells; D, *viride*; E, leaf of *D. spurium*; F, cells of leaf with papillae; G, capsule; H, teeth of peristome.

Figure 111 *Dicranum flagellare*. A, plant with capsule and flagella; B, cells from upper part of leaf; C, apex of leaf; D, base of leaf showing relative width of midrib.

Small plants, 1.5-3.5 cm tall, in dull, dark-green to yellowish-green tufts, on the bark of trees, decaying logs and sometimes on rocks, southern Canada south to North Carolina, Tennessee, and Arkansas.

Plants in dense tufts, 0.5-5.5 cm tall, dull green to brownish-green, on rotten logs, stumps, or occasionally on humus or thin soil over rocks, throughout North America from southern Canada, southern Montana, South Dakota, Minnesota, Arkansas, Tennessee and North Carolina.

3a   Plants without flagelliform branches. .. 4

3b   Plants with flagelliform branches. Fig. 54C-D; 111. ....
.... *Dicranum flagellare* Hedw.

4a   Plants small, 0.5-1 mm tall, leaves papillose on back 1/2 way down, crisped when dry, not secund; capsules erect and straight. Fig. 112A-C. ....
.... *Dicranum montanum* Hedw.

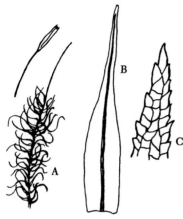

Figure 112

Figure 112 *Dicranum montanum*. A, plant; B, leaf; C, apex of leaf; D, shoot of *D. rhabdocarpum*.

In dense dark-green to brown or yellow-brown tufts, on rotten logs, stumps, tree trunks and bases and thin soil over rocks, British Columbia, Arizona, Minnesota, Iowa, Illinois, Michigan, Ontario and Quebec south to Arkansas, Tennessee and North Carolina.

4b Plants larger, usually 2 cm or more tall, leaves smooth on back to slightly rough near apex (may be strongly papillose in *D. fuscescens*), crispate to falcate-secund when dry. 5

5a Leaves when dry straight to slightly flexuous, often strongly undulate. 6

5b Leaves when dry crispate to falcate or falcate-secund. 9

6a Leaves strongly undulate in upper portion. 7

6b Leaves not to only slightly undulate in upper portion. 8

7a Upper leaf cells 1-2:1, walls not or with only a few pits. Fig. 115D-E. .... *Dicranum undulatum* Brid.

Large, 3-15 cm tall plants in dense, yellow-green to brownish-green tufts, in open, wet habitats around lake margins, bogs, and swamps, Alaska, British Columbia, Alberta, Colorado, Manitoba, Wisconsin, Michigan, Ontario, Quebec south to New York and New Jersey.

7b Upper leaf cells linear, walls pitted. Fig. 113F. .... *Dicranum polysetum* Sw.

Plants large, 4-15 cm tall, in loose, light green to yellowish green tufts, on humus, soil over rocks, logs, and in bogs, Alaska, British Columbia to Newfoundland and Nova Scotia south to Washington, Montana, Wyoming, South Dakota, Missouri, Illinois, Pennsylvania, Tennessee, North Carolina and Georgia.

8a Leaves on the back of costa, often slightly undulate, eastern North America. Fig. 114E. .... *Dicranum bonjeanii* De Not *ex* Lisa

Plants in loose, tufts, up to 10 cm tall, glossy yellow green, on soil or thin soil over rocks in wet places, Labrador to North Carolina, Pennsylvania, Illinois.

8b Leaves not at all undulate, western United States. Fig. 112D. .... *Dicranum rhabdocarpun* Sull.

Plants in dense, green to yellow green tufts, 2-6 cm high, on soil, soil over rock, and rotten wood, Wyoming, Colorado, New Mexico and Arizona.

9a Upper leaf cells pitted. 10

9b Upper leaf cells not pitted. 11

10a Leaves 10-15 mm long, setae often aggregate, plants up to 12 cm tall. Fig. 113A-D. .... *Dicranum majus* Sm.

10b Leaves usually 10 mm or less long, setae single, plants 2-8 cm tall. Fig. 114A-D. .... *Dicranum scoparium* Hedw.

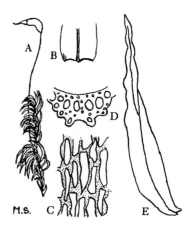

Figure 114

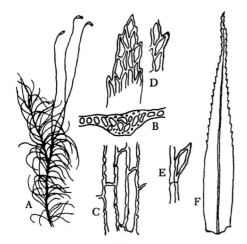

Figure 113

Figure 114 *Dicranum scoparium*. A, plant; B, base of leaf showing width of midrib; C, upper cells of leaf; D, cross section of midrib; E, leaf of *D. bonjeani*.

Plants in loose to dense, green to yellow green tufts, 2-8 cm high, on soil, humus, soil over rocks, rotten wood, and tree bases, Alaska to Newfoundland south to Oregon and California and Florida, Alabama, and Louisiana.

Figure 113 *Dicranum majus*. A, plant; B, section of midrib; C, lower cells; D, apex and tooth of leaf; E, tooth of leaf, and F, leaf of *D. polysetum*.

Plants very large, 3-12 cm tall or higher, in green to yellow-green tufts, on humus, soil or soil over rocks, rotten logs, and tree bases, Alaska, British Columbia, Washington, Ontario, Labrador, Quebec, Pennsylvania and New England.

11a Plants green to yellow green. 12

11b Plants fulvous to brownish-green; eastern North America. Fig. 110A-C. .... *Dicranum fulvum* Hook.

Plants to 5 cm tall, in extensive mats, mostly on shaded rocks, Eastern Canada south to Georgia, Alabama, Arkansas, and Minnesota.

12a Leaves undulate or rugose. 13

12b Leaves not undulate or rugose. 15

13a Leaves obtuse. Fig. 115D-E. ....
.... (p. 85) *Dicranum undulatum* Brid.

13b Leaves acute. 14

14a Leaves ovate to ovate-lanceolate, widest a little below the middle, finely papillose over all of the upper (dorsal) surface. Fig. 110 E-H. ....
.... *Dicranum spurium* Hedw.

Coarse plants in loose, dull yellow green to yellow brown tufts, 0.5-7 cm tall, Minnesota east to Newfoundland south to Arkansas, Tennessee, and North Carolina.

14b Leaves oblong-lanceolate, broadest at or near base, more or less roughened with coarse papillae on upper surface. Fig. 115A-C. ....
.... *Dicranum drummondii* C. Muell.

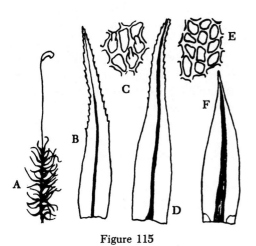

Figure 115

Figure 115 *Dicranum drummondii*. A, plant; B, leaf; C, upper cells with papillae; D, leaf of *D. undulatum;* E, cells of same; F, leaf of *D. muhlenbeckii.*

Robust plants, 4-8 cm tall, in dull, green to yellow-green tufts, in bogs, on humus, rotten logs, Manitoba and Montana to Labrador south to North Carolina.

15a Leaves strongly falcate-secund to crisped when dry. Fig. 116A-D. ....
.... *Dicranum fuscescens* Turn.

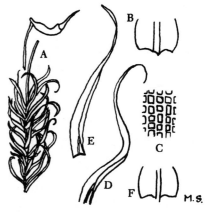

Figure 116

Figure 116 *Dicranum fuscescens*. A, plant; B, base of leaf showing relative width of midrib; C, upper cells of leaf; D, the dry curly leaf; E, dry leaf of *D. condensatum;* F, base of leaf of same.

Plants in loose, dark green to yellowish green tufts, 1-6 cm tall, on tree trunks, decaying wood, soil and humus and soil and humus over rocks, Pacific Northwest, Alaska, California, Colorado, Minnesota, Michigan, Ontario south to Georgia.

15b Leaves not secund, often straight. 16

16a Leaves subtubulose above, costa 1/5-1/6 width of leaf base. Fig. 115F. ....
.... *Dicranum muehlenbeckii* B.S.G.

Plants 3-7 cm tall, in dense green to yellowish green tufts, on soil or humus or soil over rock, Alaska, British Columbia, Colorado, North Dakota, Ontario, Quebec, New England, south to New Jersey.

16b Leaves merely channelled above, costa 1/8 width of leaf base. Fig. 54A-B; 116E-F. .... *Dicranum condensatum* Hedw.

Plants 1-5 cm tall, in yellow-brown, loose tomentose tufts, usually on light, open to shaded soil, eastern North America, Michigan to Nova Scotia south to Florida, Louisiana and Texas.

LEUCOBRYACEAE
Leucobryum

1a Leaves in cross section near the base at the middle of the costa 4-5 cells thick, i.e., with 2 layers of hyaline cells above and 1-2 layers below a single layer of green cells; leaves normally 5-8 mm long, the upper tubulose portion up to 3 times as long as the base. Fig. 45, 117E.-F. .... *Leucobryum glaucum* (Hedw.) Ångstr. *ex* Fr.

Whitish plants in hemisphaerical cushions 2-9 cm or more high, on soil, humus, rotten logs, tree bases, and moist vertical rocks, especially in forests, Newfoundland to Minnesota and south to the Gulf States.

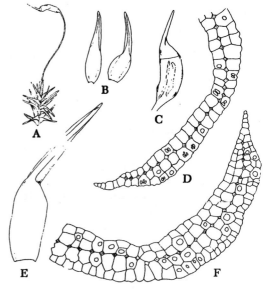

Figure 117

Figure 117 *Leucobryum albidum*. A, plant with capsules; B, leaves; C, capsule; D, cross section of leaf. *Leucobryum glaucum*. E, leaf; F, cross section of leaf.

1b Leaves in cross section near the base at the middle of the costa 3 layers thick, i.e., with 1 layer of hyaline cells above and below the single layer of green cells; leaves 2-4.5 mm long, the upper tubulose portion about the same length as the base. Fig. 117A-D. .... *Lucobryum albidum* (Brid. *ex* P.-Beauv.) Lindb.

Whitish plants in hemisphaerical cushions, on soil, tree bases and logs, Florida to Texas north to Michigan and Connecticut.

Octoblepharum

Octoblepharum albidum Hedw. (Fig. 118)

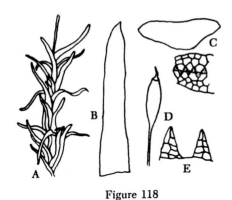

Figure 118

Figure 118 *Octoblepharum albidum* Hedw. A, plant; B, leaf; C, section of leaf; D, capsule; E, peristome.

Plants in whitish loose tufts or mats, sometimes tinged with red, on tree trunks, especially palms, and logs, common in southern Florida.

Order Pottiales

CALYMPERACEAE
Syrrhopodon

1a Leaf border doubly serrate and wing-like, cells of border not hyaline. Fig. 119C-D. ....
...... *Syrrhopodon incompletus* Schwaegr.

Plants 1-3 cm tall, in compact, dull green to brown tufts, on bark of trees, especially palms, often forming very extensive mats, Florida,

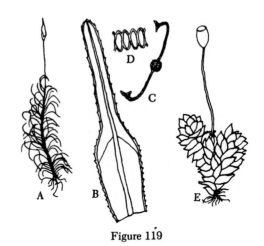

Figure 119

Figure 119 *Syrrhopodon texanus*. A, plant; B, leaf; C, cross section of leaf of *S. incompletus*; D, cells of same; E, plant of *Groutiella mucronifolia*, wet.

Musci—The Mosses

Georgia, Louisiana with a disjunct population in Long Island, New York.

1a   Leaf border composed on linear, hyaline or yellowish cells. Fig. 52, 119A-B. .... *Syrrhopodon texanus* Sull.

Plants 1-4 cm tall, in dark green or brown cushions, on soil around tree bases and shaded rocks, New York (Long Island), Missouri, Illinois, Kentucky, Tennessee, North Carolina, Mississippi, Alabama, Georgia and Florida to Texas.

Calymperes

Calymperes richardii C. Muell. (Fig. 120)

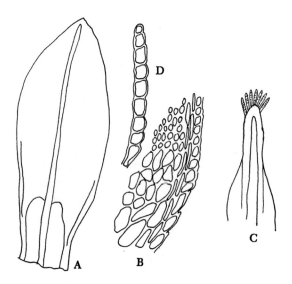

Figure 120

Figure 120 *Calymperes richardii*. A, leaf; B, leaf near base showing cancellinae and teniolae; C, leaf bearing propagula; D, propagulum.

Plants in dark, glossy green cushions or mats, 3-20 mm tall with leaves; that are involute and curved or somewhat curled when dry, the upper ones usually bearing dense clusters of filiform-clavate propagula, on dead wood, trees, or occasionally on rocks, in thickets and hammocks of southern Florida. Three other species may also be found in southern Florida.

ENCALYPTACEAE
Encalypta

There are eight species recorded for North America.

1a   Plants usually robust, 15-20 mm tall, commonly bearing abundant brood bodies, usually without sporophytes. Fig. 121A-D. .... *Encalypta procera* Bruch

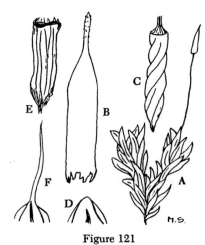

Figure 121

Figure 121 *Encalypta procera*. A, plant; B, calyptra; C, capsule; D, apex of leaf; E, capsule of *E. rhaptocarpa;* F, tip of leaf of same.

Plants in dull-green or brownish-green tufts with numerous filiform brood-bodies on simple or branched radicles on the stem, on soil or soil on rocks, Alaska to British Columbia and

Washington, Montana, South Dakota, Nevada, Colorado, Utah, Iowa, Missouri, Michigan to Nova Scotia south to North Carolina and Tennessee.

1b Plants smaller, without brood bodies, sporophytes usually present. 2

2a Leaves acute and hyaline-awned, capsules strongly ribbed. Fig. 121E-F. .... *Encalypta rhaptocarpa* Schwaegr.

Plants 5-30 mm tall in dull-green to yellow-green tufts with capsules that have straight ribs, on soil in rock crevices, on steep banks, or under overhanging ledges, Alaska south to California, Nevada, Colorado, Arizona, Utah, New Mexico, Michigan, and New York.

2b Leaves rounded-obtuse to broadly acute, capsules smooth. Fig. 122. .... *Encalypta ciliata* Hedw.

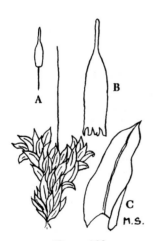

Figure 122

Figure 122 *Encalypta ciliata*. A, plant; B, calyptra; C, leaf.

Plants 5-40 mm tall, in dull-green or yellow-green dense tufts, on ledges or in crevices; of calcareous rocks, or on soil banks especially under overhanging ledges, Alaska south to California, New Mexico and Arizona, Utah, Colorado, Michigan, Iowa and Labrador south to Pennsylvania and New York.

POTTIACEAE
Anoectangium

In addition to the two more common species keyed below, five other species have been reported from North America.

1a Leaves obtuse, plants whitish in color. Fig. 123A-B. .... *Anoectangium obtusifolium* (Broth. & Paris *ex* Card.) Grout

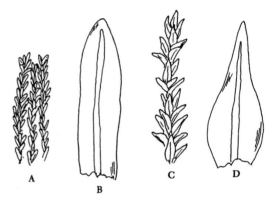

Figure 123

Figure 123 *Anoectangium obtusifolium*. A, plants; B, leaf; C, plant of *Anoectangium sendtnerianum*; D, leaf.

Plants in wide thin mats up to 5 mm tall, on rocks, Arkansas to Texas and Arizona.

1b Leaves acute, plants bluish to brownish green in color. Fig. 123C-D. .... *Anoectangium sendtnerianum* B.S.G.

Plants in dense mats up to 3 cm high, on calcareous rocks, North Carolina, Tennessee and Arkansas south to Florida and Texas.

Scopelophilia

1a   Plants usually robust, 2-5 cm high with leaves 3-5 mm long; upper leaf cells mostly hexagonal, western North America. Fig. 124. .... *Scopelophila latifolia* (Kindb.) Ren. & Card.

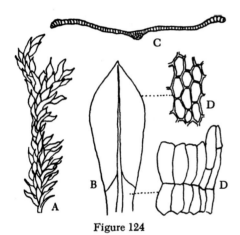

Figure 124

Figure 124 *Scopelophila latifolia*. A, plant; B, leaf; C, section of leaf; D, cells of leaf.

Plants often encrusted with lime, in compact tufts, on wet rocks or soil, British Columbia, Washington, Montana, and South Dakota south to California, Nevada, Arizona and Utah.

1b   Plants usually slender, up to 4 cm high with leaves 2-3 mm long; upper leaf cells mostly quadrate, eastern North America and in Arizona. .... *Scopelophila ligulata* (Spruce) Spruce

Plants in compact tufts, on exposed vertical rocks, Michigan and Illinois, south to Arkansas, North Carolina, Tennessee, Georgia and disjunct in Arizona. A closely related species, *Scopelophila cataractae* (Mitt.) Broth., distinguished by the presence of a red tomentum and brown lower leaves, may be encountered in North Carolina and Arizona.

Astomum

1a   Leaf margins strongly involute; setae solitary. 2

1b   Leaf margins mostly plane; setae usually clustered. .... *Astomum ludovicianum* (Sull.) Sull.

Small tufted mosses 2-3 mm high, on soil or thin soil over rocks, Arkansas to West Virginia and Virginia south to the Gulf States.

2a   Setae as long as the capsules. .... *Astomum phascoides* (Hook. ex Drumm.) Grout

Tufted mosses usually growing on bare, open soil, Saskatchewan, Minnesota, Ohio, Texas, and Arizona.

2b   Setae shorter than the capsules. Fig. 125A-B. ..... *Astomum muhlenbergianum* (Sw.) Grout

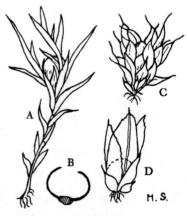

Figure 125

Figure 125 *Astomum muhlenbergianum.* A, plant; B, cross section of leaf; C, plant of *Phascum cuspidatum;* D, plant of *Acaulon muticum* var. *rufescens.*

Small plants in dense tufts, growing on open, bare clayey soil or on thin soil over rocks, Saskatchewan to Ontario south to Florida, Louisiana, Texas and Arizona.

Weissia

Eight species of this genus are known from North America, however, the following species is the one most likely to be collected.

Weissia controversa Hedw. (Fig. 44; 126)

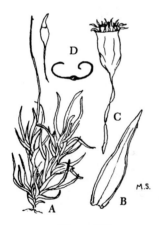

Figure 126

Figure 126 *Weisia* Hedw. A, plant; B, leaf; C, capsule and peristome; D, cross section of leaf.

Small plants, 2-5 mm tall, in dull, green, yellowish, or brownish, loose or dense tufts, on bare open soil, thin soil over rocks, in disturbed habitats, throughout North America. Without sporophytes this species cannot be distinguished from *Astomum.*

Gymnostomum

1a Leaf margins plane; upper leaf cells obscure, 5-12 μ wide. 2

1b Leaf margins recurved below on one or both margins; upper leaf cells clear, 12-15 μ wide. Fig. 127A-C. ....
.... *Gymnostomum recurvirostrum* Hedw.

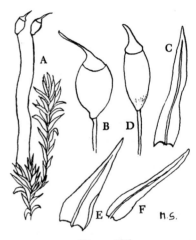

Figure 127

Figure 127 *Gymnostomum recurvirostrum.* A, plant; B, capsule; C, leaf; D, capsule and E, leaf of *G. aeruginosum.* F, leaf of *G. calcareum.*

Plants small to moderately large, 5-30 mm high, in dark-green to yellow-tufts, usually encrusted with lime, on shaded limestone rocks where seepage is common, Alaska to Nova Scotia south to California and the Gulf States.

2a Leaves oblong-lanceolate to linear-lanceolate, apex acute to blunt, upper leaf cells 7-12 μ wide. Fig. 127D-E. ....
.... *Gymnostomum aeruginosum* Sm.

Plants forming thick dense tufts, encrusted with lime, on moist rocks, in seepage areas of calcareous areas, widely distributed throughout North America.

2b Leaves lingulate, usually ending in a broad, rounded or rounded-obtuse tip; upper leaf cells 6-8 μ in diameter. Fig. 127F. .... *Gymnostomum calcareum* Nees, Hornsch. & Sturm.

Plants often forming dense tufts encrusted with lime, Colorado and California.

Husnotiella

1a Apex of leaf rounded. Fig. 128A-B. ....
.... *Husnotiella revoluta* Card.

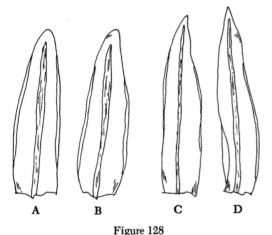

Figure 128

Figure 128 *Husnotiella revoluta*. A-B, leaves; C-D, leaves of *Husnotiella torquescens*.

Small plants, up to 6 mm high, in dense tufts, on soil and rocks, Missouri, Texas, Arizona, and California.

1b Apex of leaf acute to obtuse. Fig. 128C-D. .... *Husnotiella torquescens* (Card.) Bartr.

Plants in bright green, dense tufts, 2-10 mm high, often forming large mats over the substratum, on rocks, Arizona and Texas.

Eucladium

Eucladium verticillatum (Brid.) B.S.G. (Fig. 129)

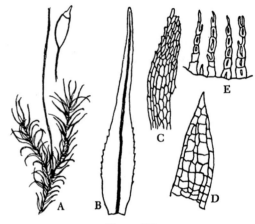

Figure 129

Figure 129 *Eucladium verticillatum*. A, plant; B, leaf; C, margin; D, apex; E, teeth of peristome.

Plants in dense, whitish tufts, 1-5 cm high, on wet, mostly calcareous rocks, Alaska, British Columbia to Ontario south to Florida, Texas, and California.

Trichostomum

Trichostomum tenuirostre (Hook. & Tayl.) Lindb. (Fig. 130)

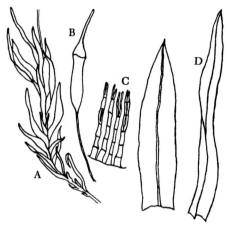

Figure 130

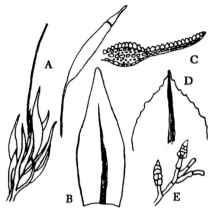

Figure 131

Figure 130 *Trichostomum tenuirostre*. A, plant; B, capsule; C, peristome; D, median and upper leaves.

Plants in loose tufts, up to 1 cm high, yellow green above or brown below, on moist rocks, soil, rock ledges or occasionally on decaying wood or trees, British Columbia to Ontario south to Washington, Arizona, Colorado, Texas, Arkansas, Tennessee and South Carolina. Four other species occur in North America.

Timmiella

Timmiella anomala (B.S.G.) Limpr. (Fig. 131A-C)

Figure 131 *Timmiella anomala*. A, plant; B, leaf; C, section of leaf; D, leaf of *Hyophila involuta*; E, gemmae.

Plants in loose mats or tufts, dirty green, up to 2 cm high, on soil, Texas and Arizona. *Timmiella crassinervis* (Hampe) L. Koch occurs in northwestern North America.

Tortella

Six species are known from North America. The following three are the ones most likely to be encountered.

1a   Leaves with nearly parallel sides to near the apex; mucronate. Fig. 132A-C. .... *Tortella humilis* (Hedw.) Jenn.

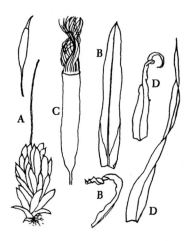

Figure 132

Figure 132 *Tortella humilis.* A, plant; B, leaf, dry and spread out; C, capsule and peristome. *Tortella tortuosa*, D, leaf, dry and spread out.

Plants in loose to dense tufts, green to yellow-green above, 1-3 cm high, on bark, tree bases, decaying wood, soil, rocks, British Columbia to New England south to Florida, Louisiana, and Texas.

1b Leaves evenly tapering from base to a very slender apex, twisted and coiled when dry. 2

2a Tips of leaves nearly all broken off. .... *Tortella fragilis* (Drumm.) Limpr.

Plants in thick, loose mats, green to yellow-green above, 1-5 cm high, on soil, tree bases and rocks, Alaska to Quebec south to Nevada, Colorado, Iowa, Tennessee and North Carolina.

2b Tips of leaves not or but little broken off. Fig. 132D. .... *Tortella tortuosa* (Hedw.) Limpr.

Plants in green to yellow-green dense, rounded tufts, 1-6 cm tall, on soil and rocks of calcareous regions, Alaska to Quebec and Nova Scotia south to Nevada, Utah, Colorado, Michigan, Tennessee and North Carolina. In the Northwestern United States *Tortella inclinata* (R. Hedw.) Limpr. may be collected. It differs from *T. tortuosa* in the leaf apex which does not taper gradually to a long-narrow point.

Pleurochaete

Pleurochaete squarrosa (Brid.) Lindb. (Fig. 133)

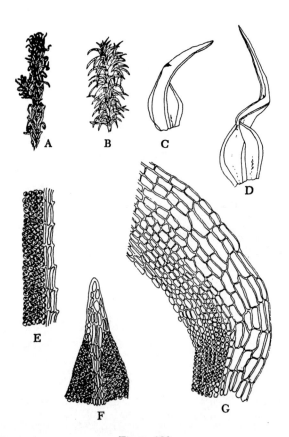

Figure 133

Figure 133 *Pleurochaete squarrosa*. A, habit of dry plant; B, habit of moist plant; C-D, leaves; E, cells of hyaline border near upper middle leaf; F, apex of leaf; G, base of leaf showing hyaline border.

Plants in loose, yellow-green tufts, 1-10 cm high, on soil and soil over calcareous rocks, especially of cedar barrens and glades, Virginia, Tennessee, Missouri, Arkansas, Oklahoma, Texas, Arizona, New Mexico.

Hyophila

Hyophila involuta (Hook.) Jaeq. & Sauerb. (Fig. 131D-E)

Plants in low, dark-green tufts up to 3 cm tall, on shaded rocks in and along streams and creeks, Michigan to Ontario south to Florida, Arkansas, Oklahoma, Texas and Arizona.

Didymodon

1a Margins of leaf unistratose, strongly decurrent, multicellular gemmae absent. Fig. 134. ....
.... *Didymodon tophaceus* (Brid.) Lisa

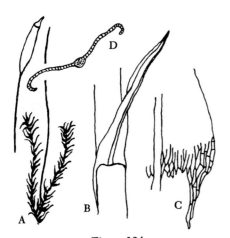

Figure 134

Figure 134 *Didymodon tophaceus*. A, plant; B, leaf; C, decurrent base; D, section of leaf.

Plants in dense, olive-green to brown tufts, up to 5 cm high, on calcareous rocks, often forming tufa, British Columbia to Ontario and south to California, Arizona, Utah, Texas, Arkansas, and Tennessee.

1b Margins of leaf 2-3 stratose in the upper part, not strongly decurrent; stems often bearing sphaerical, multicellular gemmae. Fig. 135. ....
.... *Didymodon rigidulus* Hedw.

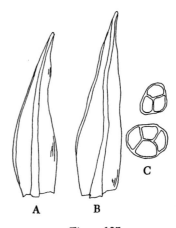

Figure 135

Figure 135 *Didymodon rigidulus*. A-B, leaves; C, gemmae.

Plants in dark green, dense mats up to 2 cm tall, on moist, usually calcareous rocks, Oregon and California to Michigan, Ontario, and in Missouri and Florida.

Bryoerythrophyllum

Bryoerythrophyllum recurvirostrum (Hedw.) Chen (Fig. 136)

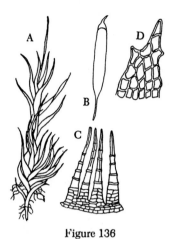

Figure 136

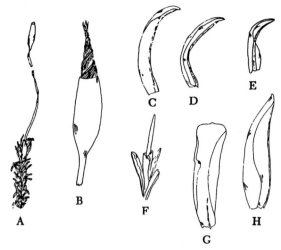

Figure 137

Figure 136 *Bryoerythrophyllum recurvirostrum*. A, plant; B, capsule; C, peristome; D, tip of leaf. Lower parts of plant reddish; sharp cell at tip of leaf is colorless.

Plants in loose or dense tufts up to 3 cm tall, green, yellow-green, brownish or red-brown, usually brick-red below, on calcareous rocks, widespread through the northern part of North America south to Oregon, the Appalachians, and the Ozarks.

Barbula

This is a rather large and complex genus with at least twenty-one valid species recognized in North America. The following species are ones most likely to be encountered. Richard Zander (Phytologia 41:11-32. 1978) has transferred many taxa in this genus to *Didymodon*.

1a Leaves oblong, ligulate or lingulate. 2

1b Leaves lanceolate to subulate, tapering gradually from an ovate base. 5

Figure 137 *Barbula convoluta*. A, plant with sporophyte; B, capsule with peristome; C-E, leaves; F, perichaetial leaves surrounding a young sporophyte; G-H, perichaetial leaves.

Small tufted plants, up to 5 (rarely more) mm high, green to yellow-green, on bare soil, usually calcareous, in disturbed places, Alaska, British Columbia to New Brunswick south to California, Oklahoma, Arkansas, and Georgia.

2a Seta yellow; perichaetial leaves convolute. Fig. 137. .... *Barbula convoluta* Hedw.

2b Seta red; perichaetial leaves not convolute. 3

3a Upper leaf-cells smooth to slightly papillose. Fig. 138. .... *Barbula ehrenbergii* (Lor.) Fleisch.

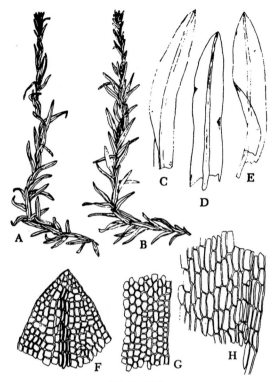

Figure 138

Figure 138 *Barbula ehrenbergii*. A-B, plants; C-E, leaves; F, apex of leaf; G, leaf cells along upper margin; H, basal leaf cells.

Plants medium, 1-5 cm high, in dense tufts, often encrusted with lime, leaves flaccid, on moist shaded calcareous rocks, Missouri, Oklahoma, Texas.

3b   Upper leaf cells densely papillose. 4

4a   Propagula present. Fig. 147D. ....
    .... *Barbula cancellata* C. Muell.

Small, loosely tufted plants, up to 2 cm tall, glaucous or yellowish-green above, on moist calcareous rocks and rock crevices, eastern United States from New Jersey south to the Gulf States, Tennessee, Missouri, and Arkansas.

4b   Propagula absent. Fig. 139. ....
    .... *Barbula unguiculata* Hedw.

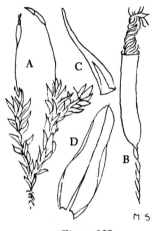

Figure 139

Figure 139 *Barbula unguiculata*. A, plant; B, capsule and peristome; C, calyptra; D, leaf.

Plants light to dirty-green, in tufts up to 2.5 cm tall, on calcareous soil and rocks, in old fields, waste places, edges of creeks and streams, throughout North America.

5a   Leaves 2.5-5 mm long. Fig. 140D. ....
    .. *Barbula cylindrica* (Tayl. *ex* Mackay) Schimp. *ex* Boul.

Plants loosely caespitose, dark green with a dull brownish or reddish tinge, 1-6 cm tall, on calcareous rocks and soil, western United States.

5b   Leaves less than 2.5 mm long. 6

6a   Cells of upper surface of costa linear. 140A-B. .... *Barbula fallax* Hedw.

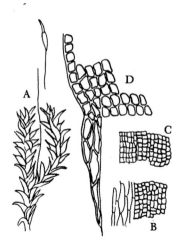

Figure 140

Figure 140 *Barbula fallax.* A, plant in wet condition; B, cells of midrib and lamina; C, cells of midrib and lamina of *B. vinealis;* D, decurrent leaf base of *B. cylindrica.*

Plants in brownish tufts, usually 5-20 mm high, on moist calcareous soil, British Columbia to Nova Scotia south to Idaho, Montana, Missouri, Ohio, Louisiana, Tennessee and North Carolina.

6b Cells of upper surface of costa quadrate. Fig. 140C. .... *Barbula vinealis* Brid.

Plants in green or greenish-brown tufts, often reddish below, 1-3 cm high, on soil and rocks, western United States.

Acaulon

Acaulon muticum (Hedw.) C. M. var. *rufescens* (Jaegr.) Crum (Fig. 125D)

Minute plants, less than 2 mm tall, sub-globose to bulbiform, yellow-green in color, on sandy or clayey soil of old field, waste lots and roadsides, usually in the spring, California to Ontario and Quebec south to Arizona, Texas, and the Gulf States. Two additional species are found in North America.

Phascum

Phascum cuspidatum Hedw. (Fig. 125C)

Plants in dense, light to yellow-green tufts, 1-2 mm high, producing sporophytes in the spring, on soil in fields, on banks, in pastures, British Columbia to Nova Scotia south to California, Arizona, Texas, Arkansas, and Tennessee. Two other rare species are found in North America.

Pottia

Pottia truncata (Hedw.) Furnr. *ex* B.S.G. (Fig. 141)

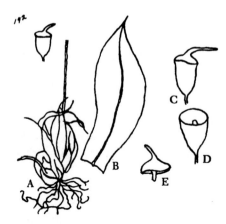

Figure 141

Figure 141 *Pottia truncata.* A, plant; B, leaf; C, capsule with operculum; D, old capsule; E, operculum with columella.

Small, 3-5 mm tall, ephemeral, occurring on bare spots of calcareous soil in fallow fields, pastures, meadows, roadsides, and ditches, British Columbia to Nova Scotia south to Michigan and Maryland. *Physcomitrium* may easily be mistaken for *Pottia,* however, unlike *Pottia, Physcomitrium* has serrate leaves with large clear cells. Eight other species of Pottia are known from North America.

Pterygoneurum

Pterygoneurum subsessile (Brid.) Jur. (Figs. 48A-B, 142A-C)

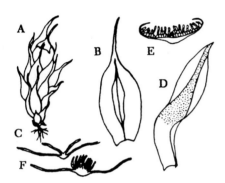

Figure 142

Figure 142 *Pterygoneurum subsessile.* A-C, (Brid.) Jur. A, plant; B, leaf; C, section of leaf; D, leaf of *Aloina rigida;* E, section of same; F, section of leaf of *Crossidium squamigerum.*

Plants 1-3 mm high, in sods, green to gray-green in color, on dry soil and rocks, British Columbia, Saskatchewan to Minnesota south to Arizona, Texas and Oklahoma. Another species, *P. ovatum* (Hedw.) Dix. is also found in the western United States and to the east in Ontario and Quebec. It is distinguished from *P. subsessile* by having a capsule that is emergent to exserted rather than immersed to emergent. Three other species are known from North America.

Crossidium

Crossidium squamiferum (Viv.) Jur. (Figs. 48D, 142F)

Small plants, up to about 6 mm high, gray-green in color, in dense tufts, on soil, walls, and rocks, California, Arizona, and Colorado. Two other species are known from the southwestern United States.

Desmatodon

Fourteen species have been recorded for North America. The following five are ones most commonly collected.

1a Capsules without a peristome. Fig. 143A-C. ....
.... *Desmatodon heimii* (Hedw.) Mitt.

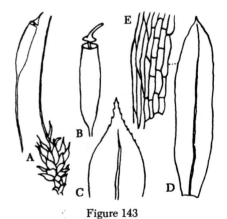

Figure 143

Figure 143 *Desmatodon heimii.* A, plant; B, capsule; C, leaf; D, leaf of *Tortula subulata;* E, margin of same.

Plants small, 2-6 mm high, in dense tufts, on sandy or alkaline soil, Western North America east to Colorado and New Mexico, Quebec and New Brunswick.

1b Capsules with a peristome. 2

2a Costa excurrent or ending in a colorless hair-point or awn. 3

2b Costa not excurrent or ending in a colorless hairpoint or awn, or may occasionally form an apiculus. 4

3a Range from western North America to Colorado and Quebec; upper leaf cells 12-21 µ wide. Fig. 144D. ....
.... *Desmatodon latifolius* (Hedw.) Brid.

Plants scattered or in tufts, 3-15 mm high, green to brownish-green, on soil in subalpine to alpine regions, western North America east to Colorado and Quebec.

3b Range in eastern United States from Missouri to Pennsylvania south to the Gulf States; upper leaf cells up to 10 µ wide. Fig. 144A-C. .... *Desmatodon plinthobius* Sull. & Lesq. *ex* Sull.

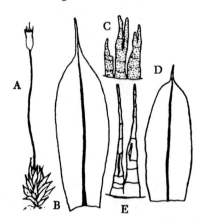

Figure 144

Figure 144 *Desmatodon plinthobius*. A, plant; B, leaf; C, peristome; D, leaf of *D. latifolius*; E, peristome of same.

Plants in dense tufts, appearing whitish because of hair-points, up to 5 mm high, on calcareous rocks, walls, pavements and soil, Missouri, Illinois to Pennsylvania south Florida, Louisiana and Texas.

4a Leaves bordered by a band of light-colored cells. Fig. 145D-E. ....
.... *Desmatodon porteri* James *ex* Sull.

Plants in compact tufts 1-3 mm tall, tawny green, on calcareous rocks, Quebec, Ontario south to Pennsylvania, Ohio, Indiana, Illinois, Missouri and Arkansas.

4b Leaves not bordered. Fig. 145A-C. ....
.... *Desmatodon obtusifolius* (Schwaegr.) Schimp.

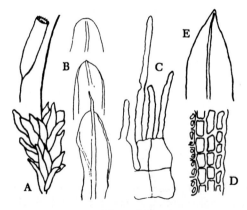

Figure 145

Figure 145 *Desmatodon obtusifolius*. A, plant; B, tips of leaves; C, peristome; D, marginal cells of *D. porteri*; E, apex of leaf of *D. porteri*.

Plants in dull-green tufts, 3-6 mm high, on damp, shaded rocks, stone walls, and rarely soil, Alaska to Quebec south to Washington; Arizona, Utah, Colorado, Oklahoma, Texas, Arkansas, Indiana, Ohio, and New Jersey.

Luisierella

Luisierella barbula (Schwaegr.) Steere (Fig. 146)

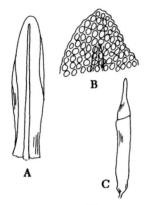

Figure 146

Figure 146 *Luisierella barbula*. A, leaf; B, apex of leaf; C, capsule.

Plants minute, up to 2 mm tall, dark green or reddish, on limestone, Florida and Texas.

Aloina

Aloina rigida (Hedw.) Limpr. (Figs. 48C, 142D-E)

Plants minute, less than 3 mm tall, single or in clusters, on calcareous soil and rocks in dry disturbed habitats, British Columbia, Alberta, California, New Mexico, Iowa.

Tortula

At least twenty-six species of this genus are found in North America.

1a Plants bearing abundant gemmae; usually growing on the trunks of living trees (or occasionally on rock walls). 2

1b Plants usually without gemmae (except *T. fragilis*); growing on soil and rocks. 4

2a Gemmae borne in the axils of leaves at the apex of the stem. Fig. 147A-B. .... *Tortula pagorum* (Milde) De Not.

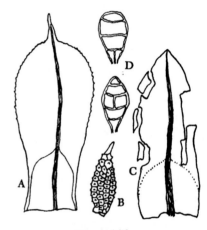

Figure 147

Figure 147 *Tortula pagorum*. A, leaf; B, gemma; C, leaf of *Tortula fragilis*; D, gemmae of *Barbula cancellata*.

Plants in small to extensive dense tufts, light-green, on tree trunks and rocks, South Dakota, Ohio and Maryland south to Georgia, Louisiana, Texas, Arizona, and California.

2b Gemmae borne on the surface of the leaf itself. 3

3a Gemmae borne only on the upper surface of the costa; eastern United States. Fig. 148A-B. .... *Tortula papillosa* Wils. *ex* Spruce

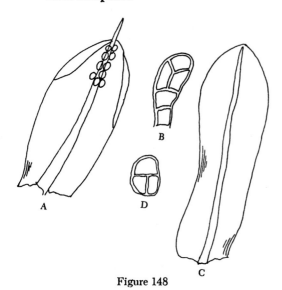

Figure 148

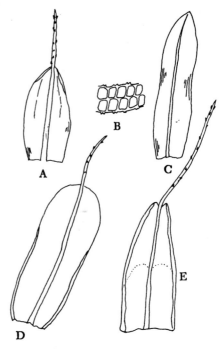

Figure 149

Figure 148 *Tortula papillosa*. A, leaf with propagula; B, propagulum. *Tortula latifolia;* C, leaf; D, propagulum.

Plants occurring singly or in small dense tufts, up to 10 mm high, dark green when moist, on tree trunks and rocks, northeastern United States south to North Carolina, Tennessee, and Missouri.

3b Gemmae borne only on the blade; Pacific Northwest. Fig. 148C-D. .... *Tortula latifolia* Bruch *ex* C. J. Hartm.

Plants in scattered, green to blackish tufts, 1-2 cm high, on trees and rock walls in wet places, British Columbia south to California.

4a Leaves bistratose (Western North America). Fig. 149A-B. .... *Tortula bistratosa* Flow.

Figure 149 *Tortula bistratosa*. A, leaf; B, cross-section of leaf. *Tortula bolanderi*. C, leaf. *Tortula princeps;* D, leaf. *Tortula obtusissima;* E, leaf.

Plants in green to brownish or reddish green tufts, 1-2 cm high, on soil or soil among rocks, often under sagebrush, Washington to Alberta south to California, Arizona, Utah, and Colorado.

4b Leaves unistratose or bistratose only along margins. 5

5a Leaves bordered by differentiated cells. 6

5b Leaves not bordered by differentiated cells. 8

6a Border of leaf composed of elongated cells, often bistratose. Fig. 143D-E. .... *Tortula subulata* Hedw.

Plants gregarious or tufted, green, 5-10 mm high, on soil at high elevations, British Columbia south to California.

6b Border of leaf not composed of elongated cells, unistratose. 7

7a Costa long excurrent as a smooth awn. Fig. 150A-C. ...... *Tortula muralis* Hedw.

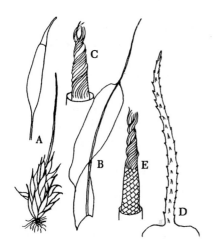

Figure 150

Figure 150 *Tortula muralis*. A, plant; B, leaf; C, peristome; D, apex of leaf of *T. ruralis*; E, peristome of same.

Plants in dense to loose cushions, glaucous-green to dark green, 3-12 mm high, on calcareous rocks, British Columbia south to California and Arizona, New England south to South Carolina, Louisiana and in Texas.

7b Costa percurrent or ending below apex. Fig. 149C. .... *Tortula bolanderi* (Lesq.) Howe

Small plants, up to 5 mm tall, gregarious or scattered, green to reddish brown, on rocks or soil over rocks, Washington, Oregon, and California.

8a Upper leaf cells smooth or only slightly papillose. Fig. 151. .... *Tortula mucronifolia* Schwaegr.

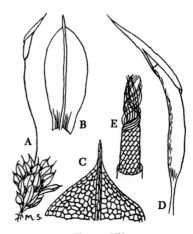

Figure 151

Figure 151 *Tortula mucronifolia*. A, plant; B, leaf; C, apex of leaf; D, capsule with calyptra; E, peristome.

Plants in tufts, scattered or gregarious, green to dark green, up to 1 cm high, on soil and rock, Alaska, British Columbia to Ontario and New England south to Arizona, Utah, Colorado, Iowa and New York.

8b Upper leaf cells strongly papillose. 9

9a Costa abruptly excurrent as a long awn that may be as long as the rest of the leaf; leaves not conspicuously broken. 10

9b Costa percurrent or excurrent as a short mucro or apiculus, never as a long awn; leaves much broken, the fragments serving as propagules. Fig. 147C. ....
.... *Tortula fragilis* Tayl.

Robust plants, 1-4 cm high, in dense cushions, dark green above, brownish-red below, on shaded rocks, southwestern United States disjunct to Missouri, Oklahoma, North Carolina, Virginia and West Virginia.

10a Excurrent costa smooth. ....
.... (p. 103) *Tortula muralis* Hedw.

10b Excurrent costa slightly to strongly serrate. 11

11a Leaves strongly squarrose-recurved when moist. Fig. 150D-E. .... *Tortula ruralis* (Hedw.) Gaertn., Meyer & Schreb.

Plants medium to large, 1-8 cm tall, in dense to loose tufts or mats, green to brownish or reddish-green, on soil and rocks, often calcareous, occasionally on trees, Pacific Northwest south to California, Nevada, Utah and Arizona, Great Lakes regions south to the Ozarks, New England south to Tennessee. A similar species, *Tortula norvegica* (Web.) Wahlenb. *ex* Lindb. has a reddish instead of whitish awn.

11b Leaves erect to wide-spreading, but not squarrose, when moist. 12

12a Margins of leaf revolute from base to apex; excurrent costa as long as rest of leaf; plants conspicuously glaucous. Fig. 149E. .... *Tortula obtusissima* (C. Muell.) Mitt.

Plants densely caespitose, bright glaucous green, up to 2 cm high, Arizona, Nevada, and Texas.

12b Leaf margins narrowly revolute in lower half or two-thirds; plants greenish to brownish-red. Fig. 149D. .... *Tortula princeps* De Not.

Plants in wide dense tufts or mats, greenish or brownish red, 2-10 cm high, on soil and rocks, British Columbia to Idaho south to California and Nevada.

Order Grimmiales

GRIMMIACEAE
Scouleria

1a Leaf margins unistratose or bistratose in spots; peristome present. Fig. 152. ....
.... *Scouleria aquatica* Hook.

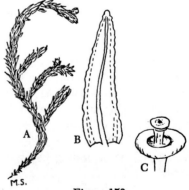

Figure 152

Figure 152 *Scouleria aquatica*. A, plant; B, leaf; C, capsule.

Large plants, stems to 15 cm long, dark green to nearly black, more or less branched, attached to wet rocks in tufts, Alaska south to California.

1b Leaf margins of 3-5 layers except near base and apex; peristome absent. .... ***Scouleria marginata*** Britt.

Stems up to 10 cm long, branched, dark green to black, attached to wet rocks in tufts, Washington, Oregon, Idaho, and California.

Grimmia

At least forty-five species of this large and difficult genus have been recorded from North America. The following key is adapted from *A Key to North America Species of Grimmia* by Geneva Sayre (Bryologist 55:251-259. 1952).

1a Upper leaves (excluding the perichaetial) without whitish hair points. 2

1b Upper leaves with distinct whitish hair points. 6

2a Apex of most upper leaves obtuse or rounded. Fig. 154G. .... ***Grimmia alpicola*** Hedw.

Plants in dark green to brown, loose to dense rigid tufts, usually 1-2 cm high, on rocks in dry exposed places, widespread in North America (absent in Florida and Louisiana).

2b Apex of upper leaves acuminate or acute. 3

3a Leaves under 1.5 mm long, linear-lanceolate, curled and twisted when dry, yellowish, translucent (glossy). Fig. 153E-F. .... ***Grimmia torquata*** Hornsch. *ex* Grev.

Plants grass green in appearance, in soft, often deep tufts, up to 4 cm tall, on rocks, soil over rocks, rock crevices, Pacific Northwest, Alaska, Nevada, Colorado and Ontario.

3b Leaves 1.5-3 mm long. 4

4a On rocks on the seashore; leaves 2 or 3 cells thick in upper part; leaves contorted when dry. Fig. 153A-B. .... ***Grimmia maritima*** Turn.

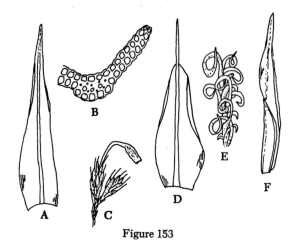

Figure 153

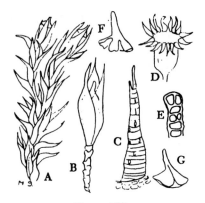

Figure 154

Figure 153 *Grimmia maritima.* A, leaf; B, cross-section of upper leaf. *G. tenerrima;* C, plant with sporophyte; D, leaf. *G. torquata;* E, plant when dry; F, leaf.

Plants in thick, robust, rigid tufts, yellowish green above, dark green or blackish below, up to 3 cm tall, on silicious and basaltic rocks near ocean, often growing within reach of salt spray, Alaska south to California, Labrador, Newfoundland, and New England.

4b   On inland rocks, leaves imbricate when dry. 5

5a   Most leaves acuminate from an ovate base; capsules oblong, peristome orange. Fig. 154A-F. .... *Grimmia apocarpa* Hedw.

Figure 154 *Grimmia apocarpa.* A, plant; B, sporophyte in perichaetial leaves; C, tooth of peristome; D, peristome; E, cross section of margin of leaf; F, calyptra; G, calyptra of *G. alpicola.*

Plants in dense, dark green or brownish tufts, 1-3.5 cm high, on rocks, usually in dry, exposed places, throughout most of North America (absent in Florida).

5b   Most leaves acute to obtuse; capsules ovate, peristome red. Fig. 154G. .... (p. 105) *Grimmia alpicola* Hedw.

6a   Leaf margins inrolled in some part of leaf or the apex channeled. 7

6b   Leaf margins plane or erect, or one or both margins revolute. 10

7a   Leaves drawn out to a long channeled acumen from an ovate base. 8

7b   Leaves lanceolate or linear-lanceolate. 9

8a Seta curved; costa mostly terete; eastern North America. Fig. 155A-C. .... *Grimmia olneyi* Sull.

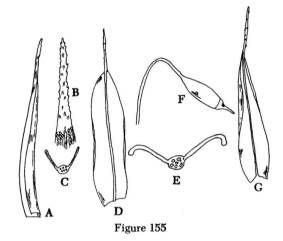

Figure 155

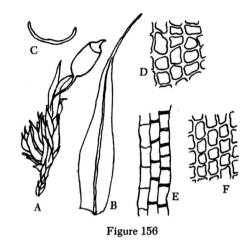

Figure 156

Figure 155 *Grimmia olneyi*. A, leaf; B, apex of leaf; C, cross section of upper leaf. *G. pulvinata;* D, leaf; E, cross of middle of leaf; F, seta and capsule. *G, donniana;* G, leaf.

Plants in loose tufts, dark green above to nearly black below, up to 2 cm high, on rocks, Nova Scotia to Michigan south to Indiana, Missouri and Georgia.

8b Seta straight; costa mostly flat; western North America. Fig. 156F. .... *Grimmia ovalis* (Hedw.) Lindb.

Plants in loose, fragile tufts, dark green above to nearly black below, on rocks, British Columbia to Montana south to Nevada and Arizona.

9a Leaves 1.5-2.5 mm long, narrow at base and tapered all the way to tip, often quite black, leaves equally distributed along the stem from base to tip; operculum long-beaked. Fig. 156A-E. .... *Grimmia montana* B.S.G.

Figure 156 *Grimmia montana*. A, plant; B, leaf; C, section of leaf; D, median cells; E, basal marginal cells; F, median cells of *G. ovalis*.

Plants in dense tufts, dark green or gray-green and hoary, up to 2 cm high, on rock and soil over rock, Pacific Northwest, California, Nevada, Utah, Colorado.

9b Leaves 1-1.5 mm long, broad at base, abruptly pointed, green, often glaucous, tending to be in terminal "cabbage-head" tufts with lower stem leaves absent; operculum short-beaked and blunt. Fig. 153C-D. .... *Grimmia tenerrima* Ren. & Card.

Plants in dense tufts or cushions, green to grayish or glaucous, up to 1 cm high, on rock, Pacific Northwest, California, Arizona, Nevada, Utah, Colorado, Quebec, Labrador.

10a Leaves not keeled, usually clearly concave. 11

10b Leaves keeled, some lying folded at least at apex. 14

11a Leaves broader above the middle or oblong, abruptly acuminate, hair point 1-1/2 times the length of the upper leaves. Fig. 157E-F. .... *Grimmia wrightii* (Sull.) Aust.

Plants in gray-green, usually hoary, compact tufts, up to 1.4 cm high, on rock, usually sandstone, Alberta, South Dakota, and Minnesota south to Arizona, Nevada, Utah, Texas, Kansas, and Missouri.

11b Leaves broader below the middle, gradually acuminate or acute, hair point shorter. 12

12a Leaf cells large and clear, mostly unistratose; hairpoint not decurrent. 13

12b Leaf cells small and obscure, mostly bistratose; hair point decurrent. Fig. 157A-D. .... *Grimmia laevigata* (Brid.) Brid.

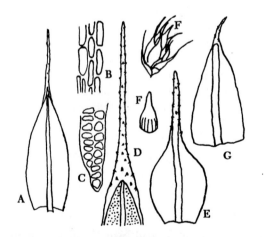

Figure 157

Figure 157 *Grimmia laevigata*. A, leaf; B, median cells; C, section of leaf; D, decurrent awn; E, leaf of *G. wrightii;* F, plant and calyptra; G, leaf of *G. rauii*.

Plants in loose fragile tufts or cushions, hoary above, dark green to almost black below, up to 1.5 cm high, on dry rocks, throughout most of the United States.

13a Leaves acute or obtuse; capsule unsymmetrical, calyptra small. Fig. 160C-D. .... *Grimmia plagiopoda* Hedw.

Plants in fragile tufts or cushions, dark green to nearly black, on rock and soil over rock, British Columbia and Alberta south to Arizona, New Mexico and Colorado, Ontario south to Iowa, Ohio, Maryland.

13b Leaves acuminate; capsule symmetrical, calyptra large, reaching below the middle of the capsule. Fig. 157G. .... *Grimmia rauii* Aust.

Plants in small, compact, gray or greenish tufts, up to 1 cm high, on rocks, usually sandstone, Minnesota and South Dakota, Colorado, Kansas, Missouri, Oklahoma, Texas.

14a Leaves distinctly oblong and acute, both margins revolute near the middle; capsule exserted on a curved seta. Fig. 155D-F. .... *Grimmia pulvinata* (Hedw.) Sm.

Plants in dense hoary cushions, dark green to nearly black below, up to 1.4 cm tall, on rocks, often siliceous, on concrete walls, western North America east to Iowa, Missouri, Arkansas, Maryland, Texas, Ontario.

14b Leaves ovate, lanceolate, or linear, usually acuminate, margins revolute throughout or one or both plane. 15

15a Most mature leaves 2 mm long or longer. 16

15b Most mature leaves 1-1.5 mm long. 21

16a Both leaf margins equally and distinctly revolute; capsule immersed in perichaetial leaves. 17

16b Margins plane or only one revolute, or one revolute and the other slightly recurved below. 18

17a Leaves narrowly acuminate from an ovate base, costa forming about 1/4 of the leaf base, lower leaf cells 3:1, the alr cells hyaline. Fig. 160E-G. .... *Grimmia pilifera* P.-Beauv.

Plants in dense, coarse, spreading, tufts, dark green to slightly hoary, 1-3 cm high, on rocks, rarely on decaying wood, eastern North America from Nova Scotia to Minnesota south to Oklahoma, Arkansas, and Georgia.

17b Leaves lanceolate, some acute, costa narrower, lower leaf cells 2:1, alar cells not differentiated. Fig. 154A-F. .... (p. 106) *Grimmia apocarpa* Hedw.

18a Calyptra covering most of the capsule; hair point 1/2 to 2/3 the length of the leaf, some leaves lanceolate to acute. Fig. 158H. .... *Grimmia calyptrata* Hook. *ex* Drumm.

Plants in tufts or cushion, hoary above, dark green below, up to 2.5 cm high, on rocks, usually in dry places, Pacific Northwest to California, Arizona, New Mexico, Nevada, Utah, Colorado.

18b Calyptra smaller; hairpoint shorter or leaves slenderly acuminate or both. 19

19a Leaves long acuminate from an ovate base, alar cells clearly quadrate with thicker cross-walls; seta straight. Fig. 158A-D. .... *Grimmia affinis* Hoppe & Hornsch. *ex* Hornsch.

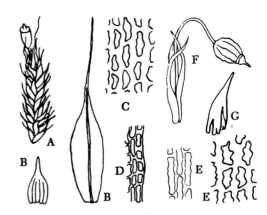

Figure 158

Figure 158 *Grimmia affinis*. A, plant; B, leaf; C, median cells of leaf; D, basal marginal cells; E, median cells of *G. trichophylla*; F, seta and capsule; G, calyptra; H, calyptra of *G. calyptrata*.

Plants in dense tufts or cushions, hoary above, dark green below, up to 2.5 cm high, on rocks, Alaska, British Columbia, Washington to Alberta, Idaho, Montana, Wyoming, New Mexico, Colorado, South Dakota, Minnesota, Ontario.

19b Leaves linear or lanceolate or the alar cells not conspicuously different; seta straight or curved. 20

20a Seta straight; leaf cells bistratose in upper 2/3, margins mostly erect, median leaf cells not sinuose. Fig. 159. .... *Grimmia montana* B.S.G.

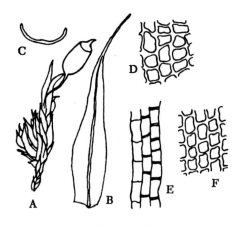

Figure 159

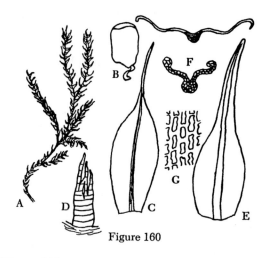

Figure 160

Figure 160 *Grimmia anodon.* A, plant; B, capsule and seta. *G. plagiopodia*; C, leaf; D, tooth; E, leaf of *G. pilifera*; F, sections of leaf; G, median cells.

Figure 159 *Grimmia montana.* A, plant; B, leaf; C, section of leaf; D, median cells; E, basal marginal cells; F, upper leaf cells.

Plants in dense, dark green or gray-green tufts, 0.5-2 cm high, on rocks and soil, Pacific Northwest, California, Nevada, Utah, and Colorado.

Plants in dense tufts or loose cushions, dark green to nearly black, up to 1.2 cm high, on soil or soil over rocks, Yukon south to California, Arizona, Nevada, North Dakota, Colorado, New Mexico.

20b Seta curved; leaf cells unistratose, one or both margins recurved, median leaf cells sinuose. Fig. 158E-G. .... *Grimmia trichophylla* Grev.

21b Leaves keeled, linear or lanceolate or ovate-lanceolate; peristome present. .. 22

22a Alar cells differentiated, usually hyaline. Fig. 155G. .... *Grimmia donniana* Sm.

Plants in tufts or mats, green to yellow green, often black below, 1-3 cm high, on rocks, western North America east to Idaho, Montana and Wyoming.

Plants in dense tufts, dark green to hoary, often nearly black below, up to 2 cm high, on rocks, arctic-alpine, Alaska, Oregon, New England.

21a Leaves concave, broadly oblong-lanceolate; peristome absent. Fig. 160A-D. .... *Grimmia anodon* B.S.G.

22b Alar cells not differentiated. Fig. 154A-F. .... (p. 106) *Grimmia apocarpa* Hedw.

Rhacomitrium

Ten species occur in North America.

1a Leaves without hyaline hair points. 2

1b Leaves with hyaline hair points. 3

2a Apex broad and round. Fig. 161B. .... *Rhacomitrium aciculare* (Hedw.) Brid.

Plants in green to dark green mats or tufts, often nearly black with age, 2-5 cm high or higher, on rocks in or near streams, Alaska south to California, Michigan, Ontario, Quebec to Newfoundland south to Georgia.

2b Apex acute or occasionally narrowly obtuse. Fig. 163F. .... *Rhacomitrium fasciculare* (Hedw.) Brid.

Plants usually yellow-green, dark or brownish with age, in tufts or mats, 3-10 cm tall, stems often prostrate, on rocks, Alaska to Washington, Alberta, Montana, Minnesota, Colorado, Quebec, Ontario, Newfoundland, New England and New York.

3a Leaf cells smooth. 4

3b Leaf cells papillose. 5

4a Hyaline hair point smooth to merely toothed. Fig. 161C-E. ...... *Rhacomitrium heterostichum* (Hedw.) Brid.

Plants often branched, with or without short tuft-like branches, 2-6 cm long, yellow-green to dark green, occasionally gray-green or hoary, on rocks, rotten wood, tree trunks and rail fences, Alaska south to California, Minnesota, Michigan, Ontario, Labrador, Quebec south to Georgia.

4b Hyaline hair point dentate with large papillae along the margin. Fig. 161A. .... *Rhacomitrium lanuginosum* (Hedw.) Brid.

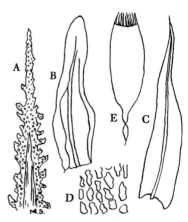

Figure 161

Figure 161 *Rhacomitrium lanuginosum*. A, apex of leaf; B, leaf of *R. aciculare;* C, leaf of *R. heterostichum;* D, cells of lower part of leaf, and E, capsule of the last.

Plants in green, gray-green or brownish tufts or mats, 4-12 cm long, stems with numerous short tuft-like branches, on rocks in rather dry places, Alaska south to Oregon and Idaho, Nova Scotia and Newfoundland south to Quebec and New England.

5a Papillae high. Fig. 162. .... *Rhacomitrium canescens* (Hedw.) Brid.

Musci—The Mosses   111

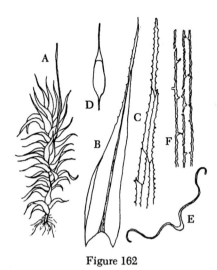

Figure 162

5a Pappilae low, Pacific Northwest. Fig. 163. .... *Rhacomitrium varium* (Mitt.) Jaeg. & Saverb.

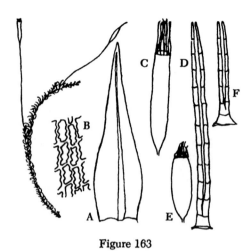

Figure 163

Figure 162 *Rhacomitrium canescens.* A, shoot; B, leaf; C, apex of leaf; D, capsule; E, section of leaf; F, median cells.

Plants in dense to loose, green, yellowish, or brownish mats, stems curved ascending with few to many short, lateral tuft like branches, 2-8 cm long, a pioneer on sand and gravel, on rocks and soil, in rather dry places, circumpolar, Alaska to Greenland south to California, Colorado, Michigan, New York.

Figure 163 *Rhacomitrium varium.* A, leaf; B, upper cells; C, capsule; D, peristome; E, capsule; F, peristome of *R. fasciculare.*

Plants in large, green to yellow-green mats, often with many long branches, 4-10 cm long, on rocks, soil, or logs, endemic to the Pacific Northwest.

Order Funariales

DISCELIACEAE
Discelium

Discelium nudum (Dicks.) Brid. (Fig. 164)

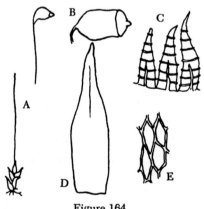

Figure 164

112   Musci—The Mosses

Figure 164 *Discelium nudum.* A, plant; B, capsule; C, teeth of peristome; D, leaf, with faint midrib in upper part; E, cells of leaf.

Minute plants with persistent protonema, almost stemless, on bare clay in spring, British Columbia, Northwest Territories, Illinois, New York, New Jersey, Pennsylvania, Ohio.

EPHEMERACEAE
Ephemerum

The following key is adapted from "The Ephemeraceae in North America by Virginia S. Bryan and Lewis E. Anderson (Bryologist 60:-67-102, 1957).

1a   Upper leaf essentially without a costa. Fig. 46. .... *Ephemerum serratum* (Hedw.) Hampe

Minute plants, practically stemless, growing on soil from persistent, much branched protonema, late fall to early spring, California, Nova Scotia, Saskatchewan, Ohio, Massachusetts, New York, New Jersey.

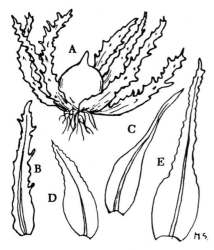

Figure 165

Figure 165 *Ephemerum spinulosum.* A, plant; B, leaf; C, leaf of *E. cohaerens;* D, leaf of *E. crassinervium.*

1b   Upper leaf with a costa. 2

2a   Cells in middle third of leaf smooth, in diagonal rows from costa to margin. Fig. 165D. .... *Ephemerum cohaerens* (Hedw.) Hampe

Plants green or brownish green, ephemeral, on bare soil in wet places, along streams, in pastures, marshes and meadows, Quebec south to Florida, Louisiana, west to Iowa, Missouri, Texas.

2b   Cells in middle third of leaf papillose above, if smooth, not in diagonal rows. 3

3a   Upper leaves with spines mostly recurved at an angle of 45° or more; leaf cells at middle of leaf about 4-8:1. Fig. 165A-B. .... *Ephemerum spinulosum* Bruch & Schimp. *ex* Schimp.

Arises from unusually abundant and persistent, felt-like protonema, on bare soil following flooding, Minnesota and Ontario south to the Gulf States and Texas.

3b   Upper leaves almost entire to strongly toothed, most of the teeth extending at an angle less than 45°; leaf cells at middle of leaf usually less than 4:1. Fig. 165D, 166E. .... *Ephemerum crassinervium* (Schwaegr.) Hampe

Plants arising from uncrowded protonema, on moist soil in disturbed places, especially old fields, on river banks, late summer to early spring, Ontario to Connecticut, Michigan, Missouri, Kansas, Florida, Louisiana, and Texas.

Micromitrium

Micromitrium austinii Aust. (Fig. 166A-D)

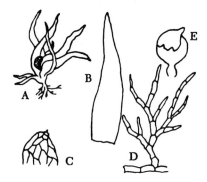

Figure 166

Figure 166 *Micromitrium austinii*, A-D, Aust. A, plant; B, leaf; C, calyptra on capsule; D, protonema; E, calyptra and capsule of *Ephemerum crassinervium*.

Stemless or nearly stemless plants developing from persistent protonema, on moist soil, on soil of flood plains, Gulf States north to Illinois and Connecticut. Four other species occur in North America.

FUNARIACEAE
Physcomitrella

Physcomitriella patens (Hedw.) B.S.G. (Figs. 47A, 167, 168E)

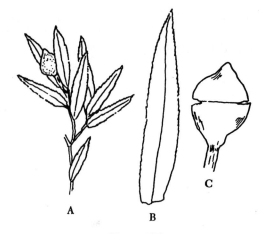

Figure 167

Figure 167 *Physcomitrella patens*. A, plant; B, leaf; C, capsule.

Small, inconspicuous plants with immersed, inoperculate capsules, on alluvial soil around lakes, ponds, streams, Quebec, Ohio, Minnesota to Missouri and South Carolina. *P. californica* Crum and Anderson occurs in California.

Aphanorrhegma

Aphanorrhegma serratum (J. Hook. & Wils. ex Drumm.) Sull. (Figs. 47B, 168)

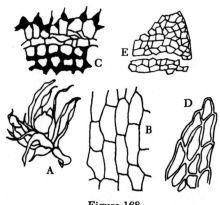

Figure 168

114   Musci—The Mosses

Figure 168 *Aphanorhegma serratum*. A, plant; B, cells of leaf; C, cells of outer wall of capsule, with thickened walls; D, apex of leaf; E, line of dehiscence of *Physcomitrella patens*.

Minute plants growing of alluvial soils in late fall and early spring, eastern North America from southern Canada to the Gulf States.

Physcomitrium

Eight species are known from North America.

1a   Seta not longer than perichaetial leaves. Fig. 169. ....
.... *Physcomitrium immersum* Sull.

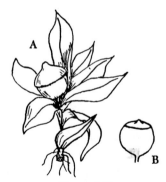

Figure 169

Figure 169 *Physcomitrium immersum*. A, plant; B, capsule just mature; C, old capsule.

Often growing with and confused with *Aphanorrhegma serratum;* plants usually somewhat larger, up to 8 mm, and its capsules are turbinate when dry, on damp soil often along streams, British Columbia, Minnesota, Wisconsin, Quebec, New England south to Oregon, Iowa, Pennsylvania.

1b   Seta longer than perichaetial leaves. 2

2a   Seta 5-15 mm long; mouth of capsule bordered by 8-10 rows of flattened cells. Fig. 170A-D. .... *Physcomitrium pyriforme* (Hedw.) Hampe

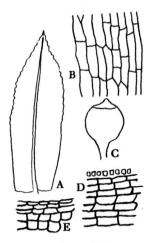

Figure 170

Figure 170 *Physcomitrium pyriforme*. A, leaf; B, cells of leaf; C, capsule just mature; D, cells at mouth of capsule; E, cells at mouth of capsule of *P. hookeri*.

Small, light green, tufted plants, on soil in pastures, lawns, waste places, throughout eastern North America.

2b   Seta 2-4 mm long; mouth of capsule bordered by 3-5 rows of flattened cells; annulus large, rolling back when the capsule opens. Fig. 170E. ....
.... *Physcomitrium hookeri* Hampe

Small plants, 1-2 mm high, densely clustered, light to brownish green, on soil, Montana, Wyoming, Manitoba, Utah, Wisconsin, Nebraska, Iowa.

Pyrimidula

Pyramidula tetragona (Brid.) Brid. (Fig. 171)

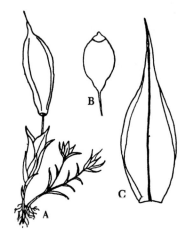

Figure 171

Figure 171 *Pyramidula tetragona.* A, plant; B, capsule; C, leaf.

Small annual plants in dense to loose tufts or mats, up to 1 mm high, calyptra of capsule 4-angled, on soil, Colorado, Minnesota, Nebraska, Iowa, Texas.

Funaria

Six species are reported from North America.

1a Annulus present and conspicuous. 2

1b Annulus absent. 4

2a Segment of inner peristome obtuse, less than 1/2 length of outer teeth; spores 20-30 μ in diameter. Fig. 172F. .... *Funaria flavicans* Michx.

Plants in loose, light green tufts, 1.5-5 mm high, on soil often mixed with *F. hygrometrica,* Iowa, Indiana, southern Ontario to Connecticut south to the Gulf States.

2b Segments of inner peristome lanceolate, slenderly pointed, at least 3/4 length of outer teeth; spores 12-18 μ in diameter. 3

3a Capsule horizontal to pendant, turgid, strongly arcuate. Fig. 172A-D. .... *Funaria hygrometrica* Hedw. var. *hygrometrica*

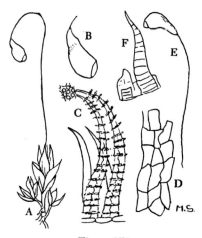

Figure 172

Figure 172 *Funaria hygrometrica.* A, plant; B, capsule; C, peristome; D, cells of leaf. *F. hygrometrica* var. *calvescens.* E, erect seta with the capsule inclined; *F. flavicans.* F, peristome. segments short and truncate.

Plants in loose to dense tufts, often in large patches, light green to pale yellowish green, 3-10 mm high, in waste places and on bare soil, often where fires have occurred, throughout North America.

3b Capsule slender, merely inclined, less arcuate. Fig. 172E. ... *Funaria hygrometrica* var. *calvescens* (Schwaegr.) Mont.

On soil, Iowa, Tennessee, and North Carolina south to the Gulf States and Texas.

4a Leaves entire, costa excurrent. Fig. 173A-C. .... *Funaria americana* Lindb.

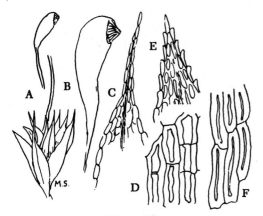

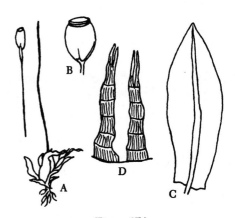

Figure 173

Figure 174

Figure 173 *Funaria americana*. A, plant; B, capsule; C, apex of leaf and margin; D, upper cells of capsule of *F. serrata* from a furrow; E, apex and margin of leaf of same; F, upper cells of capsule from a ridge.

Plants light green to yellowish green, in tufts up to 8 mm tall, on moist soil or humus, Washington, Canadian Rockies, Montana, Wisconsin, Illinois, Ohio, Pennsylvania south in Missouri, Oklahoma, Tennessee and Texas.

4b Leaves serrate, costa ending below apex in most leaves. Fig. 173E-F. .... *Funaria serrata* Brid.

Plants gregarious, up to 6 mm high, on bare soil, Georgia, Louisiana, Arkansas, Oklahoma, Texas.

Entosthodon

Entosthodon drummondii Sull. (Fig. 174)

Figure 174 *Entosthodon drummondii* Sull. A, plant; B, ripe capsule; C, leaf; D, peristome. Stems 1-4 mm tall; seta 10-15 mm. Without annulus.

Plants in loose tufts, yellowish-green, 1-4 mm high, on sandy soil, Georgia, Florida, Louisiana, Texas. Twelve species are known from North America.

OEDIPODIACEAE
Oedipodium

Oedipodium griffithianum (Dicks.) Schwaegr. (Figs. 49D, 175)

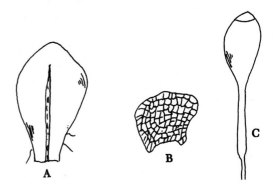

Figure 175

Musci—The Mosses   117

Figure 175 *Oedipodium griffithianum.* A, leaf; B, multicellular gemmae; C, capsule.

Small, glaucous green, gregarious plants 2-5 mm high, on soil in rock crevices, Alaska, Washington.

SPLACHNACEAE
Voitia

Voitia nivalis Hornsch. (Fig. 176)

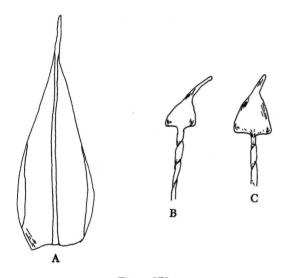

Figure 176

Figure 176 *Voitia nivalis.* A, leaf; B-C, dry capsules.

Plants in light green to yellow green tufts, 1-6 cm high, on peaty soil at high elevations, Alaska, Alberta and Colorado.

Tayloria

Tayloria serrata (Hedw.) B.S.G. (Figs. 49C, 177A-D)

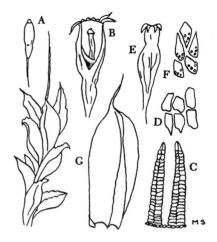

Figure 177

Figure 177 *Tayloria serrata.* A, plant; B, capsule; C, peristome; D, cells of leaf, E, capsule of *Tetraplodon mnioides;* F, cells of leaf; G, leaf.

Tufted plants, clear green above, brown below, 0.5-3 cm high, on humus or animal excrement, Alaska, Alberta south to Oregon and Montana, Nova Scotia, Quebec and New England. Five other species are known from North America.

Tetraplodon

Tetraplodon mnioides (Hedw.) B.S.G. (Figs. 49B, 177E-G)

Plants in dense green tufts, brown below, 2-8 cm high, on rich organic soil or dung, Alaska, British Columbia to Ontario, Nova Scotia, Labrador south to Washington, Colorado, New England, New York. Two other species of *Tetraplodon* are found in North America.

Splachnum

Six species are found in North America.

1a Leaves narrowly lanceolate; hypothesis of capsule narrower or only slightly wider than the capsule. .... *Splachnum pennsylvanicum* (Brid.) Grout *ex* Crum

Plants 1-3 cm high, leaves distant on stem, green above, in loose tufts on soils rich in organic matter, Atlantic Coast, Nova Scotia to Florida, Louisiana, Texas.

1b Leaves not narrowly lanceolate; hypothesis broader than capsule. 2

2a Hypothesis more or less globose, rugose when dry. Fig. 178. .... *Splachnum ampullaceum* Hedw.

Figure 178 *Splachnum ampullaceum*. A, plant; B, capsule; C, peristome; D, cells of leaf.

Plants in pale green tufts, 1-2.5 cm high, growing on soil rich in organic matter, Alaska to British Columbia, Michigan, Ontario, Quebec, New England south to New Jersey and Pennsylvania.

2b Hypothesis umbrella shaped, smooth. .... *Splachnum luteum* Hedw.

Plants loosely tufts, 2-3 cm high, on dung in woods, swamps, and bogs, Northwest Territories, New Brunswick. This species like several others of this genus is striking in the appearance of it large hypothesis. Many species are said to emit a manurial odor and thus attract flies which carry the spores.

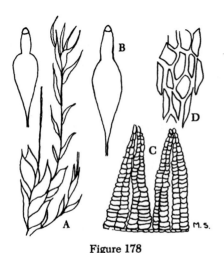

Figure 178

Order Schistostegales

SCHISTOSTEGACEAE
Schistostega

Schistostega pennata (Hedw.) Web. & Mohr.

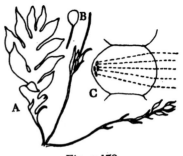

Figure 179

Figure 179 *Schistostega pennata.* The Luminous Moss. A, the "leaf" or sterile shoot; B, capsule; C, one cell of protonema.

Plants glaucous green or brown, 2-10 mm high, leaves distichous, on damp soil, wood, or upturned stumps, in caves or dark places in woods, British Columbia and Alberta south to Washington, Montana, Michigan, Wisconsin, Ontario, Quebec, New Brunswick, Nova Scotia south to Ohio, New England and New York. The cells of protonema are so shaped as to focus incoming light on chloroplats at the inner side of the cell cavity. This light is reflected back, causing a greenish golden glow.

Order Eubryales

BRYACEAE
Mielichhofera

Mielichhoferia mielichhoferi (Funck *ex* Hook.) Loeske (Fig. 180)

Figure 180 *Mielichhoferia mielichhoferi.* A, plant with sporophyte; B-C, leaves; D, apex of leaf; E, portion of peristome; F, annulus.

Plants in erect, dense, yellowish green tufts 1-2 cm high, on rocks, Colorado, California, Michigan, North Carolina, Tennessee. The presence of this moss usually indicates that copper is contained in the rock. One other species, *M. macrocarpa* (Hook. *ex* Drumm.) Bruch & Schimp. *ex* Jaeg. & Sauerb. is found in western North America.

Pohlia

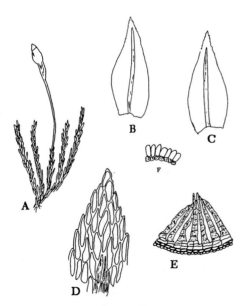

Figure 180

1a	Plants producing gemmae in the axils of leaves. 2
1b	Plants not producing gemmae. 5
2a	Gemmae large, rounded, reddish, occurring singly in leaf axils. Fig. 181A. .... *Pohlia rothii* (Corr. *ex* Limpr.) Broth.

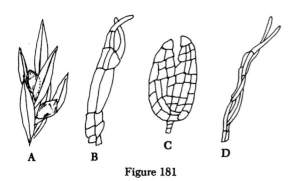

Figure 181

Figure 181 Gemmae. A, *Pohlia rothii;* B, *P. proligera;* C, *P. annotina* var. *annotina;* D, *P. annotina* var. *decipiens.*

Plants in loose to dense tufts, yellow-green with a decided luster, up to 2 cm high, on moist sand, Northeast North America.

2b Gemmae smaller, not reddish, usually more than one in a leaf axis. 3

3a Plants glossy with broad leaves. Fig. 181B. .... *Pohlia proligera* (Kindb. *ex* Limpr.) Lindb. *ex* Arnell

Plants up to 2 cm high in glossy, yellow-green tufts, on moist, often sandy, soil, British Columbia, Alaska, Wyoming, Idaho, Quebec, south to Washington, Nevada, Minnesota, Michigan.

3b Leaves narrower, not glossy. 4

4a Gemmae short, ovoid to wedge-shaped. Fig. 181C. .... *Pohlia annotina* (Hedw.) Lindb. var. *annotina*

Plants in yellow-green tufts 1-2 cm high, on damp soil, Alaska south to California and Colorado, may occur in the eastern United States.

4b Gemmae elongated and twisted. Figs. 181D, 184E. ..... *Pohlia annotina* (Hedw). Lindb. var. *decipiens* Loeske

On damp soil, British Columbia, Washington, Iowa, Missouri, Arkansas, New England to South Carolina.

5a Leaves glossy, with a metallic luster. Fig. 182A-B. .... *Pohlia cruda* (Hedw.) Lindb.

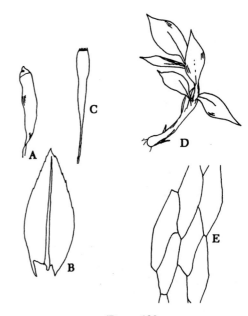

Figure 182

Figure 182 *Pohlia cruda.* A, capsule; B, leaf. *P. elongata;* C, capsule. *Epipterygium tozeri;* D, plant; E, cells of leaf.

Plants in loose, soft, shiny ligh or whitish green, yellowish or bluish tufts, 1-5 cm high, on soil or humus in shaded places, especially clifts or banks protected by overhanging turf, Alaska, Pacific Northwest to Quebec and New Brunswick south to California, Arizona, Colorado,

Minnesota, Nebraska, Iowa, North Carolina and Tennessee.

5b Leaves not glossy. 6

6a Plants whitish green in color, leaves widely spaced. Fig. 183. .... *Pohlia wahlenbergii* (Web. & Mohr.) Andr.

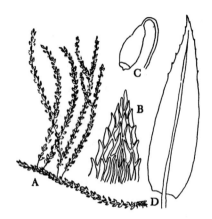

Figure 183

Figure 183 *Pohlia wahlenbergii*. A, plant; B, apex of leaf; C, capsule; D, leaf.

Plants in loose tufts or mats, light green to whitish green, 1-5 cm high, on wet soil near streams, throughout North America.

6b Plants not as above. 7

7a Neck as long or longer than rest of capsule; alpine in distribution. Fig. 182C. .... *Pohlia elongata* Hedw.

Plants tufted, green to yellow green, 1-3 cm high, on soil among rocks usually alpine, British Columbia, Colorado, Minnesota, New England, North Carolina, Tennessee.

7b Neck shorter than rest of capsule; widespread and common. Fig. 184A-D. .... *Pohlia nutans* (Hedw.) Lindb.

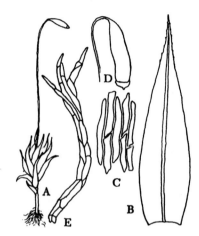

Figure 184

Figure 184 *Pohlia nutans*. A, plant; B, leaf; C, cells of leaf; D, capsule; E, gemma of *Pohlia annotina* var. *decipiens*.

Plants in tufts, green to yellow-green, up to about 1 cm high, on wet to dry soil, decaying wood, rock crevices, cosmopolitan, throughout most of North America, absent in the Gulf States.

Epipterygium

Epipterygium tozeri (Grev.) Lindb. (Fig. 182D-E)

Plants scattered or in loose tufts, green to reddish, 5-10 mm high, on wet clay and sandy banks of lowlands, British Columbia, to California.

Brachymenium

Four species of *Brachymenium* are known from North America.

1a Costa excurrent as a long slender hyaline hair-point; leaves overlapping when dry. Fig. 185A. ...... *Brachymenium systylium* (C. Muell.) Jaeg. & Sauerb.

Leptobryum

Leptobryum pyriforme (Hedw.) Wils. (Fig. 186)

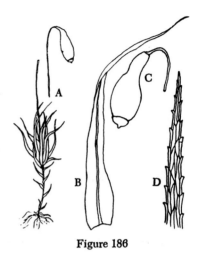

Figure 186

Figure 186 *Leptobryum pyriforme*. A, plant; B, leaf; C, capsule; D, apex of leaf.

Slender plants, 0.5-1.5 cm high, in loose or dense shiny, light or yellow-green tufts, on soil, rotten wood, humus, or rocks, in moist places, especially on disturbed soil, throughout eastern North America, in the west from Alaska south to California, Arizona, New Mexico, Colorado, Utah.

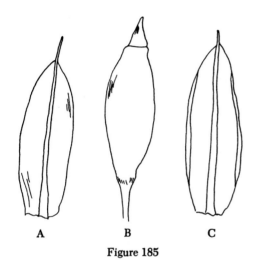

Figure 185

Figure 185 *Brachymenium systylium*. A, leaf; *B. macrocarpum*; B, capsule; C, leaf.

Densely tufted plants, dark green to reddish, tomentose below, on sandy calcareous soil, Florida, North Carolina, New Mexico and Arizona.

1b Costa short-excurrent, not hyaline; leaves spirally twisted when dry. Fig. 185B-C. .... *Brachymenium macrocarpum* Card.

Tufted, dark green to brownish plants, tomentose below, on trees, logs, rocks, Florida, Louisiana, Arizona, New Mexico.

Bryum

Forty-eight species of this difficult genus are found in North America.

1a Plants silvery to whitish green in color. Fig. 187. .... *Bryum argenteum* Hedw.

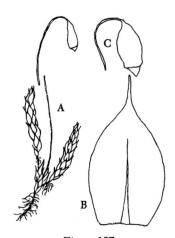

Figure 187

Figure 187 *Bryum argenteum*. A, plant; B, leaf; C, capsule.

Small, silvery to whitish green tufted plant, on soil, rock, brick walls, sidewalks, shingle roofs, cosmopolitan.

1b   Plants not silvery or whitish green in color. 2

2a   Leaf cells large, 150 x 50 μ, western North America. Fig. 188A-C. .... *Bryum sandbergi* Holz.

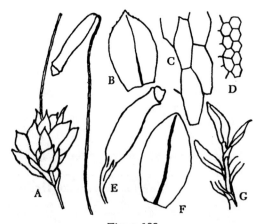

Figure 188

Figure 188 *Bryum sandbergii*. A, plant; B, leaf; C, cells of leaf; D, cells of *Mnium cuspidatum* on same scale; E, capsule; F, leaf of *B. miniatum;* G, stem leaves of *B. weigelii*.

Plants large, scattered in loose, light green tufts, often glossy, up to 2.5 cm high, on soil and humus in wood, Pacific Northwest, California, Colorado. Many authors place this species in the genus *Roellia* in the Mniaceae.

2b   Leaf cells smaller, less than 100 $\times$ 35 μ. 3

3a   Leaves far apart with strongly decurrent bases, border absent or indistinct, Fig. 188G. .... *Bryum weigelii* Spreng.

Plants large, 3-10 cm high, in soft, loose, light green tufts, in wet places, Alaska to California, Utah, Colorado, Montana, Wyoming, Ontario, Quebec, Nova Scotia to Michigan, New England.

3b   Leaves not as above. 4

4a   Upper leaves obtuse to rounded at apex, not decurrent, border of well defined cells absent, costa usually ending below apex. Fig. 188F. ... *Bryum minatum* Lesq.

Medium to large plants in red to reddish brown or green tufts, up to 5 cm high, on moist rocks and soil, British Columbia to California, inland to Montana and Utah and disjunct to south-central Missouri.

4b   Upper leaves not as above or if obtuse then slightly decurrent. 5

5a Upper leaves 1.5-2.5 mm long, not long acuminate, apex acute to obtuse, margins entire and not bordered; gemmae often present in axis of leaves. Fig. 193D. ....
.... **Bryum gemmiparum De Not**

Plants in green tufts, often whitish due to deposits of calcium carbonate, 1-2 cm high, on calcareous soil and rocks, Washington, Oregon, Montana, California, Oklahoma, Missouri, Arizona, Ontario south to Tennessee and North Carolina.

5b Plants not as above. 6

6a Small plants up to 0.5 cm (rarely 1 cm) high, gemmae often present in axils of leaves. Fig. 189A-C. ....
.... **Bryum bicolor Dicks.**

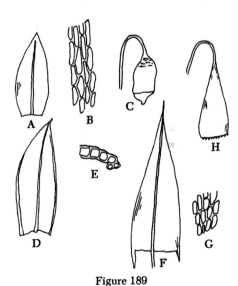

Figure 189

Figure 189 *Bryum bicolor.* A, leaf; B, median leaf cells; C, dried capsule. *B. pallens;* D, leaf; E, cross edge of leaf. *B. turbinatum;* F, leaf; G, median leaf cells; H, dried capsule.

Small tufted plants, green above, brown below, on moist soil, often in disturbed areas, British Columbia, Washington, Missouri, Indiana, New England to Florida.

6b Plants larger. 7

7a Median leaf cells long, 6-7:1; leaves imbricate; costa excurrent. Fig. 190 A-D. ....
.... **Bryum caespiticium Hedw.**

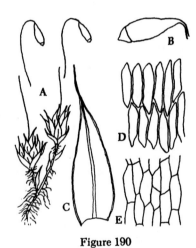

Figure 190

Figure 190 *Bryum caespiticium.* A, plant; B, capsule; C, leaf; D, cells of leaf; E, leaf cells of *B. capillare.*

Plants caespitose, green above, stems red, up to 1 cm high, on soil often in disturbed places, cosmopolitan.

7b Median leaf cells shorter, 2-3:1. 8

Musci—The Mosses

8a   Exostome joined to the endostome thus giving the peristome a chambered appearance (as seen under the microscope before complete soaking), capsules usually pendant. Fig. 191E. ....
.. ***Bryum algovicum* Sendtn. *ex* C. Muell.**

Plants densely tufted, green or yellow green above, radiculose below, up to 1 cm high, on soil, rocks, sand or gravel, Alaska to Quebec south to California, Nevada, Colorado, Kansas, Missouri, Indiana, Ohio, District of Columbia.

8b   Peristome not as above. 9

9a   Leaves with a distinct border, bistratose in part. 10

9b   Leaves with a unistratose border. 11

10a Plants dioicous; spores 13-20 μ in diameter. Fig. 189D-E. ....
...... ***Bryum pallens* (Brid.) Sw. *ex* Roehl.**

Plants in loose tufts, green to reddish, up to 1 cm high, on wet soil, British Columbia to Oregon and Wyoming, Quebec and Newfoundland south to New York.

10b Autoicous; spore 23-30 μ in diameter. Fig. 191A-D. ....
.... ***Bryum uliginosum* (Brid.) B.S.G.**

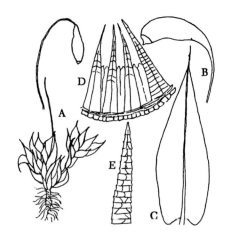

Figure 191

Figure 191 *Bryum uliginosum*. A, plant; B, capsule; C, leaf; D, peristome; E, tooth of *B. algovicum*.

Plants in loose or dense, green, yellowish or brownish tufts, up to 2 cm high, on peaty or sandy soil in wet places, British Columbia, Alberta, Montana, Colorado, Minnesota, Michigan, Labrador south to Illinois and New York, Texas and New Mexico.

11a Leaves obovate, twisted and contorted when dry. Fig. 190E. ....
.... ***Bryum capillare* Hedw.**

Plants usually in dark green or brownish tufts, up to 1 cm high, papillose gemmae often clustered in leaf axils, on rock, soil or humus, tree bases or in crotches or drainage channels of tree trunks, cosmopolitan.

11b Leaves not as above. 12

12a Costa percurrent to short excurrent; leaf cells with pitted walls. 13

12b Costa distinctly excurrent at least in upper leaves; cell walls of leaf hardly or not pitted. 15

13a Capsule brown; leaves somewhat contorted when dry, decurrent. 14

13b Capsule red; leaves imbricate, not or scarcely decurrent. Fig. 192D. ....
.... *Bryum pseudotriquetrum* (Hedw.) Gaertn., Meyer & Schreb. var. *crassirameum* (Ren. & Card.) Lawt.

Tufted green to reddish brown plants, up to 8 cm high, on wet soil and soil over rocks, often near streams, British Columbia to Montana and Calif.

14a Dioicous. Fig. 192A-C. ....
.... *Bryum pseudotriquetrum* (Hedw.) Gaertn., Meyer & Schreb. var. *pseudotriquetrum*

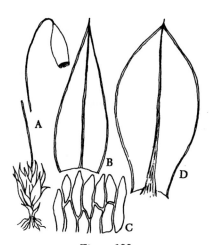

Figure 192

Figure 192 *Bryum pseudotriquetrum*. A, plant; B, leaf; C, cells of leaf; D, leaf of *B. pseudotriquetrum* var. *crassirameum*.

Tufted green to reddish brown plants, up to 8 cm high, on wet soil, soil over rocks, humus, decaying wood, Pacific Northwest south to California, Colorado, throughout the midwest and eastern North America south to North Carolina, Tennessee, Arkansas, Oklahoma, Texas.

14b Synoicous. ...... *Bryum pseudotriquetrum* (Hedw.) Gaertn., Meyer & Anderson var. *bimum* (Schreb.) Lilj.

On soil and rocks, British Columbia, Washington.

15a Leaf border indistinct; dioicous; outer peristome teeth light yellow. Fig. 189F-H. ....
Bryum turbinatum (Hedw.) Turn.

Plants in loose to dense, yellow green or reddish tufts, up to 4 cm high, on wet soil in mountains, Western North America, Yukon, British Columbia, Wyoming and Montana to California, Utah, Colorado and Arizona.

15b Leaf border distinct; autoicous or synoicous; outer peristome teeth darker, at least below. 16

16a Synoicous; leaves imbricate to slightly contorted when dry. Fig. 193A-B. ....
.... *Bryum creberrimum* Tayl.

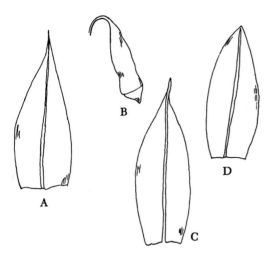

Figure 193

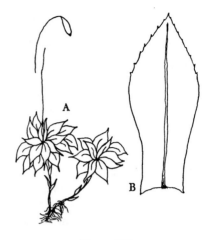

Figure 194

Figure 193 *Bryum creberrimum.* A, leaf; B, dry capsule. *B. pallescens;* C, leaf. *B. gemmiparum;* D, leaf.

Plants in green tufts, up to 1 cm high, on soil or soil over rocks, cosmopolitan.

16b Autoicous; leaves somewhat contorted when dry; in dense cushions, western North America. Fig. 193C. ....
Bryum pallescens Schleich. *ex* Schwaegr.

Plants in dense, deep tufts, green to yellow-green, up to 4 cm high, on moist soil or soil over rock, Pacific Northwest, California, Nevada, Colorado.

Rhodobryum

Rhodobryum roseum (Hedw.) Limpr. (Fig. 194)

Figure 194 *Rhodobryum roseum.* A, plant; B, leaf.

Plant large, up to 5 cm high, dark green and often tinged with red, secondary stems erect, bearing rosettes of leaves up to 1.5 cm across, in woods on soil and soil over rocks, widely distributed in North America south North Carolina, Arkansas and Arizona.

MNIACEAE
Leucolepis

Leucolepis menziesii (Hook.) Steere *ex* L. Koch (Fig. 195)

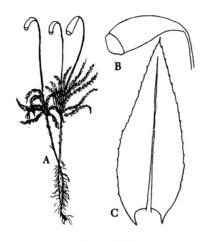

Figure 195

Figure 195 *Leucolepis menziesii*. A, plant; B, capsule; C, leaf.

Plants dendroid, large, 4-8 cm high, reddish brown to almost black, on soil, logs, tree trunks, endemic to western North America, Alaska to California, Idaho.

Mnium

Twenty-eight species of *Mnium* are reported from North America. Recent authors, notably Lawton (1971), divide this genus into three genera, *Mnium, Rhizomnium* and *Plagiomnium*.

1a Margins of leaves, entire, thickened. 2

1b Margins of leaves toothed. 3

2a Leaf apex rounded at apex; cell walls thick and pitted. Fig. 196A-B. .... *Mnium glabrescens* Kindb.

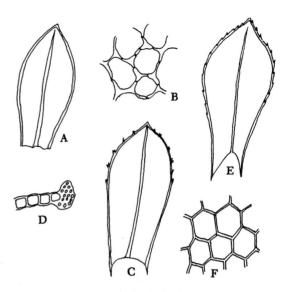

Figure 196

Figure 196 *Mnium glabrescens*. A, leaf; B, median leaf cells. *M. spinulosum;* C, leaf; D, cross section of edge of leaf showing stereid cells. *M. drummondii;* E, leaf; F, median leaf cells.

Plants scattered or tufted, green, up to 3 cm high, well developed leaves clustered near top of erect, naked stems, on humus, soil over rocks, decaying logs, endemic to western North America, Alaska to Idaho and California.

2b Leaf apex broadly rounded to emarginate. Fig. 197A-B. .... *Mnium punctatum* Hedw. var. *punctatum*

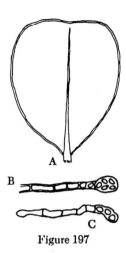

Figure 197

Figure 197 *Mnium punctatum*. A, leaf; B, section of border; C, section of border of leaf of *Cinclidium stygium*.

Moderate to large plants in loose light to dark green tufts 5-6 cm tall, on wet soil, humus, logs and rocks, usually in swamps, along streams and near springs, throughout eastern North America south to Gulf States. *M. punctatum* var. *elatum* Schimp. is a very large (up to 12 cm tall) form with leaves that lack an apiculus and a costa that ends below the apex.

3a   Leaves without a distinct border of elongated cells. Fig. 198. .... *Mnium stellare* Hedw.

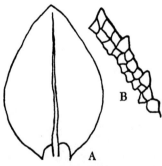

Figure 198

Figure 198 *Mnium stellare*. A, leaf; B, apex and margin of leaf.

Small plants in loose tufts up to 5 cm tall, dark green or red brown, leaf cells with pointed ends, on shaded, wet soil, humus, logs or rocks, Saskatchewan, Ontario to New Brunswick south to North Carolina, Indiana, Illinois, Arkansas.

3b   Leaves with a border of elongated cells. 4

4a   Marginal teeth double. 5

4b   Marginal teeth single. 8

5a   Leaves long and narrow, the costa ending below the apex, toothed on back near apex. Fig. 199. .... *Mnium hornum* Hedw.

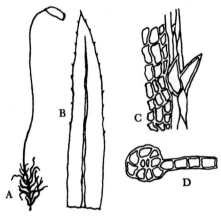

Figure 199

Figure 199 *Mnium hornum*. A, plant; B, leaf; C, cells and margin with paired teeth; D, section of margin.

Plants, green, in close to loose tufts, up to 7-8 cm high, on humus soil over rocks in wet places and along streams, Ontario to Labrador south to Georgia, Alabama, Arkansas.

5b Leaves obovate to ovate-lanceolate, costa reaching apex to excurrent, not toothed on back. 6

6a Leaf cells 21-30 µ, occasionally larger, in diameter; plants; synoicous or paroicous. 7

6b Leaf cells 15-21 µ in diameter; plants dioicous. Fig. 200A-C. .... *Mnium thomsonii* Schimp.

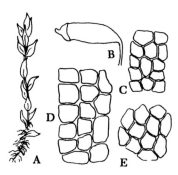

Figure 200

Figure 200 *Mnium thomsonii*. A, plant; B, capsule and operculum; C, cells of leaf; D, cells of *M. marginatum*; E, cells of *M. spinulosum*.

Plants up to 2 cm high, in green with reddish tinged tufts, on damp shaded soil, humus or soil over rocks, tree bases along stream banks and clifts, Alaska and British Columbia to Ontario, Quebec and Nova Scotia south to Oregon, Montana, Wyoming, Colorado, Arkansas, Tennessee and North Carolina.

7a Cell walls of leaves not thickened at the corners, border with inner cells stereid; synoicous. Fig. 196C-D. .... *Mnium spinulosum* B.S.G.

Plants up to 2.5 cm high, in loose deep green tufts, often tinged with red, on soil, humus, logs, and rocks of coniferous forests, Alaska to Ontario south to Oregon, Idaho, Colorado, Minnesota, Michigan, New England, New York.

7b Cell walls of leaves thickened at the corners, leaf border without stereid cells; paroicous. Fig. 200D. .... *Mnium marginatum* (With) Brid. *ex* P. -Beauv.

Slender plants in green or brownish tufts, 1-4 cm high, on moist, usually calcareous soil and rocks in shaded places, Alaska to Ontario and Quebec south to California, Arizona, Colorado, Iowa, Arkansas, Tennessee, and North Carolina.

8a All stems erect, without runners; capsule with a brown warty neck. Fig. 201E. .... *Mnium venustum* Mitt.

Plants loosely tufted, 2-4 cm high, green to yellowish green, on soil, rotten logs and tree trunks in woods, Alaska to California, Idaho, Montana.

8b Erect fruiting stems arising from horizontal runners. 9

9a Leaf cells hexagonal, cell walls thin; plants glaucous. Fig. 196E-F. .... *Mnium drummondii* Bruch & Schimp.

Plant glaucous in tufts or mats, 2-4 cm high, on moist soil and decaying wood, British Columbia, Northwest Territory to Ontario, Quebec and Nova Scotia south to Minnesota and New England.

9b Leaf cells longer in one axis than another, not precisely hexagonal. 10

10a Stems winged by the broadly decurrent leaves; dioicous. .... *Mnium insigne* Mitt.

Plants in green to yellow-green scattered or matted tufts, 3-8 cm high, on moist soil in woods, Alaska to California, Montana, Idaho.

10b Leaves much less broadly decurrent or not at all. 11

11a Leaves serrate only in upper 1/2 or 2/3. Fig. 201A-B. .... *Mnium cuspidatum* Hedw.

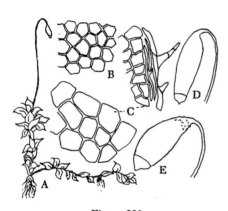

Figure 201

Figure 201 *Mnium cuspidatum*. A, plant; B, cells of leaf; C, cells of *M. affine* var. *ciliare* with 3-celled tooth on border; D, capsule of same; E, capsule of *M. venustum*.

Plants in dark green to yellow-green mats with fertile stems erect, 1-2 cm high, on shaded soil and rocks, British Columbia to Oregon and Montana in the west, widespread elsewhere in North America.

11b Leaves serrate nearly to base. 12

12a Teeth of leaves 1-3 cells long; apex obtuse or mucronate; dioicous. Fig. 201C-D. .... *Mnium affine* Bland. *ex* Funck. var. *ciliare* C. M.

Plants in green to yellow-green loose or dense tufts 3-9 cm high, on moist humus, soil, rocks, rotten wood, cosmopolitan.

12b Teeth of leaves one-cell long, rarely more; apex acute to acuminate; synoicous. .... *Mnium medium* B.S.G.

Plants in loose tufts, light to yellow-green, 2-7 cm high, on humus, soil, rock, bases of trees in wet woods, Alaska to Ontario and Nova Scotia south to California, Nevada, Utah, Colorado, Arkansas and Maryland.

Cinclidium

Cinclidium stygium Sw. (Fig. 197C)

Plants large, 2-10 cm high, in deep tufts, green at tips, reddish brown to nearly black below, in logs and wet places, Alaska to Labrador and Newfoundland south to British Columbia, Ontario, Michigan. Three other species are known from North America.

RHIZOGONIACEAE
Rhizogonium

Rhizogonium spiniforme (Hedw.) Bruch *ex* Kraus (Fig. 202)

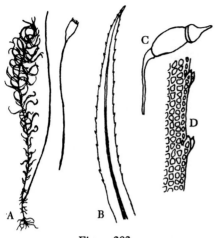

Figure 202

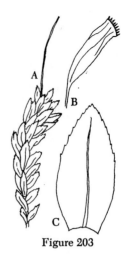

Figure 203

Figure 202 *Rhizogonium spiniforme*. A, a twig; B, leaf; C, capsule; D, margin of leaf.

Plant large, up to 8 cm high, in large loose tufts, deep yellowish or brownish green, on sandy soil or roots in wet places, especially seepage springs, Florida, Georgia, Louisiana. A tropical genus and family.

AULACOMNIACEAE
Aulacomnium

Five species are known from North America.

1a Leaves concave, broadly oblong or oblong-ovate, coarsely serrate in the upper half, scarcely contorted when dry, cells smooth. Fig. 203. .... *Aulacomnium heterostichum* (Hedw.) B.S.G.

Figure 203 *Aulacomnium heterostichum*. A, plant; B, capsule; C, leaf.

Plants in loose, green or yellow-brown tufts, 1.5-4 cm high, on soil, tree bases, and stumps in woods, widespread in eastern and midwest North America.

1b Leaves keeled, oblong-lanceolate, serrulate near apex, loosely erect and curved or contorted when dry; cells unipapillose on both surfaces. 2

2a Plants fairly robust, 3-9 cm high; leaves contorted when dry; alar cells swollen and usually brownish in several rows; sterile stems bearing terminal clusters of triangular gemmae at the end a seta-like extension of stem and also a few along the extension. Fig. 204 A-D. .... *Aulacomnium palustre* (Hedw.) Schwaegr.

Musci—The Mosses 133

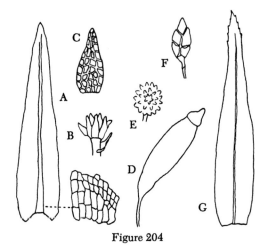

Figure 204

Figure 204 *Aulacomnium palustre*. A, leaf with swollen basal cells; B, cluster of gemmae; C, gemma; D, capsule; E, cluster of gemmae of *A. androgynum*; F, one gemma; G, leaf.

Large plants in loose to dense, yellow, yellow-green or yellow-brown tufts, often conspicuously tomentose below, on moist humus or soil in swamps, fens, bogs (burned-over) and moist rock ledges, cosmopolitan.

2b Plants smaller, 1.5-3 cm high, leaves not usually contorted when dry; alar cells not or only slightly differentiated; sterile stems bearing terminal clusters with very small fusiform gemmae at end of seta-like extension of stem but not along its length. Fig. 204E-G. .... *Aulacomnium androgynum* (Hedw.) Schwaegr. ....

Small plants in dense dark green to sometimes yellow-brown tufts, on soil or humus of banks in woods; occasionally on rotten wood, Alaska to Newfoundland south to California, Wyoming, the Great Lakes and West Virginia.

MEESIACEAE
Amblyodon

Amblydon dealbatus (Hedw.) B.S.G. (Fig. 205)

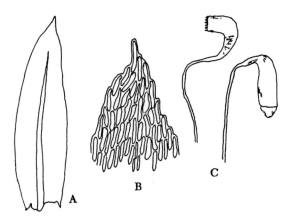

Figure 205

Figure 205 *Amblyodon dealbatus*. A, leaf; B, cells of leaf apex; C, capsules.

Plants up to 1 cm high (rarely larger), tufted or gregarious, shiny pale or bright green or blue-green tinged with red, on peaty soil or logs in wet seepy places such as bogs and river banks, Northwest Territories to Ontario, Quebec, and Nova Scotia south to British Columbia, Colorado, Minnesota, Wisconsin, Michigan.

Meesia

Three species are reported from North America.

1a Plants small, 1-4 cm high, leaves erect and not 3-ranked, margins strongly recurved and entire. Fig. 206A, E, F. .... *Meesia uliginosa* Hedw.

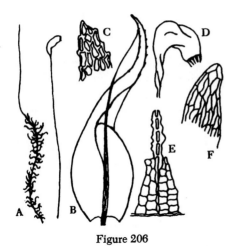

Figure 206

Figure 206 *Meesia uliginosa*. A, plant and capsule. E, peristome; F, apex of leaf; B, leaf of *M. triquetra*; C, apex of leaf; D, dry capsule.

Plants in dense sordid green or yellowish tufts, brown or blackish below, leaves contorted when dry, on wet humus or peaty soil, occasionally on logs, in marly fens, bogs, or rock crevices, Alaska to Ontario, Labrador and Newfoundland south to California, Nevada, Colorado, Michigan and New York.

1b   Plants larger, 2-14 cm high, leaves 3-ranked, usually strongly contorted when dry, margins scarcely revolute, serrulate above the shoulders. Fig. 206B-D. .... *Meesia triquetra* (Richt.) Ångstr.

Plants tufted, dark green to yellow-green, on calcareous wet soil or humus in bogs or swamp forests, Alaska to Ontario, Newfoundland and Nova Scotia south to California, Nevada, Michigan, New England, New York.

CATOSCOPIACEAE
Catoscopium

Catoscopium nigritum (Hedw.) Brid. (Fig. 207)

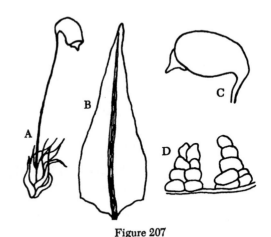

Figure 207

Figure 207 *Catoscopium nigritum*. A, plant; B, leaf; C, capsule; D, tooth of peristome.

Small tufted plant, 3-35 mm high, dark green to brownish green, on moist soil, Alaska to Newfoundland and New Brunswick south to British Columbia, Montana, Iowa, Michigan, Ontario.

BARTRAMIACEAE
Plagiopus

Plagiopus oederiana (Sw.) Limpr. (Fig. 208)

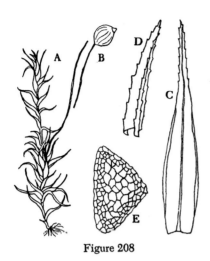

Figure 208

Figure 208 *Plagiopus oederiana.* A, shoot; B, capsule; C, leaf; D, apex of leaf; E, section of stem.

Plants in dark green dense tufts, 2-4 cm high, on shaded calcareous rocks and clifts, Alaska to Nova Scotia south to Oregon, Idaho, Montana, Colorado, Iowa, Michigan, New York and Virginia.

Anacolia

Anacolia menziessi (Turn.) Paris (Fig. 209)

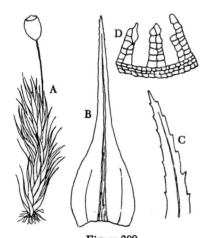

Figure 209

Figure 209 *Anacolia menziesii.* A, plant; B, leaf; C, apex of leaf; D, peristome.

Plants in loose to dense yellow-green tufts, 2-6 cm high, normally with conspicuous clusters of rhizoids on lower part of stem, on shaded soils and rocks, British Columbia to Montana, Wyoming and California. One other species, *A. laevisphaera* (Tayl.) Flow., occurs in Arizona.

Bartramia

Seven species of *Bartramia* occur in North America.

1a Leaves evenly tapering from base to apex. 2

1b Leaves broad and clasping at base, abruptly narrowed to the slender apex. Fig. 210D. .... *Bartramia ithyphylla* Brid.

Green to glaucous green plants in dense tufts, 0.5-2.5 cm high, (sometimes higher), on soil over rocks and in crevices, Alaska to Quebec and Labrador south to California, Colorado, Michigan and Pennsylvania.

2a Leaves crisped and contorted when dry, spreading. Fig. 210A-E. .... *Bartramia pomiformis* Hedw.

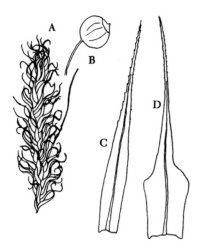

Figure 210

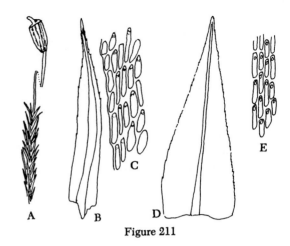

Figure 211

Figure 210 *Bartramia pomiformis*. A, plant; B, capsule; C, leaf; D, leaf of *B. ithyphylla*.

Plants in dense, green or glaucous green tufts, 2-10 cm high, on shaded soil, soil over rocks, and in crevices of rocks, of steep banks; ledges, and faces of clifts, Alaska to Labrador south to Oregon, Idaho, Montana, Colorado and the Gulf States.

2b   Leaves erect, slender, brittle. .... ***Bartramia stricta*** Brid.

Plants in dense green tufts, brownish below, 0.7-3 cm high, on soil over rocks, California, Idaho, Montana, and Colorado.

Conostomum

Conostomum tetragonum (Hedw.) Lindb. (Fig. 211A-C)

Figure 211 *Conostomum tetragonum*. A, plant with sporophyte; B, leaf; C, median leaf cells. *Philonotis marchica;* D, leaf; E, median leaf cells.

Plants small, 0.5-3 cm high, in glaucous-green tufts, on soil in rock crevices, Alaska to Labrador south to British Columbia, Montana, Quebec, New York, New England.

Philonotis

Nine species are found in North America. The following key is based upon a monograph of the genus *Philonotis* in North America by W. M. Zales, (unpublished dissertation, Univ. of British Columbia, Can., 1976).

1a   Leaf cells papillose at the upper ends throughout. 2

1b   Leaf cells papillose at the lower ends at least in lower portion of leaf. 5

2a   Leaf cells elongate, 6-20:1, in more or less vertical rows. 3

2b Leaf cells quadrate, 1:1, to rectangular, 4:1, in less obvious rows. 4

3a Leaves long triangular-lanceolate, straight or falcate and flat, cells long and narrow, 9-15:1, papillae pointed at the extreme upper cell ends, monoicous. Fig. 212A-D. ....
.... *Philonotis longiseta* (Michx.) Britt.

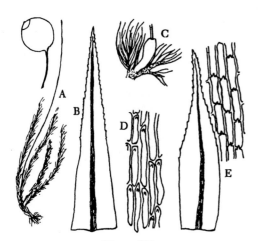

Figure 212

Figure 212 *Philonotis longiseta*. A, plant; B, leaf; C, antheridial bud; D, median cells; E, leaf and cells of *P. sphaericarpa*.

Plants in loose to dense, green to yellow-green tufts, 1-2.5 cm high, on moist rocks and soil, Iowa, Nebraska, Indiana, Pennsylvania south to the Gulf States.

3b Leaves triangular to slightly ovate-lanceolate, carinate, cells rectangular (less than 9:1), papillae round near the upper cell ends; dioicous. Fig. 211D-E. ....
.... *Philonotis marchica* (Hedw.) Brid.

Plants in dense to loose tufts, bright green, 2-8 cm high, British Columbia to Ontario and Labrador south to Arizona, New Mexico, and the Gulf States.

4a Papillae distinct and numerous, margins rolled and strongly scabrous, costa long excurrent, Florida. Fig. 212E. ....
Philonotis sphaerocarpa (Hedw.) Brid.

Plants tufted, light green, on moist clay soils, Florida.

4b Papillae faint, costa faint or ending short of the apex, margins composed of rounded papillose cells appearing doubly serrate, southeastern United States. ....
Philonotis glaucescens (Hornsch.) Broth.

Plants small, tufted, yellowish to brownish with age, on moist soil and calcareous rock, southeastern United States west to Texas.

5a Leaves ovate-lanceolate more or less falcate, rarely with shallow plicae on each side of costa, stem tips not twisted. Fig. 213. .... *Philonotis fontana* (Hedw.) Brid. var. *fontana*

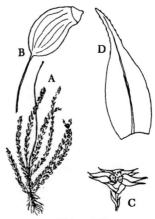

Figure 213

Figure 213 *Philonotis fontana.* A, plant; B, capsule; C, antheridial head; D, leaf.

Plants in dense tufts, bright green or yellowish-green, 3-16 cm high, on soil and rock in wet seepy places, cosmopolitan (absent in Florida).

5b Leaves broadly ovate and short lanceolate, widely spaced with several plicae on each side of the costa, stem tips twisted, western. .... *Philonotis fontana* var. *americana* (Dism.) Flow. *ex* Crum

Large plants, 3-10 cm high, in dense green or yellow-green tufts, on soil and rocks in wet places, Alaska to California, Wyoming, Montana.

TIMMIACEAE
Timmia

Three species are reported from North America, all easily recognized in the field by the calyptra standing erect at the bend of the seta.

1a Leaf sheath hyaline or yellow when old. 2

1b Leaf sheath orange to brown. .... *Timmia austriaca* Hedw.

Plants in robust, tomentose, reddish brown tufts, up to 15 cm high, on soil or rocks, Pacific Northwest to Nevada, Utah, Colorado, South Dakota and Michigan.

2a Leaf cells 9-18 μ in diameter, the walls thick. Fig. 214. .... *Timma megapolitana* Hedw. var. *megapolitina*

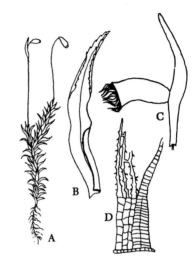

Figure 214

Figure 214 *Timmia megapolitana.* A, plant; B, leaf; C, calyptra and capsule; D, peristome.

Plants up to 6 cm high, in loose, dirty green tufts, on moist to wet shaded humus or soil, decaying wood, especially along streams, Alaska to Quebec south to California, Nebraska, Arkansas, North Carolina and Tennessee.

2b Leaf cells 6-9 μ in diameter, rarely larger, the walls thin. .... *Timmia megapolitana* var. *barvarica* (Hessl.) Brid.

Mainly in western North America.

Order Isobryales

ERPODIACEAE
Erpodium

Erpodium biseriatum (Aust.) Aust. (Fig. 63)

Green, yellowish, or brownish plants in loose to dense flat mats, on trunk of smooth bark trees, Georgia, Florida, Louisiana. *Erpodium acrifolium* Purs. has been found on the Edward's Plateau of Texas. This species has terete branches, is loosely erect when dry with leaves ending in a hyaline awn.

PTYCHOMITRIACEAE
Ptychomitrium

1a Leaves entire or nearly so. 2

1b Leaves serrate in upper half. 3

2a Leaves 1-2 mm long; seta about 2 mm long; peristome teeth separate to base. Fig. 13. .... *Ptychomitrium incurvum* (Schwaegr.) Spruce

Plants in small, 5-7 mm high, brownish or blackish-green tufts, on rocks, Minnesota to New England south to Florida and Texas.

2b Leaves 3.5-5 mm long; seta about 5 mm long; peristome teeth joined at base. .... *Ptychomitrium leibergii* Best.

Plants in loose tufts, up to 1 cm high, on rock and logs, Missouri, Texas, Arizona.

3a Leaves 4-6 mm long; setae 1-2 cm long; peristome teeth separate to base; Pacific Northwest. .... *Ptychomitrium gardneri* Lesq.

Plants in dark green to black tufts or cushions, 2-4 cm high, on rocks, Washington, Oregon, Idaho, California. A closely related species, *P. serratum* B.S.G. is reported from West Texas and Tennessee.

3b Leaves 1-1.5 mm long; setae 2-3 long; peristome teeth 2-3 cleft above the middle. .... *Ptychomitrium drummondii* (Wils.) Sull.

Plants in dense, green tufts, on trees, Missouri, Oklahoma, Tennessee, Virginia, North Carolina south to the Gulf States.

ORTHOTRICHACEAE
Zygodon

Zygodon viridissimus (Dicks.) Brid. var. *rupestris* Lindb. *ex* C. J. Hartm. (Fig. 215)

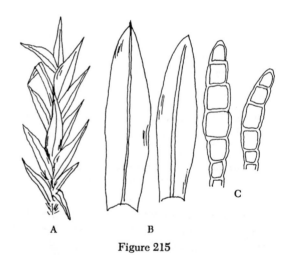

Figure 215

Figure 215 *Zygodon viridissimus* var. *rupestris*. A, plant; B, leaves; C, gemmae.

Plants scattered or in loose, small tufts, 0.5-1.5 cm high, green to brownish green, leaves crisped and contorted when dry, 3-6 celled frusiform, ellipsoidal to subclavate gemmae in axils of leaves, on rocks and trees Alaska to Ontario south to California, Arizona, New Mexico, Wisconsin, Tennessee and North Carolina. Four other species are reported from North America.

Amphidium

1a Leaf margins plane or nearly so; upper leaf cells with large, conspicuous papillae. Fig. 216A-D. ....
.... *Amphidium lapponicum* (Hedw.) Schimp.

1b Leaf margins normally recurved; upper leaf cells not strongly papillose. 2

2a Leaves entire or minutely serrate near apex. Fig. 216E. ....
Amphidium mougeotii (B.S.G.) Schimp.

Plants in green to yellow-green tufts, 1-8 cm high, on rocks, Alaska to Newfoundland, Labrador and Nova Scotia south to Oregon, Montana, Ohio, Georgia.

2b Leaves toothed near apex; western North America. ....
Amphidium californicum (Hampe *ex* C. Muell.) Broth.

Tufted plants 2-7 cm high, on soil in crevices of rocks, British Columbia to California.

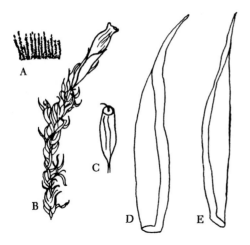

Figure 216

Figure 216 *Amphidium lapponicum*. A, cluster; B, one plant; C, capsule; D, perichaetial leaf; E, perichaetial leaf of *A. mougeotii*.

Plants in dense green to yellow-green tufts, 1-3 cm high, on rocks, Alaska to Quebec, Newfoundland, Nova Scotia south to Arizona, Nevada, Colorado, Minnesota, Michigan, Pennsylvania and New Jersey.

Orthotrichum

This is a large (35 species in North America), and complex genus. In many species, good capsules are required for accurate determination. The following key is based upon a recent treatment of this genus by Dale Vitt (A revision of the genus *Orthotrichum* in North America, north of Mexico, 1973, Bryophytorum Bibliotheca Band I, J. Cramer).

1a Leaves ending in a serrate hyaline awn. Fig. 217A-B. ....
.... *Orthotrichum diaphanum* Brid.

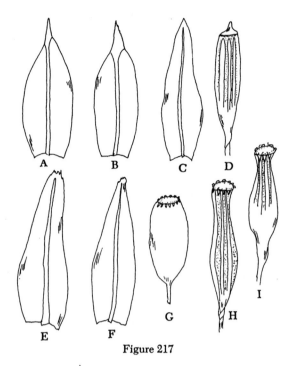

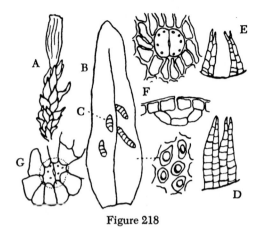

Figure 217 *Orthotrichum diaphanum.* A-B, leaves. *O. pulchellum;* C, leaf; D, capsule with operculum. *O. rivulare;* E, leaf. *O. pusillum;* F, leaf; G, dry mature capsule. *O. affine;* H, dry mature capsule. *O. sordidum;* I, dry mature capsule.

Plants scattered or in loose tufts, 2-7 mm high, light-green, dark-green or brownish, usually on bark of trees in dry areas, occasionally on rocks, California, Arizona, New Mexico, Texas, Colorado, Nebraska, Missouri to Louisiana.

1b Leaves not ending in a serrate hyaline awn. 2

2a Leaf margins erect incurved with broadly obtuse to blunt apex; usually with abundant gemmae. Fig. 218. .... ***Orthotrichum obtusifolium* Brid.**

Figure 218 *Orthotrichum obtusifolium.* A, plant; B, leaf; C, gemma; D, tooth; E, segments; F, superficial stoma, face view and section; G, immersed stoma.

Plants in dull orange-yellow to olive-brown tufts, 3-14 mm high, on deciduous trees in open areas, especially common on poplar trees, Alaska to Quebec south to California, Arizona, New Mexico, Nebraska, Minnesota, Michigan, Pennsylvania, North Carolina.

2b Leaf margins plane, recurved, revolute or thickened; gemmae absent or not abundant. 3

3a Leaves crisped-flexuose when dry; endostome segments as long as exostome, 8 or 16; stomates of capsule immersed; Pacific Northwest and Alaska. 4

3b Leaves erect-appressed when dry; endostome segments shorter than exostome, usually 8; stomates superficial or immersed; throughout North America. 5

4a Plants 0.8-2.0 cm high; capsules long exserted; peristome teeth 8, light brown. Fig. 222E. .... ***Orthotrichum consimile* Mitt.**

Plants yellow-green above, olive-green or brownish below, in coarse, loose tufts, on tree trunks and branches in coniferous forests, occasionally on shaded boulders, British Columbia to California.

4b   Plants usually less than 1.3 cm high; capsules barely exserted; peristome teeth 16, red. Fig. 217C-D. ....
.. *Orthotrichum pulchellum* Brunt. *in* Sm.

Plants in delicate, loose, light- to yellow-green tufts, on trunks and twigs of trees, Alaska to Washington.

5a   Stomates superficial; basal leaf cells elongate, ± nodose and thick-walled. .. 6

5b   Stomates immersed; basal leaf cells rectangular, not nodose, thin-walled. 10

6a   Endostome segments lacking or rare and rudimentary; almost always on rock. 7

6b   Endostome segments well developed; usually on trees. 8

7a   Capsules cylindric, fully exserted; endostome rudimentary or lacking. Fig. 220C & F. .... *Orthotrichum laevigatum* Zett.

Plants in dirty-green dense tufts or cushions, 0.5-4.0 cm high, on non-calcareous moist boulders usually in open coniferous forests, British Columbia south to California, Nevada, Colorado, South Dakota.

7b   Capsules emergent, globose-ovate to short-oblong, if cylindric then strongly 8-ribbed and endostome well-developed. 17

8a   Plants dioecious; exostome teeth acuminate; capsules long-cylindric and 8-ribbed 1/2 to entire length; gemmae sometimes present on leaves. Fig. 219A-B. ....
.... *Orthotrichium lyellii* Hook. & Tayl.

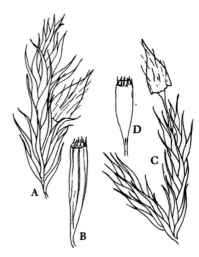

Figure 219

Figure 219 *Orthotrichum lyellii*. A, plant; B, 8-ribbed capsule and seta; C, plant of *O. speciosum* var. *speciosum*; D, smooth capsule of same.

Plants in loose to dense yellow-green to dark green tufts, up to 13 cm high, usually on trees, along the Pacific Coast from Alaska to California.

8b   Plants goniautoecious; exostome teeth acute or truncate; capsule cylindric to ovate-cylindric, lightly ribbed 1-1/2 its length to smooth. 9

9a   Capsules lightly 8-ribbed. Fig. 219C-D. .. .. *Orthotrichum speciosum* Nees. *in* Sturm. var. *speciosum*

Plants in robust, loose, yellow-brown, olive-green or brown mats, 0.8-5.0 cm high, usually on tree trunks, Alaska to Quebec and Newfoundland south to Oregon, Idaho, Michigan.

9b   Capsules smooth. Fig. 220A-B. .... *Orthotrichum speciosum* var. *elegans* (Schwaegr. *ex* Hook. & Grev.) Warnst.

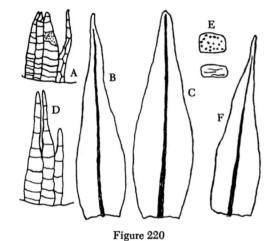

Figure 220

Figure 220   *Orthotrichum speciosum* var. *elegans*. A, tooth and segment of peristome; B, leaf. C & F, leaf of *O. laevigatum*; D, tooth and segment; E, upper and lower parts of tooth.

Plants 5-15 mm tall, in dark-green to brownish tufts, usually on trunks and branches of trees, Northwest Territories to Quebec and Newfoundland, south to Alberta, Wisconsin, Michigan and New Jersey.

10a   Plants on rocks or at the base of trees near streams, margins of leaves coarsely dentate near the apex; Pacific Northwest. Fig. 217E. .... *Orthotrichum rivulare* Turn.

Plants in dark-green, brown or blackish tufts, 1-5 cm high, British Columbia to California and Montana.

10b   Plants growing on trunks of trees or dry rock faces. 11

11a   Exostome teeth erect or spreading, striate, reticulate, or papillose-striate; endostome absent; on rocks. 12

11b   Exostome teeth reflexed or recurved, papillose, rarely, striate at tips endostome usually present; on trees. 15

12a   Capsules fully exserted, cylindrical, usually with 8 long and 8 short ribs. Fig. 221A-C. .... *Orthotrichum anomalum* Hedw.

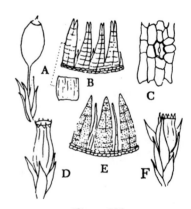

Figure 221

Figure 221   *Orthotrichum anomalum*. A, capsule and seta; B, peristome; C, stoma; D, capsule and seta of *O. strangulatum*; E, peristome of same; F, capsule and seta of *O. cupulatum*.

Plants in dark-green, olive-green or brown tufts, 0.6-2-5 mm high, on rocks, especially limestone, Alaska to Quebec south to Arizona,

New Mexico, South Dakota, Iowa, Illinois, Ohio, Pennsylvania and Virginia.

12b Capsules immersed or emergent, ovate, or oblong, usually with 8 or 16 ± uniform ribs. 13

13a Capsule with 16 ribs. Fig. 221F. .... *Orthotrichum cupulatum* Brid.

Plants in light-green, olive-green or brown cushions, up to 1.2 cm high, British Columbia south to California, Arizona, Montana, South Dakota, Colorado and disjunt to central Texas.

13b Capsules with 8 ribs. 14

14a Exostome teeth finely papillose or finely reticulate-papillose, yellowish; eastern North America. Fig. 221D-E. .... *Orthotrichum strangulatum* P.-Beauv.

Plants in dull, olive-green, dark-green or blackish tufts, 4-10 mm high, on dry calcareous or dolomitic rocks and bluff faces; Manitoba, Ontario and New England south to South Dakota, Texas, Arkansas, Alabama and Georgia.

14b Exostome teeth ridged-striate, coarsely striate-reticulate or striate-papillose, white to clear, western North America. Fig. 222A-D. .... *Orthotrichum hallii* Sull. & Lesq. *in* Sull.

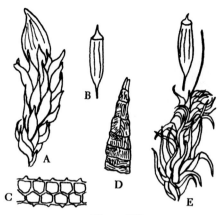

Figure 222

Figure 222 *Orthotrichum hallii*. A, plant; B, capsule; C, section of leaf; D, peristome. Rocky Mountains. E, plant of *O. consimile*, dry.

Plants in dull olive-green or brownish tufts or cushions, up to 2.5 cm high, usually on rocks, British Columbia, Idaho, California, Nevada, Arizona, Montana, Colorado, New Mexico.

15a Capsule smooth when dry; exostome of 16 teeth, erect to recurved teeth; endostome absent. Fig. 217F-G. .... *Orthotrichum pusillum* Mitt.

Plants soft, dark-green, scattered or in tufts, 2-4 mm high, on trunks of deciduous trees, Kansas, Missouri, Illinois, Ohio to New England south to Texas, Alabama, and Georgia.

15b Capsule 8-ribbed when dry; exostome teeth 8, or 8 splitting to 16; endostome usually present. 16

16a Exostome teeth recurved. Fig. 223A-E. .. .. *Orthotrichum pumilum* Sw.

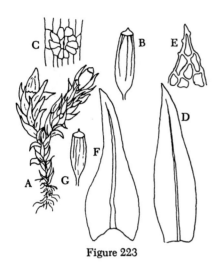

Figure 223

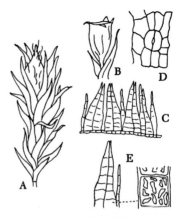

Figure 224

Figure 223 *Orthotrichum pumilum*. A, plant; B, capsule; C, stoma; D, leaf; E, apex of leaf; F, leaf of *O. ohioense*; G. capsule of same.

Plants in soft, dark- to black-green tufts up to 5 mm high, usually on deciduous trees, Alberta, Washington, Idaho, to Michigan, New York south to Arizona, New Mexico, Oklahoma, Missouri, Kentucky and Virginia.

16b Exostome teeth reflexed. Fig. 223F-G. .. .. *Orthotrichum ohioense* Sull. & Lesq. *ex* Aust.

Plants in yellow-green to dark-brown tufts, 0.4-1.4 cm high, on deciduous trees, Wisconsin to New England south to Oklahoma, Arkansas, and the Gulf States.

17a Exostome teeth erect; endostome never present. Fig. 224. .... *Orthotrichum rupestre* Schleich. *ex* Schwaegr.

Figure 224 *Orthotrichum rupestre*. A, plant; B, capsule and seta; C, and 3, peristome; D, stoma.

Plants in loose tufts or cushions, 1-12.5 cm high, yellow-green to olive or dark-green above, on moist non-calcareous boulders and cliff faces at higher elevations, Rocky Mountains to the West Coast.

17b Exostome teeth reflexed; endostome usually present. 18

18a Capsules greater than 1.5 mm long, ribbed the entire length; western North America. Fig. 217. .... *Orthotrichum affine* Brid

Plants in olive-green, dark-green, or brownish tufts, up to 3.0 cm high, on trunks and bases of deciduous trees, rarely on shaded rocks, British Columbia, Alberta, south to California, Montana, Utah.

18b Capsules less than 1.5 mm long, ribbed for ± 1/2 the length of the capsule; Great Lakes, Northeastern North America, and Alaska. Fig. 217I. ....: *Orthotrichum sordidum* Sull. & Lesq. *in* Aust.

Plants in dark- to yellow-green tufts or cushions, on trunks of both deciduous and coniferous trees; Alaska, Ontario to Quebec and Newfoundland south to Wisconsin, Michigan, Pennsylvania, and New York.

Ulota

Ten species of *Ulota* are reported from North America. Rev. W. R. Megaw of Belfast, Ireland, has written a novel entitled "Ulota"; it tells of the loves, human and vegetal, of a bryologist.

1a Leaves nearly straight and erect when dry. Fig. 225A-B. .... *Ulota hutchinsiae* (Sm.) Hamm.

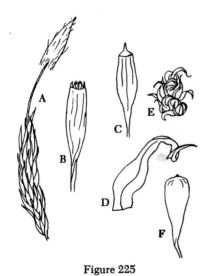

Figure 225

Figure 225 *Ulota hutchinsiae*. A, plant; B, capsule; C, mature capsule of *U. crispa*; D, dry leaf; E, a dry shoot; F, capsule of *U. coarcta*.

Plants are rigid tufts, 1-1.5 cm high, dull yellow-brown to dark-green above, blackish below, on acidic rocks, rarely on tree trunks, Alaska, southern Ontario to Newfoundland south to Oklahoma, Arkansas, Alabama and Georgia.

1b Leaves very curly when dry. 2

2a Leaves with clusters of septate gemmae at tips; dioicous and only rarely fruiting. Fig. 226A-D. .... *Ulota phyllantha* Brid.

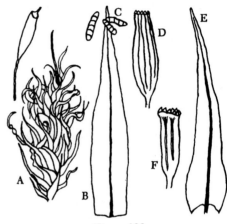

Figure 226

Figure 226 *Ulota phyllantha*. A, plant; B, leaf; C, gemma; D, capsule; E, leaf of *U. obtusiuscula*; F, capsule of same.

Plants in dense, yellow-green to brownish tufts, up to 3 cm high, on trees and rocks near coastal areas, Alaska to Oregon, Quebec, New Brunswick, Nova Scotia and Newfoundland south to New England.

2b Gemmae absent; monoicous. 3

3a Capsule club-shaped, smooth except near mouth rounded to the mouth. Fig. 225F. .... *Ulota coarcta* (P.-Beauv.) Hamm.

Plants in yellow-green to brown tufts, 5-10 mm high, on bark of trees in moist woods, Ontario to Newfoundland south to Wisconsin, Michigan, Tennessee and North Carolina.

3b Capsule 8-ribbed, oblong-cylindric to fusiform. 4

4a Capsule widest at mouth, evenly tapering to base; Pacific Northwest. Fig. 226E-F. .... *Ulota obtusiuscula* C. Muell. & Kindb. *ex* Macoun & Kindb.

Plants in dense cushions, 1-5 cm high, green to yellow-green, on trees, Alaska to Oregon.

4b Capsule contracted below the mouth; Alaska and eastern North America. Fig. 225C-E. ...... *Ulota crispa* (Hedw.) Brid.

Plants in green, yellow-green or yellow-brown tufts, 1-3 cm high, on bark of trees in moist woods, Alaska, Quebec to Newfoundland south to Wisconsin, Indiana, Ohio, Pennsylvania, and Georgia.

Drummondia

Drummondia prorepens (Hedw.) Britt. (Fig. 227A, C, D)

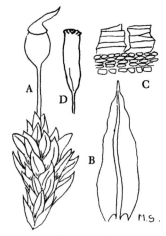

Figure 227

Figure 227 *Drummondia prorepens*. A, plant; B, leaf; C, peristome; D, capsule of *Schlotheimia rugifolia*.

Plants in dark-green thin mats, produced by numerous erect-ascending branches arising from a creeping stem up to 10 cm or more long, on bark of trees, Ontario to New England south to Oklahoma, Arkansas, and Georgia.

Groutiella

Groutiella mucronifolia (Hook. & Grev.) Crum & Steere (Fig. 119E)

Plants in dark greenish-black with numerous short secondary branches, on trees, Florida *G. tomentosa* (Hornsch.) Wijk & Marg. is also reported from Florida and is distinguished by the fragile leaf tips of upper leaves and the dark-purple, oblong-cylindric capsules.

Schlothemia

Schlothemia rugifolia (Hook.) Schwaegr. (Fig. 227B).

Plants dark-green to reddish-brown, in spreading mats, on trees and logs, Virginia, Tennessee and North Carolina south to the Gulf States and Texas. *S. lancifolia* Bartr. is an endemic species in the vicinity of Highlands, North Carolina.

FONTINALACEAE

The following keys are based upon "A Monograph of the Fontinalaceae" by Winona H. Welch (Martinus Nijhoff, 1960).

Fontinalis

Sixteen species of this aquatic genus are reported from North America.

1a Leaves usually carinate or carinate-conduplicate. 2

1b Leaves concave or plane. 3

2a Ends of foliated stems and branches conspicuously elongated, triangular pyramidal in shape; keel only slightly curved; leaf apices acute. Fig. 228A. ....
...... *Fontanilis neomexicana* Sull. & Lesq.

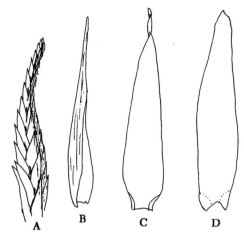

Figure 228

Figure 228 *Fontinalis neomexicana*. A, branch; *F. filiformis;* B, leaf *F. sullivantii;* C, leaf *F. falccida;* D, leaf.

Plants pale green, yellowish-green, golden-brown or brown, stems rigid up to 56 cm long, Alaska to Montana, Washington, California, Idaho and in Michigan.

2b Ends of foliated stems and branches not conspicuously elongated, triangular pyramidal in shape; keel moderately to strongly curved; leaf apices subobtuse to broadly abuse. Fig. 229. ....
.... *Fontinalis antipyretica* Hedw.

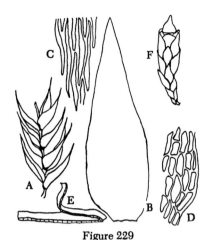

Figure 229

Figure 229 *Fontinalis antipyretica*. A, shoot; B, leaf; C, median cells; D, alar cells; E, section of leaf; F, capsule with operculum.

A highly variable species that includes a number of varieties. Plants range from slender to robust with slightly rigid to somewhat flaccid stems, up to 80 cm in length, Alaska to Labrador, south to California, New Mexico, Minnesota, Wisconsin, Michigan, Pennsylvania, New York. The name *antipyretica* (against fire) has reference to the old Germanic custom of filling cracks of the log houses with this moss, whereby, according to popular belief, conflagration was supposed to be prevented.

3a Leaves usually concave. 4

3b Leaves usually plane. 7

4a Leaves erect to slightly spreading, appearing to be appressed. Fig. 230. .... *Fontinalis dalecarlica* Schimp. ex B.S.G. ....

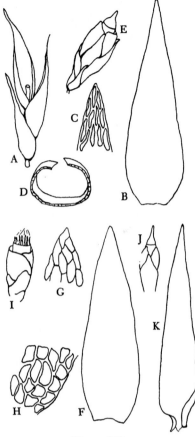

Figure 230

Figure 230 *Fontinalis dalecarlica*. A, shoot; B, leaf; C, apex of leaf; D, section of leaf; E, perichaetium and capsule; *Fontinalis hypnoides* var. *duriaei*; F, leaf; G, apex of leaf; H, alar cells; I, perichaetium and capsule; *F. novae-angliae*; J, leaf; K, perichaetium and capsule.

Plants slender, yellowish-green to dark-green, stems up to 90 cm long, Ontario, Quebec to Newfoundland south, Minnesota, Wisconsin, Indiana, Georgia and Florida.

4b Leaves erect-spreading, not appearing to be appressed. 5

5a Leaves ovate lanceolate, margins usually involute. Fig. 230J-K. .... *Fontinalis novae-angliae* Sull.

Plants slender to robust, color variable from copper color to green, stems up to 40 cm long, Ontario, Quebec to Newfoundland south to Oklahoma, Arkansas, Alabama, Mississippi and Florida.

5b Leaves narrowly lanceolate, margins not involute. 6

6a Leaf blades concave throughout or subconcave at base and plane above; 0.75-1.5 mm wide. Fig. 228C. .... *Fontinalis sullivantii* Lindb.

Plants slender, yellowish-green, green to brownish-green, stems up to 25 cm long, Indiana, Michigan, Pennsylvania to New York and Massachusetts south to Louisiana and Florida.

6b Leaf blade concave or deeply concave to convolute tubulose; 0.35-0.5 mm wide. Fig. 228F. .... *Fontinalis filiformis* Sull. & Lesq. *ex* Aust.

Plants very slender, yellowish, to green, to reddish-brown, stems up to 20 cm long, Illinois, Michigan to New Jersey south to Texas, Louisiana, and Florida.

7a Leaves generally broadly ovate-lanceolate or oval lanceolate, 1-2.5 mm wide; margins tapering from approximately middle into apex; apices short and broadly acuminate. Fig. 230F-I. .... *Fontinalis hypnoides* C. J. Hartm. var. *duriaei* (Schimp.) Husn.

Plants slender to medium in size, stems flaccid up to 30 cm long, British Columbia to Ontario and New Brunswick south, California, Arizona, Colorado, Minnesota, Texas, Missouri, Indiana, Ohio, Pennsylvania, New York.

7b Leaves generally narrowly ovate-lanceolate or lanceolate, 0.5-1.75 mm wide; margins tapering from basal fourth or half into apex; apices long acuminate. .. 8

8a Apices gradually narrowed; tips commonly acute and entire, auricles usually absent, occasionally slight. .... *Fontinalis hypnoides* C. J. Hartm. var. *hypnoides*

Plants slender and delicate, pale green, green to brown, up to 30 cm long, Northwest Territories, British Columbia to Ontario, Quebec and Nova Scotia south to Oregon, Montana, Colorado, Missouri, Michigan and Connecticut.

8b Apices frequently abruptly narrowed; tips commonly obtuse to truncate and serrulate; auricles very conspicuous. Fig. 228D. .... *Fontinalis flaccida* Ren. & Card.

Plants slender, yellowish to green to brownish green, up to 40 cm long, Ontario to Quebec south to Georgia, Louisiana and Florida.

Brachelyma

Brachyelyma subulatum (P.-Beauv.) Schimp. *ex* Card. (Fig. 231)

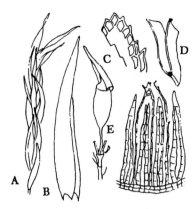

Figure 231

Figure 231 *Brachelyma subulatum*. A, shoot; B, leaf; C, apex of leaf; D, section of leaf; E, seta, capsule and calyptra; F, peristome.

Plants slender medium in size, yellow-green, green, or brown, up to 20 cm long, floating or submerged, Illinois, Texas, Missouri, Virginia south to Florida.

Dichelyma

1a Costa subpercurrent to briefly excurrent. 2

1b Costa shortly excurrent. 3

2a Costa shortly excurrent. ....
 .... *Dichelyma falcatum* (Hedw.) Myr.

Plants medium to robust, yellowish-green, green, to copper-colored, golden brown or blackish green, up to 15 cm in length, Alaska to Quebec and Newfoundland south to Utah, Colorado, Minnesota, Wisconsin, Michigan, Maine and New Hampshire.

2b Costa not excurrent. ....
 .... *Dichelyma pallescens* B.S.G.

Plants slender, yellow-green, green, or brownish, up to 10.5 cm long, Ontario, Quebec to Nova Scotia and Newfoundland south to Minnesota, Wisconsin, Michigan, Pennsylvania, New York and Massachusetts.

3a Median stem leaves commonly falcate, 4-5 mm long, Northwest North America.
 .... *Dichelyma uncinatum* Mitt.

Plants slender, yellowish-green, green to brown, up to 12 cm long, British Columbia to Oregon, Wyoming, and Montana.

3b **Median stem leaves commonly erect-ascending, straight to moderately curved along keel, 5-7 mm long, eastern North America. Fig. 232.** ....
 .... *Dichelyma capillaceum* (With.) Myr.

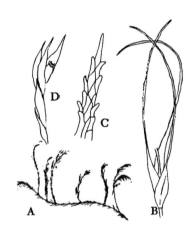

Figure 232

Figure 232 *Dichelyma capillaceum*. A, shoot; B, leaves; C, apex of leaf; D, capsule and perichaetium.

Plants slender, yellowish-green, green, or brownish-green, stems up to 20 cm in length, Ontario, Nova Scotia and Newfoundland south to Wisconsin, Michigan, Indiana, Louisiana, Florida.

CLIMACIACEAE
Climacium

The following key is based upon a recent study of this genus in North America by Diana Horton and Dale Vitt. (Reproduced by permission of the National Research Council of Canada from the Canadian Journal of Botany, Volume 54, pp. 1872-1883, 1976.)

1a   Apex of stem leaves gradually to rapidly narrowed, acuminate; upper leaf cells of branch leaves 3-8:1, auricles of branch leaves generally project well below level of insertion; capsules long cylindric, 3-6 mm. Fig. 233A-C, F. .... *Climacium americanum* Brid.

1b   Apex of stem leaves sharply contracted, obtuse, and apiculate, upper leaf cells of branch leaves 7-13:1, auricles of branch leaves project only slightly below level of insertion, capsules short-cylindric, 1.5-4 mm. Fig. 233D-E. .... *Climacium dendroides* (Hedw.) Web. & Mohr.

Plants yellow or yellow-green, dendroid, on soil and humus in shady, swamp places, Alaska, Northwest Territories to Quebec and Newfoundland south to Washington, New Mexico and Arizona, Minnesota, Wisconsin, Michigan, Pennsylvania, Virginia, New York, Vermont, and New Hampshire.

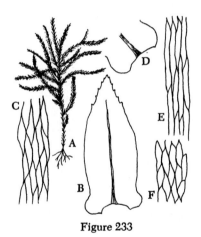

Figure 233

Figure 233 *Climacium americanum*. A, plant; B, leaf; C, upper median cells of leaf; D, base of leaf of *C. dendroides*; E, cells of same; F, leaf cells of *C. americanum*.

Plants dull yellowish-green to dark green, branches typically erect and regularly branched, usually tree-like, arising from a creeping stem, some forms that grow along stream banks are not regularly branched or erect, on shaded soil and rock ledges of woods, often in swamps, Ontario, Quebec to Nova Scotia south to Texas, Louisiana, and Florida. This and the following species are known as the "tree moss."

HEDWIGIACEAE
Hedwigia

Hedwigia ciliata (Hedw.) P.-Beauv. (Figs. 55B, 234)

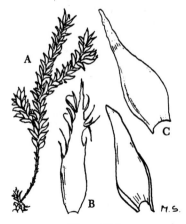

Figure 234

Musci—The Mosses   153

Figure 234 *Hedwigia ciliata*. A, plant; B, leaf of perichaetium; C, leaves with and without awn point.

Plants coarse and usually robust, in dull mats, stems branched, erect-ascending or creeping, green, gray-green or brownish or glaucous when dry, leaves typically with a white awn, on dry rocks or rarely tree bases, cosmopolitan.

Pseudobraunia

Pseudobraunia californica (Lesq.) Broth. (Fig. 235A-C)

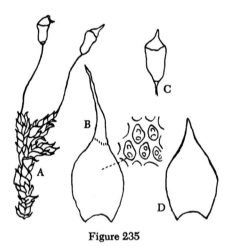

Figure 235

Figure 235 *Pseudobraunia californica*. A, plant; B, leaf; C, capsule; D, leaf of *Braunia secunda*.

Plants in large mats, yellow-green, brownish, or gray-green, with ascending or creeping stems to about 8 cm long, on dry rocks, British Columbia to Idaho and California.

Braunia

Braunia secunda (Hook.) B.S.G. (Fig. 235D)

Plants yellow-green, in mats, stems creeping or ascending, 4-6 cm long, on rocks, West Texas, Arizona.

CRYPHAEACEAE
Cryphaea

Four species of *Cryphaea* have been reported from North America.

1a Leaves ovate, with a short acumination, costa ending near middle of leaf. Fig. 236. ....
.... *Cryphaea glomerata* B.S.G. *ex* Sull.

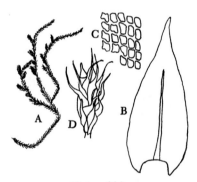

Figure 236

Figure 236 *Cryphaea glomerata*. A, plant; B, leaf; C, cells of leaf near base; D, branch with capsule.

Plants in thin mats, light gray-green, stems slender, julaceous, on trees and shrubs especially near streams, Oklahoma, Missouri, New England south to the Gulf States and Texas.

1b Leaves ovate-lanceolate, with long acumination, costa ending in the apex. ....
.. *Cryphaea nervosa* (Drumm.) C. Muell.

On tree trunks, Virginia and Tennessee south to Gulf States.

Alsia

Alsia californica (Hook. & Arnott) Sull. (Fig. 237)

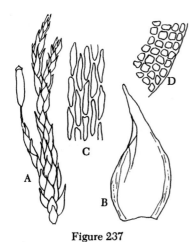

Figure 237 *Alsia californica*. A, twig; B, leaf; C, median cells; D, alar cells.

Plants in mats, green to yellow-green, stems up to 8 cm long with numerous short branches, on trees usually near coast, British Columbia to California.

Forsstroemia

1a Secondary stems profusely branched; leaves plicate; costa thin, reaching middle of leaf to short and double. Fig. 238. .... *Forsstroemia trichomitria* (Hedw.) Lindb.

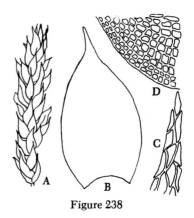

Figure 238 *Forsstroemia trichomitria*. A, shoot; B, leaf; C, apex of leaf; D, alar cells.

Plants in dense to loose tufts or mats, yellowish to dark green, on trees and rock faces, southern Canada to the Gulf States, and Oklahoma.

1b Secondary stems sparingly branched; leaves not plicate; costa strong and protruding on dorsal side. .... *Forsstroemia ohioense* (Sull.) **Lindb.**

Plants slender, green to dark green, on tree trunks, Missouri, Ohio, Virginia.

Dendroalsia

Dendroalsia abietina (Hook.) Britt. (Fig. 239)

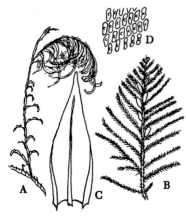

Figure 239

Figure 239 *Dendroalsia abietina*. A, a dry plant; B, branch with capsules; C, leaf; D, cells of leaf.

Plants dendroid, dark green, erect stems up to 11 cm high, horizontal stem up to 15 cm long, on rocks and trunks of trees, British Columbia to California.

LEUCODONTACEAE
Leucodon

1a Leaves ovate-elliptical abruptly short acuminate, scarcely plicate when dry. Fig. 241E. .... *Leucodon julaceus* (Hedw.) Sull.

Plants small to robust, dark-green to yellow-brown, secondary stems julaceous, straight or curved when dry, on tree trunks, logs, stumps, & rocks, Eastern North America from southern Canada to the Gulf States, west to Kansas, Oklahoma and Texas.

1b Leaves ovate to ovate-acuminate, plicate when dry. 2

2a Secondary stems well developed, rarely with clusters of flexuose branchlets in the axils of some leaves; seta shorter than perichaetial leaves; leaves gradually acuminate. Fig. 240. .... *Leucodon brachypus* Brid. var. *brachypus*

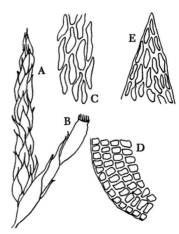

Figure 240

Figure 240 *Leucodon brachypus* var. *brachypus*. A, shoot; B, sporophyte and perichaetium; C, median cells of leaf; D, basal cells of leaf; E, apex of leaf.

Plants in loose curley tufts or patches, branches julaceus, on trees and rocks, Northeastern North America south to Appalachians, Gulf States, west to Kansas, Oklahoma, and in Arizona.

2b Secondary stems with flexuose branchlets regularly produced in the axils of some leaves; seta longer than perichaetium; leaves slenderly acuminate, strongly plicate wet or dry. Fig. 241A-D. .... *Leucodon brachypus* Brid. var. *andrewsianus* Crum & Andr.

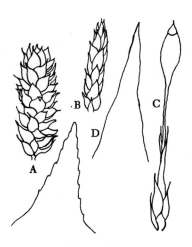

Figure 241

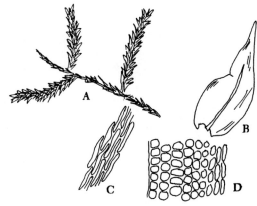

Figure 242

Figure 241 *Leucodon brachypus* var. *andrewsianus*. A, wet shoot; B, dry shoot; C, sporophyte and perichaetium; D, apex of leaf; E, apex of leaf of *L. julaceus*.

Plants moderately coarse and robust, dull dirty-green to brownish, secondary stems julaceous, usually curved when dry, on tree trunks, rarely on limestone rocks, Northeastern North America south to Georgia and Arkansas.

Leucodontopsis

Leucodontopsis geniculata (Mitt.) Crum & Steere (Fig. 242)

Figure 242 *Leucodontopsis geniculata*. A, habit of plant; B, branch leaf; C, median leaf cells; D, quadrate alar cells.

Plants in loose mats, primary stem creeping, secondary stems short, straight or curved, julaceus; septate gemmae sometimes in leaf axils, on tree trunks and fallen limbs, Florida.

Antitrichia

1a Branches not julaceous; leaves with supplementary costal up to 1/3 the length of the leaf; cell walls strongly pitted. Fig. 243A-F. ....
Antitrichia curtipendula (Hedw.) Brid

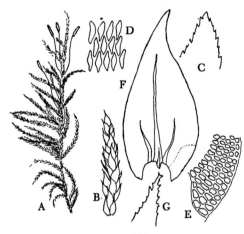

Figure 243

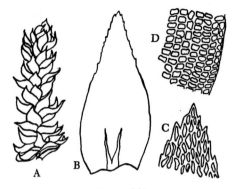

Figure 244

Figure 243 *Antitrichia curtipendula*. A, shoot; B, branch; C, apex of leaf; D, median cells; E, alar cells; F, whole leaf with accessory ribs; G, apex of leaf of *A. californica*.

Plants dark green to yellow-green or yellow-brown, in loose mats, stems up to 15 cm long, branches often pendant and flagelliform, on trees and logs, rarely on rocks, Alaska south to California, Montana, Newfoundland.

1b Branches julaceous; leaves without or with only weak secondary costae; cells wall not or only inconspicuously pitted. Fig. 243G. .... **Antitrichia californica** Sull. *ex* Lesq.

Plants in green to yellow-green or yellow-brown mats, stem stiff, 5-10 cm long with numerous julaceus branches, on trees, logs or rocks, British Columbia to California, Idaho, Montana, Nevada.

Pterogonium

Pterogonium gracile (Hedw.) Sm. (Fig. 244)

Figure 244 *Pterogonium gracile*. A, shoot; B, leaf; C, apex of leaf; D, cells near basal angles.

Plants in loose mats, dark green above, brownish below, primary stem creeping, secondary stems ascending, up to 5 cm or more long, strongly julaceous, often all turned to one side, on rocks and trees, British Columbia to California.

PTEROBRYACEAE
Jaegerina

Jaegerina scariosa (Lor.) Arzeni (Fig. 245A-B)

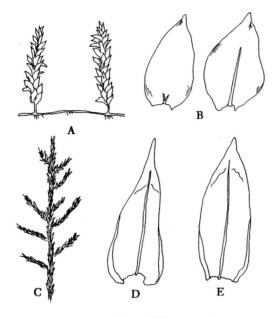

Figure 245

Figure 245 *Jaegerina scariosa*. A, habit of plant; B, branch leaves. *Pireella pohlii*; C, habit of plant; D, leaf. *P. cymbifolia*; E, leaf.

Plants in tufts developing from naked, slender creeping stems, secondary stems, few uncrowded, erect, simple, leaves with a costa that is short and double or single, on bark of hardwood trees, Florida.

Pireella

1a Leaf base auriculate; subquadrate cells in alar region few and inconspicuous. 245C-D. .... *Pireella pohlii* (Schaegr.) Card.

Plants arising from leafless creeping stems, secondary stems, robust, fern like or dendroid 3-5 cm long, dull or shiny yellow-green, on hardwood trees of mesic forests, Florida, Louisiana.

1b Leaf base not auriculate; subquadrate cells in alar region, numerous. Fig. 245E. .... *Pireella cymbifolia* (Sull.) Card.

Plants similar to *P. pohlii* but smaller, secondary stems 2-3 cm long, on hardwood trees of mesic forests, central to southern Florida and Louisiana.

METEORIACEAE
Papillaria

Papillaria nigrescens (Hedw.) Jaegr. & Sauerb. (Fig. 246A-C)

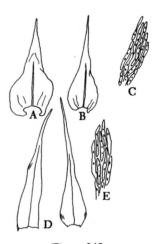

Figure 246

Figure 246 *Papillaria nigrescens*. A, stem leaf; B, branch leaf, C, median leaf cells. *Barbella pendula*; D, leaves; E, median leaf cells.

Plants with freely branching stems forming interwining mats, dull to light-greenish yellow in color, leaves decurrent and conspicuously auriculate, leaf cells with a single row of papillae, on trees and shrubs, occasionally on limestone, Florida, Louisiana.

Barbella

Barbella pendula (Sull.) Fleisch. (Fig. 246D-E).

Plants with elongated, thread-like branches, dull yellowish-green, leaves lanceolate with thread-like tips, leaf cells with a single row of papillae, on trees and shrubs, southern Florida, Louisiana, Mississippi.

NECKERACEAE
Neckera

1a Leaves rounded and usually apiculate at the apex, not strongly undulate. Fig. 247. ...... *Neckera complanta* (Hedw.) Hueb.

1b Leaves acute to acuminate, strongly undulate. 2

2a Leaves entire or slightly denticulate above. Fig. 248. ....
.... *Neckera pennata* Hedw.

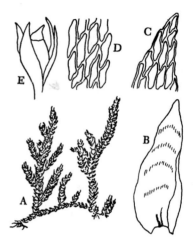

Figure 248

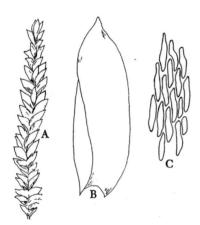

Figure 247

Figure 247 *Neckera complanata*. A, branch of plant; B, leaf; C, median leaf cells.

Plants 2-4 cm long, stems complanate-foliate, green to pale green, in loose mats, on rocks and trunks of trees, Labrador, southern Canada south to Tennessee, North Carolina and in Arkansas.

Figure 248 *Neckera pennata*. A, shoot; B, leaf; C, apex of leaf; D, median cells; E, capsule.

Plants large, stems 5-10 cm long, bright to yellowish green, complanate-foliate, mats hanging from tree trunks or limbs, occasionally on rock, Alaska to British Columbia, Arizona and New Mexico, Ontario to Newfoundland south to North Carolina.

2b Leaves with numerous sharp teeth above; western United States. Fig. 249. ..
.... *Neckera douglasii* Hook.

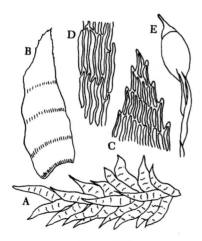

Figure 249

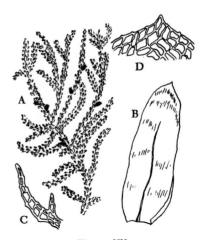

Figure 250

Figure 249 *Neckera douglasii*. A, shoot; B, leaf; C, apex of leaf; D, median cells; E, capsule.

Plants large, stems 5-10 cm long, green to yellow-green, complanate foliate, on trunks and branches of trees, Alaska to California and Idaho.

Metaneckera

Metaneckera menziesii (Hook. *ex* Drumm.) Steere (Fig. 250)

Figure 250 *Metaneckera menziesii*. A, shoot with capsules; B, leaf; C, paraphyllium; D, apex of leaf.

Plants small to large, stems 5-20 cm long, branches often flagelliform, branches and stems complanate and undulate, on tree trunks and branches, occasionally on rocks, Alaska and Alberta south to Idaho, Montana, South Dakota, Nevada and California.

Neckeropsis

1a Leaves basically flat when wet. Fig. 251A. ....
 .... **Neckeropsis disticha** (Hedw.) **Kindb.**

Musci—The Mosses 161

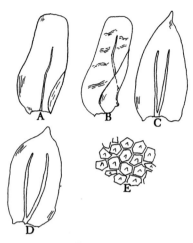

Figure 251

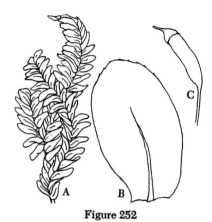

Figure 252

Figure 251 *Neckeropsis disticha* A, leaf; *N. undulata*; B, leaf; *Cyclodictyon varians*; C, leaf; *Callicostella pallida*; D, leaf; E, median leaf cells.

Plants robust, in loose mats, secondary stems to 5 cm, strongly complanate, on trees, stumps and limestone, peninsular Florida.

1b Leaves transversely undulate when wet. Fig. 251B. .... *Neckeropsis undulata* (Hedw.) Reich.

Plants similar to above, leaves strongly undulate, on trees, stumps, and limestone, peninsular Florida, Texas.

Homalia

Homalia trichomanoides (Hedw.) B.S.G. (Fig. 252)

Figure 252 *Homalia trichomanoides*. A, shoot; B, leaf; C, capsule with operculum.

Plants in green to dark green, glossy mats, shoots complanate foliate, often ending in slender, terete-foliate branches, leaves asymmetrical, on rocks, rarely tree trunks, British Columbia and Washington, Wisconsin, Michigan, Ontario, Quebec to Newfoundland south to North Carolina, Tennessee and Arkansas.

Thamnobryum

Thamnobryum alleghaniense (C. Muell.) Nieuwl. (Fig. 253)

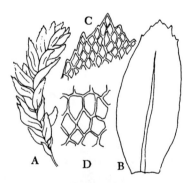

Figure 253

Figure 253 *Thamnobryum alleghaniense.* A, shoot; B, leaf; C, apex and serration of leaf; D, upper median cells.

Plants robust, conspicuously dendroid, light to dark green, on humus, soil, moist rock faces, boulders, rarely on wood, Northeastern Canada south to Alabama and Arkansas. *T. neckeroides* (Hook.) Lawton is found in western North America.

Order Hookeriales

HOOKERIACEAE
Hookeria

Hookeria acutifolia Hook. & Grev. (Fig. 254A-B)

south to Arkansas, Alabama and Georgia. Plants with obtuse leaves are designated *H. lucens* (Hedw.) Sm. (Fig. 254C-D) and occur from Alaska south to Idaho and California.

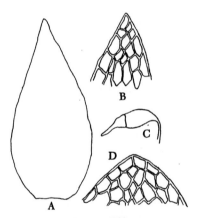

Figure 254

Figure 254 *Hookeria acutifolia.* A, leaf; B, cells of apex of leaf; C, capsule of *H. lucens;* D, apex of leaf of same.

Pale watery mosses in soft mats with acute leaves, on dripping sandstone ledges, British Columbia to Oregon, Indiana to New England

Cyclodictyon

Cyclodictyon varians (Sull.) Kuntze (Fig. 251C)

Plants delicate pale green with bluish iridescence, in branching, flatten mats, cells of leaf smooth, on moist soil or rotten wood in rich forests, Florida, Louisiana.

Callicostella

Callicostella pallida (Hornsch.) Ångstr. (Fig. 251D-E)

Plants pale, dull, whitish green or brownish, in mats with flattened stems, cells of leaf papillose, on moist soil or limestone along streams or in springs, Florida, Louisiana.

Order Hypnobryales

THELIACEAE
Thelia

1a Papillae unbranched, curved toward the apex of the leaf. Fig. 255. .... *Thelia hirtella* (Hedw.) Sull.

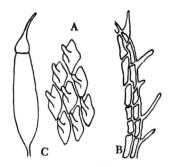

Figure 255

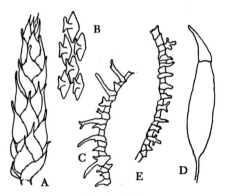

Figure 256

Figure 255 *Thelia hirtella.* A, papillae; B, margin of leaf; C, capsule.

Stems creeping, closely 1-pinnate, green, yellow-brown, or grayish-yellow, usually on tree trunks, rarely on rock, throughout eastern North America from southern Canada to the Gulf States west to Nebraska, Kansas, Oklahoma and Texas.

1b Papillae forked, 3 (2-4)-pointed. 2

2a Stems creeping, radiculose, 1-pinnately branched; leaf margins ciliate papillose above, long ciliate below. Fig. 256A-D. .. .. *Thelia asprella* Sull.

Figure 256 *Thelia asprella.* A, shoot; B, papillae; C, margin of leaf; D, capsule; E, margin of leaf of *T. lescurii.*

Plants green to glaucous-green, usually in thick mats or cushions, on bark at base of trees, sometimes on rock and sterile soil in dry open woods, Minnesota and southern Canada south to the Gulf States.

2b Stems crowded and ascending, not to only slightly radiculose, irregularly branched, margins not long-ciliate below. Fig. 256E. .... *Thelia lescurii* Sull.

Similar to the preceeding species, on rocks and sterile soils, Wisconsin, southern Canada, New England south to the Gulf States.

Myurella

1a Leaves obtuse or sometimes with a small apiculus. .... *Myurella julacea* (Schwaegr.) B.S.G.

Plants glaucous, in tufts or mats, stems 1-3 cm long, often branched, both stems and branches julaceous, on soil in rock crevices, occasionally on logs, Alaska, Ontario, New England south to Washington, Idaho, Colorado, Michigan.

1b Leaves acuminate to apiculate. 2

2a Leaf cells with a single large papilla on the dorsal surface. Fig. 257. .... *Myurella sibirica* (C. Muell.) Reim.

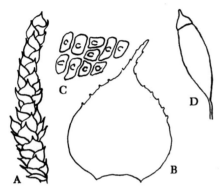

Figure 257 *Myurella sibirica*. A, shoot; B, leaf; C, papillae; D, capsule.

Plants slender, in thin, glossy light to dark green mats or tufts, stem freely branching, julaceous, leaves loosely imbricate, on thin soil of shaded rocks and rock crevices, Alaska to Nova Scotia south to Arkansas, Georgia.

2b Leaf cells smooth to faintly papillose with a single low papilla on dorsal surface. .... *Myurella tenerrima* (Brid.) Lindb.

Similar to the preceeding species, on soil in rock crevices, arctic-alpine, Alaska, British Columbia, Montana.

FABRONIACEAE
Fabronia

This genus consists of several intergrating species that are at times difficult to separate.

1a Peristome teeth single, broad and obtuse. 2

1b Peristome teeth imperfect; teeth of leaf often of more than one cell. .... *Fabronia gymnostoma* Sull. & Lesq. *ex* Sull.

Plants in thin mats or patches, apparently rare, Oklahoma, New Mexico.

2a Leaves ovate lanceolate. 3

2b Leaves lanceolate, teeth composed of single cells; Texas and Arizona. .... *Fabronia wrightii* Sull.

Similar to *F. ciliaris*, on rock, soil, and trees, Texas and Arizona.

3a Teeth of leaf often composed of more than one cell. Fig. 258E. .... *Fabronia pusilla* Raddi

Similar to *F. ciliaris*, on rocks and tree trunks, British Columbia to Idaho, California, Arizona, Colorado.

3b Leaves with teeth composed of single large cells or entire or nearly so. 4

4a Leaves entire or nearly so. .... *Fabronia ravenelii* Sull.

Similar to *F. wrightii*, on bark of trees and decaying logs, Pennsylvania to Tennessee, Florida, and in Arkansas.

4b Leaves irregularly serrate dentate. Fig. 258A-D. ....
Fabronia ciliaris (Brid.) Brid.

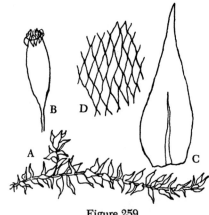

Figure 259

Figure 259 *Anacamptodon splachnoides*. A, plant; B, open capsule with peristome; C, leaf; D, median cells.

Plants in dark-green or yellowish mats, stems creeping, on bark of trees, especially in crotches, knot-holes, fissures, occasionally on log and stumps, and woody fungi, Illinois to Maine south to Florida, Gulf States and Texas.

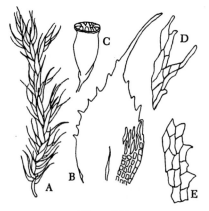

Figure 258

Figure 258 *Fabronia ciliaris*. A, shoot; B, leaf; C, capsule; D, margin of leaf; F, margin of leaf of *F. pusilla*.

Plants small, in thin light-green mats or patches, on tree trunks, rocks, Oregon, South Dakota, Minnesota, Michigan, Pennsylvania, to Texas, Arkansas, Florida.

Anacamptodon

Anacamptodon splachnoides (Froel. *ex* Brid.) Brid. (Fig. 259)

Schwetschkeopsis

Schwetschkeopsis fabronia (Schwaegr.) Broth. (Fig. 260)

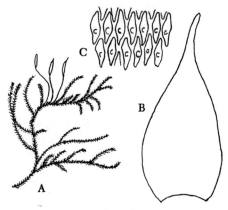

Figure 260

166   Musci—The Mosses

Figure 260 *Schwetschkeopsis fabronia*. A, plant; B, leaf; C, cells of leaf.

Plants light olive to yellow-green, glossy, in crowded mats, stems slender, julaceous; leaf cells unipapillose, on tree trunks, and rocks, Missouri, Illinois, Indiana, Pennsylvania to Connecticut south to Florida and the Gulf States.

Clasmatodon

Clasmatodon parvulus (Hampe) Hook. & Wils. *ex* Sull. (Fig. 261)

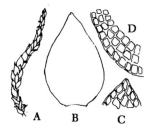

Figure 261

Figure 261 *Clasmatodon parvulus*. A, shoot; B, leaf; C, apex of leaf; D, alar cells.

Plants in light to dark green mats, stems julaceous, on tree trunks and rocks, Oklahoma, Missouri, to Indiana, Pennsylvania south to the Gulf States and Texas.

LESKEACEAE
Lindbergia

Lindbergia brachyptera (Mitt.) Kindb. (Fig. 262)

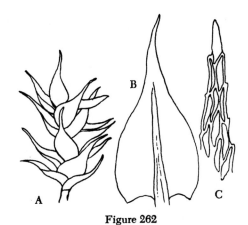

Figure 262

Figure 262 *Lindbergia brachyptera*. A, a wet branch; B, leaf; C, apex of leaf.

Plants creeping, slender, in loose scattered or small mats, dark-green, yellowish or brownish, leaves squarrose when moist and tipped with short hair points, on trunks of trees, occasionally on logs, rarely on rock, Minnesota to Quebec south to Texas, Arizona, Louisiana, South Carolina. *L. mexicana* (Besch.) Card. *ex* Pringle, occurs in the southwest United States in Texas and Arizona and has smooth leaf cells.

Leskea

1a Leaves small, averaging less than 0.6 mm long, median leaf cells 5-7 μ wide. .... *Leskea australis* Sharp

Plants in spreading tufts green to reddish-brown or blackish, on tree trunks in lowlands, Missouri, Tennessee, North Carolina south to the Gulf States and Texas.

1b Leaves larger, averaging more than 0.7 mm long; median leaf cells 7-10 μ wide. 2

2a Stem leaves symmetric, biplicate, one or both margins usually recurved, apices gradually acute, obtuse, or blunt-pointed. 3

2b Stem leaves asymmetric, not plicate, margins not revolute, apices rounded-obtuse to sub-acute. Fig. 263D. .... *Leskea obscura* Hedw.

Plants in loose, spreading, dark-green to olive-green tufts, on trees and bases, soil, and rocks, widespread throughout the eastern United States from Minnesota and Ontario south to the Gulf States and Texas.

3a Stem leaves averaging less than twice as long as wide. Fig. 263A-C. .... *Leskea gracilescens* Hedw.

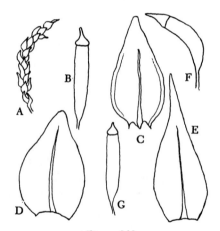

Figure 263

Figure 263 *Leskea gracilescens*. A, shoot; B, capsule; C, leaf; D, leaf of *L. obscura*; E, leaf of *L. polycarpa*; F, G, capsules of same.

Plants in dark green mats with creeping, pinnately branched stems, on tree bases and trunks, rotten log, soil and rocks, throughout eastern North America.

3b Stem leaves averaging more than twice as long as wide. Fig. 263E-F. .... *Leskea polycarpa* Hedw.

Similar to the preceeding species, on trunks and bases of trees, British Columbia to Eastern Canada south to Oregon, Idaho, Montana, South Dakota, Missouri, Tennessee and North Carolina.

Leskeella

Leskeella nervosa (Brid.) Loeske (Fig. 264)

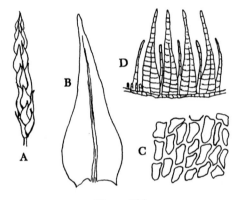

Figure 264

Figure 264 *Leskeella nervosa*. A, shoot; B, leaf; C, median cells of leaf; D, peristome.

Plants in loose to dense, rigid mats, dark-green to brown, stems creeping, subpinnately branched; nearly always producing dense, axillary clusters of brood-branchlets, on bark of trees and rocks, Alaska to Quebec, Nova Scotia and Labrador south to Arizona, Utah, Colorado, Minnesota, Wisconsin, Michigan, Pennsylvania, North Carolina.

Pseudoleskeella

Pseudoleskeella tectorum (Funck *ex* Brid.) Kindb. *ex* Broth (Fig. 265)

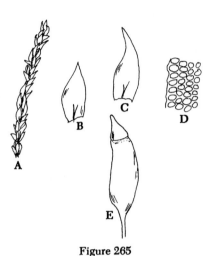

Figure 265

Figure 265 *Pseudoleskeella tectorum*. A, branch; B-C, branch leaves; D, upper leaf cells; E, capsule.

Plants in small, green, yellow-green, or brownish mats, branches often flagelliform, costa of leaves variable usually ending at middle of leaf but often nearly lacking to extending to 2/3 the length of leaf, on rock, soil, or occasionally on logs, British Columbia to Ontario south to Arizona, New Mexico, Idaho, Wyoming, Colorado, Utah, Montana, Michigan. Three other species of *Pseudoleskeella* are known from North America.

Pseudoleskea
(Lescuraea of some authors)

Eight species of *Pseudoleskea* are reported from North America.

1a   Leaf cells with large papillae on the lumen, usually central. Fig. 266. ....
.... *Pseudoleskea patens* (Lindb.) Kindb.

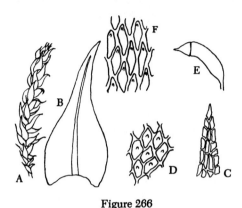

Figure 266

Figure 266 *Pseudoleskea patens*. A, plant; B, leaf; C, apex of leaf; D, median cells; E, capsule.

Plants in dark-green, yellow-green, or brownish mats, stems 1-6 cm long, branching irregularly, paraphyllia numerous, on rocks or a occasionally rotting wood, usually alpine, Alaska to Nova Scotia and Newfoundland south to California, Nevada, Montana, Colorado, Utah, Michigan, New England.

1b   Leaf cells with small papillae at cell ends or papillae absent. 2

2a   Median leaf cells of stem leaves usually 1-2:1. Fig. 267F. ....
Pseudoleskea incurvata (Hedw.) Loeske

Plants in small to large mats, yellow-green to brownish when old, stem very short to 10 cm long, simple or branching irregularly, paraphyllia numerous on stems, on rocks or occasionally on rotting wood, Alaska south to Oregon, Idaho, Nevada, Colorado, Montana.

2b Median leaf cells of stem leaves usually 3-4:1 or longer. 3

3a Leaves gradually long-acuminate, acumen more than half the length of the leaf; papillae over the cell cavity at the upper end. ...... *Pseudoleskea stenophylla* Ren. & Card. *ex* Roell

Plants in loose mats, green to yellow-green, stems to 5 cm long, branching irregularly, paraphyllia numerous, on branches of living trees, Alaska to Oregon, Idaho and Montana.

3b Leaves with a shorter acumen, papillae when present formed by projecting cell ends. Fig. 267A-E. .... *Pseudoleskea radicosa* (Mitt.) Macoun & Kindb.

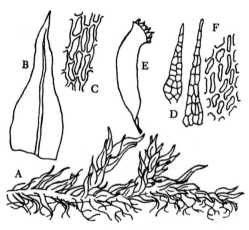

Figure 267

Figure 267 *Pseudoleskea radicosa*. A, shoot; B, leaf; C, median cells; D, paraphyllia; E, capsule; F, median cells of *P. incurvata*.

Plants in dark green to yellow-green mats, stems up to 10 cm long, simple or branching irregularly, branches often hooked at ends, paraphyllia usually numerous, on rock, soil, or wood, Alaska to Quebec south to California, Arizona, Utah, Colorado, Michigan, New England.

Pterigynandrum

Pterigynandrum filiforme Hedw. (Fig. 268)

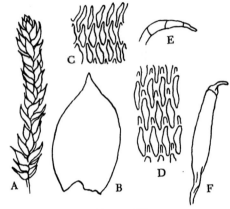

Figure 268

Figure 268 *Pterigynandrum filiforme*. A, shoot; B, leaf; C, median cells; D, upper cells with papillae; E, paraphyllium; F, capsule.

Small, slender plants in dark-green to yellowish or brownish mats, stems creeping, terete or julaceous, frequently with slender, nearly filiform gemmae in the leaf axils, on rocks and trunks of trees, Alaska to Ontario, Quebec, Newfoundland and New Hampshire south California, Colorado, Missouri, Michigan, Virginia, Tennessee and North Carolina. *P. sharpii* Crum & Anderson is known from Tennessee and North Carolina, lacks axillary gemmae and pseudoparaphyllia and has less crowded and concave leaves.

THUIDIACEAE
Heterocladium

Heterocladium macounii Best (Fig. 269)

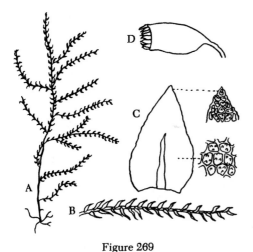

Figure 269

Figure 269 *Heterocladium macounii*. A, plant; B, shoot, wet; C, leaf; D, capsule.

Plants in small, thin to dense spreading mats, stems papillose, irregularly to more or less regularly branching with simple, serrate paraphyllia with papillose cells, on moist rocks, rarely on wood, Alaska to Oregon and Idaho, North and South Carolina, Tennessee. *H. dimorphum* (Brid.) B.S.G. has papillose leaves but smooth stems and is found in the Pacific Northwest east to Michigan, Quebec, New England, Labrador, Nova Scotia and Newfoundland. *H. procurrens* (Mitt.) Rav & Herv. has smooth stems and leaves and is found from British Columbia to Oregon, Idaho, and Montana.

Haplohymenium

Haplohymenium triste (Ces. *ex* De Not.) Kindb. (Fig. 270)

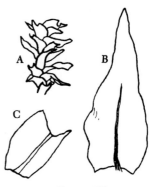

Figure 270

Figure 270 *Haplohymenium triste*. A, shoot; B, a whole leaf; C, the usual broken leaf.

Plants in slender, dull green, mats, stems prostrate, irregularly to subpinnately branching, older leaves broken off at tips, on tree trunks, decaying wood, and rocks, Ontario to Nova Scotia south to Oklahoma, Louisiana, Tennessee, Georgia.

Anomodon

1a Plants with leaves ending in a hyaline hair-point. Fig. 271. ....
.. *Anomodon rostratus* (Hedw.) Schimp.

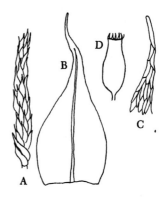

Figure 271

Musci—The Mosses   171

Figure 271 *Anomodon rostratus*. A, shoot; B, leaf; C, apex of leaf; D, capsule.

Plants in yellowish to dark green dense mats with branches terete, on tree bases, soil, rock, southern Canada south to Arizona, Texas, Louisiana, Florida.

1b Plants with leaves not ending in a hyaline hair-point. 2

2a Secondary stems much branched, many attenuate to flagelliform branches evident; leaves often apiculate or toothed at apex. Fig. 272A-C. ....
.. *Anomodon attenuatus* (Hedw.) Hueb.

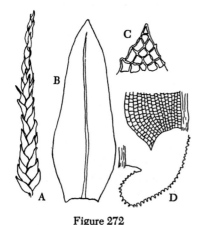

Figure 272

Figure 272 *Anomodon attenuatus*. A, shoot with flagelliform end; B, leaf; C, apex and base of leaf; D, base of leaf of *A. rugelii*.

Plants in loose, dark green or yellow to brown mats, on tree trunks and bases, stumps, logs, rocks, occasionally soil, southern Canada south to Arizona, Texas, Louisiana and Florida.

2b Secondary stems sparingly branched, attenuate or flagelliform branches absent. 3

3a Leaves erect or imbricate, scarcely contorted when dry. Fig. 273A-D. ....
.... *Anomodon minor* (Hedw.) Fuernr.

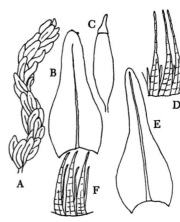

Figure 273

Figure 273 *Anomodon minor*. A, a leafy shoot; B, detached leaf; C, capsule; D, peristome; E, leaf of *A. viticulosus;* F, peristome of same.

Coarse plants in loose dark or glaucous green to yellowish or brownish-green mats, leaves broadly lingulate and rounded obtuse, on trunks and bases of trees, logs, rocks, South Dakota, southern Canada south to Arizona, Texas, Louisiana, Florida.

3b Leaves erect, incurved-contorted when dry. 4

4a Leaves scarcely tapered from shoulders to the rounded apex, rounded-auriculate at base. Fig. 272D. ....
.... *Anomodon rugelii* (C. Muell.) Keissl.

Coarse plants in dark green, yellow-green or brownish mats, on bark of trees, stumps, rocks, Michigan, Ontario to New England south to Missouri, Georgia.

4b Leaves tapered from the shoulders to the narrow, obtuse or rounded-obtuse apex, bases broadly decurrent. Fig. 273E-F. .... *Anomodon viticulosus* (Hedw.) Hook. & Tayl.

Plants robust and coarse, in dark-green to yellowish or brownish-green, dense mats, Michigan to Nova Scotia south to Arkansas and Tennessee.

Herpetineuron

Herpetineuron toccoae (Sull. & Lesq. *ex* Sull.) Card (Fig. 274)

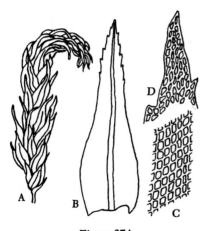

Figure 274

Figure 274 *Herpetineuron toccoae*. A, plant; B, leaf; C, median cells; D, apex of leaf with teeth.

Plants loosely caespitose, in dark green to yellowish mats, secondary stems simple, densely foliate, subcircinate, paraphyllia absent, on tree trunks, and boulders, Arkansas to Virginia, south to Florida, Georgia and Louisiana.

Claopodium

Four species are present in North America.

1a Leaves without hairpoints. Fig. 275A-C. .... *Claopodium whippleanum* (Sull.) Ren. & Card.

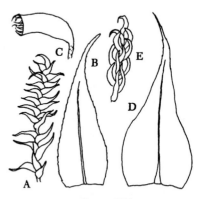

Figure 275

Figure 275 *Claopodium whippleanum*. A, shoot; B, leaf; C, capsule; D, leaf of *C. crispifolium*; E, dry shoot of same.

Plants small, stems creeping, often stoloniferous, 2-4 cm long, green to yellow-green, on soil, logs, roots, British Columbia to California.

1b Leaves with a hair point. Fig. 275D-E. .... *Claopodium crispifolium* (Hook.) Ren. & Card.

Plants large, in wide mats, green to yellow-green, stems pinnately branching, 4-8 cm long, leaf cells with a single large papilla on each surface, on logs, tree bases, and soil, Alaska to Idaho and California. A similar species, *C. bolanderi* Best has 2 or more papillae on each surface is also found in the Pacific Northwest.

Haplocladium

1a Stem leaves roundish-ovate, scarcely plicate, abruptly acuminate to a broad, oblong point. Fig. 276 & 279D. .... *Haplocladium virginianum* (Brid.) Broth.

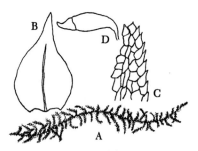

Figure 276

Figure 276 *Haplocladium virginianum*. A, plant; B, leaf; C, apical cells of leaf; D, capsule.

Plants in tangled mats, dark to dirty green, stems pinnately branched, 2-4 cm long, branches rigid, julaceous and with numerous paraphyllia, on soil, occasionally on rocks and soil, Minnesota to New England south to Arizona, Texas, Louisiana, Georgia.

1b Stem leaves gradually long-acuminate, biplicate. .... *Haplocladium microphyllum* (Brid.) Broth.

Plants medium in size in thin spreading mats, light to yellowish green, stems 3-5 cm long, stems and branches loosely foliate, on soil, humus, and decaying wood, Wisconsin, Ontario to New England south to Florida, Louisiana, Texas and Arizona.

Thuidium

Nine species are known from North America.

1a Plants small; leaf cells minutely pluripapillose. 2

1b Plants large to robust; leaf cells stoutly unipapillose. 4

2a Branches smooth. 3

2b Branches papillose. Fig. 277D. .... *Thuidium pygmaeum* B.S.G.

Small, dark green plants, in this mats, on moist rocks near streams and waterfalls, widely distributed in eastern North America.

3a Plants 1-pinnate; branches terete when dry; leaves imbricate when dry, erect spreading when dry; leaf cells papillose on back. .... *Thuidium scitum* (P.-Beauv.) Aust.

Small, rigid, dark-green, yellow-green or brown plants, stems creeping, on bark usually at base of trees, on logs, stumps or rocks, Michigan to New Brunswick, south to northern Missouri and North Carolina.

3b Plants 1-2 pinnate; branches not terete; leaves incurved-catenulate when dry, wide spreading when moist; leaf cells papillose on both surfaces. Fig. 277A-C. .. *Thuidium minutulum* (Hedw.) B.S.G.

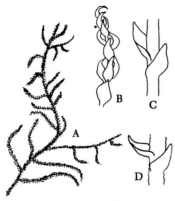

Figure 277

Figure 277 *Thuidium minutulum*. A, plant; B, dry branch; C, twig; D, branch of *T. pygmaeum*.

Small, dark green plants, stems creeping, 2-4 cm long, on decaying wood, soil and rocks, Minnesota to New Brunswick south to Gulf Coast.

4a   Stems 1-pinnate, sub-erect. Fig. 278. ......
        .... *Thuidium abietinum* (Brid.) B.S.G.

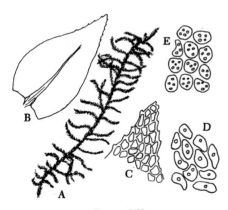

Figure 278

Figure 278 *Thuidium abietinum*. A, plant; B, leaf; C, apical cells of leaf; D, median cells.

Robust, rigid, dark-green, yellowish, or dark-brown plants, stems up to 12 cm high, erect ascending, on soil and rocks, usually calcareous, Alaska to Nova Scotia, Newfoundland south to Arizona, Utah, Colorado, Iowa, Indiana and Virginia.

4b   Stems 2-3 pinnate, spreading or curved-ascending. 5

5a   Stem leaves spreading-recurved, papillae on paraphyllia near the upper end of the cell; perichaetial leaves without cilia. Fig. 279A-B. ....
       .. *Thuidium recognitum* (Hedw.) Lindb.

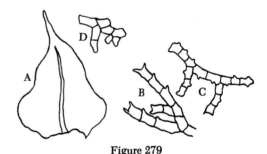

Figure 279

Figure 279 *Thuidium recognitum*. A, stem leaf; B, paraphyllium; C, paraphyllium of *T. delicatulum*; D, paraphyllium of *Haplocladium virginianum*.

Robust plants in rigid light-to-yellow-green or yellow-brown mats, stems 4-9 cm long, on soil, humus, logs, tree bases in wet to dry woods, Alaska to Nova Scotia south to British Columbia, Oklahoma, Arkansas, Georgia.

5b   Stem leaves erect, papillae on paraphyllia near middle of cells; perichaetial leaves with long cilia on the margin or cilia absent. 6

Musci—The Mosses   175

6a Stem leaves, not ending in a capillary point; perichaetial leaves ciliate. 7

6b Stem leaves ending in a capillary point formed by 2-7 or more uniseriate cells; perichaetial leaves not or rarely ciliate. .. .. *Thuidium delicatulum* var. *radicans* (Kindb.) Crum, Steere & Anderson

Similar to the next species, on soil, humus, on logs in wet woods, Alaska to Quebec and northern Ontario south to British Columbia, Alberta, Iowa, Michigan and Virginia.

7a Stems regularly bi- or tripinnately branched; inner perichaetial ending in a hair point. Fig. 280. .... *Thuidium delicatulum* (Hedw.) B.S.G. var. *delicatulum*

7b Stems less regularly pinnate; inner perichaetial leaves not ending in a hair point. .... *Thuidium alleni* Aust.

Plants in loose mats, leaf cells with papillae on each surface, on soil and decaying wood, Pennsylvania to Florida.

Helodium

1a Leaf cells smooth or weakly papillose because of projecting upper angles. Fig. 281. .... *Helodium paludosum* (Sull.) Aust.

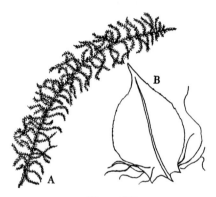

Figure 281

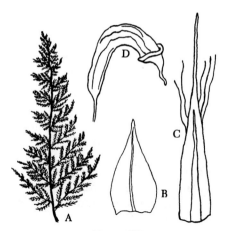

Figure 280

Figure 280 *Thuidium delicatulum*. A, plant; B, stem leaf; C, perichaetial leaf; D, capsule.

Plants in robust green or yellow-green mats, stem 3-8 cm long, spreading or arched-ascending, on soil, humus, decaying wood, rocks, or tree bases, in moist areas, Alaska to Labrador south to Arizona, in the east south to the Gulf States.

Figure 281 *Helodium paludosum*. A, plant; B, leaf with paraphyllia.

Plants in yellowish-green deep, soft, tufts, branching irregularly pinnate, on soil in wet open places such as bogs and pastures, Iowa, Illinois to Ontario south to Missouri, Tennessee and North Carolina.

1b Leaf cells strongly unipapillose at back near upper ends or centrally. .... *Helodium blandowii* (Web. & Mohr.) Warnst.

Plants in dense, yellow to light green mats, closely 1-pinnate, on soil and humus in wooded to open bogs or swamps, along streams and in ditches, Alaska to Quebec, Nova Scotia, Newfoundland south to Arizona, Colorado, Illinois, Ohio and New York.

AMBLYSTEGIACEAE
Cratoneuron

Four species are reported from North America.

1a Leaves not plicate nor circinate. Fig. 282A-C. ....
Cratoneuron filicinum (Hedw.) Spruce

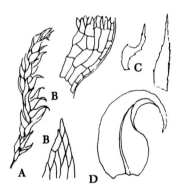

Figure 282

Figure 282 *Cratoneuron filicinum*. A, branch; B, leaf base and apex; C, paraphyllia; D, leaf of *C. commutatum*.

Plants in green, yellowish, or brownish cushions up to 12 cm high, stems spreading to erect ascending, paraphyllia none to numerous, on wet calcareous soil or rocks in seepage areas or edge of spring branches or streams, Alaska to Newfoundland south to California, New Mexico, Texas, Alabama, Georgia.

1b Leaves plicate and circinate. Fig. 282D. .... *Cratoneuron commutatum* (Hedw.) Roth

Plants in mats or tufts, stem spreading, 5-15 cm long, branching regularly or irregularly, paraphyllia numerous, on soil and rocks near water, Alaska and Yukon to Quebec and Newfoundland south to Nevada, Utah, Colorado, Michigan, New York.

Campylium

Eight species are reported from North America.

1a Costa absent or very short and double. 2

1b Costa single, ending near or above the middle of the leaf. 3

2a Slender plants with creeping stems; leaves finely serrate from base to apex; alar cells small. Fig. 283. ...... *Campylium hispidulum* (Brid.) Mitt.

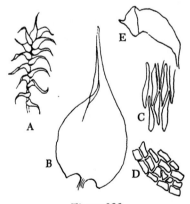

Figure 283

Musci—The Mosses   177

Figure 283 *Campylium hispidulum*. A, shoot; B, leaf; C, median cells; D, alar region; E, capsule.

Plants in bright-green or yellowish mats, stems creeping, leaves squarrose, on soil, rocks, logs, tree bases, Alaska to Newfoundland south British Columbia, Montana, Utah, Colorado, Texas, Louisiana, Alabama and Georgia.

2b   Moderately robust plants with erect-ascending stems; leaves entire; alar cells inflated. Fig. 284. ....
Campylium stellatum (Hedw.) C. Jens.

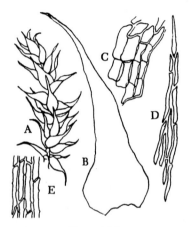

Figure 284

Figure 284 *Campylium stellatum*. A, shoot; B, leaf; C, alar region; D, apex of leaf; E, cells from lower part of leaf.

Plants in loose to dense, yellow, golden, or brownish mats, on soil or humus in wet, calcareous habitats, Alaska to Quebec, Newfoundland and Nova Scotia south to British Columbia, Wyoming, New Mexico, Colorado, Iowa, Michigan, Ohio, Pennsylvania, New York and New England.

3a   Leaves erect-spreading, gradually acuminate; alar cells enlarged and inflated. Fig. 285. .... *Campylium polygamum* (B.S.G.) C. Jens.

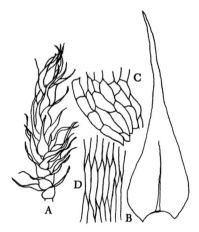

Figure 285

Figure 285 *Campylium polygamum*. A, shoot; B, leaf; C, alar region; D, median cells.

Moderately robust plants in loose to dense, yellow-green to golden-brown mats, on soil and humus of meadows, swamps, or bogs, British Columbia to Labrador south to the Gulf States in eastern North America.

3b   Leaves wide-spreading to squarrose, abruptly acuminate; alar cells not to only slightly inflated. 4

4a   Leaves crowded, often secund at tips; upper leaf cells linear. Fig. 286. ....
.... *Campylium chrysophyllum* (Brid.) J. Lange

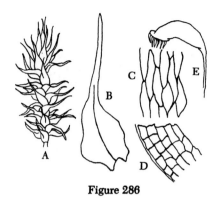

Figure 286

Figure 286 *Campylium chrysophyllum.* A, shoot; B, leaf; C, median cells; D, alar region; E, capsule.

Plants in green, yellow-green, or yellow-brown mats, stems creeping, freely and irregularly branched, on soil, humus, rocks, tree bases, British Columbia to Newfoundland south to Oregon, Arizona, Montana, Colorado, Texas and the Gulf States.

4b   Leaves not crowded or secund; upper leaf cells short-rhomboidal (4-6:1). .... *Campylium radicle* (P.-Beauv.) Grout

Slender plants in soft, loose, light-green or yellowish mats, on decaying leaves, twigs and mucky soil in swampy habitats, Ontario south to Florida.

Leptodictyum

A highly variable complex of species. Many are aquatic. At least seven species are described from North America.

1a   Median leaf-cells averaging 3-10:1. 2

1b   Median leaf-cells averaging 8-16:1. .... *Leptodictyum riparium* (Hedw.) Warnst.

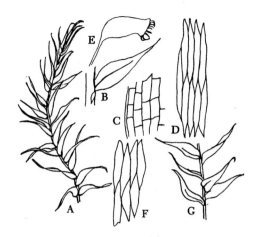

Figure 287

Figure 287 *Leptodictyum riparium.* A, shoot; B, leaf with portion of stem; C, alar cells; D, median cells; E, capsule; F, median cells of *L. laxirete;* G, shoot of *f. fluitans.*

A number of poorly defined forms of this species are recognized and may be separated by the following characters suggested by Conard (Bryologist 62:96-104. 1959).

     a.   Apex of leaf acute, ending in mostly 2 cells. .... *L. riparium* fo. *obtusum* (Grout) Grout

     a.   Apex acuminate to slenderly acuminate, ending mostly in 1 cell. .... b

     b.   Plants of wet habitats, but not floating or submerged. Fig. 287 A-E. .... *L. riparium* fo. *riparium*

     b.   Plants floating or submerged. c

     c.   Plants usually floating; leaves 3.5-5 mm long, acumination, very long filiform. .. d

     c.   Plants submerged-floating; apices long and filiform. Fig. 287G. .... *L. riparium* fo. *fluitans* (Lesq. & James) Grout

  d. Leaves crowded, imbricate. ....
   .... *L. riparium* fo. *elongatum* **Moenk.**

  d. Leaves distant. .... *L. riparium* fo. *longifolium* (**Schultz.**) **Grout**

Plants medium to robust, green, yellowish, to brownish plants in loose, flat mats, on humus, twigs, branches, roots and rocks in wet places inundated by flowing water, throughout North America.

2a Costa of stem leaves ending near the middle of the leaf. 3

2b Costa of stem leaves subpercurrent, percurrent to excurrent. 5

3a Plants robust, somewhat complanate, distinctly aquatic, usually growing in swiftly flowing water. Fig. 287F. ....
  .... *Leptodictyum laxirete* (**Card. & Thér.**) **Broth.**

Robust, dull green plants with soft, flaccid leaves, stems up to 10 cm long, on calcareous soil and rocks, Texas, Missouri, Indiana and Georgia.

3b Plants small, leaves not complanate, on shaded soil. 4

4a Leaves distant, separated by nearly their own width. .... *Leptodictyum sipho* (**P.-Beauv.**) **Broth.**

Plants light green, small, stems 1-2 cm long, on humus and decaying wood, Georgia, Florida, Louisiana and Texas.

4b Leaves not distant. Fig. 288. ....
  .... *Leptodictyum brevipes* (**Card. & Thér.** *ex* **Holz.**) **Broth.**

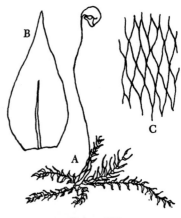

Figure 288

Figure 288 *Leptodictyum brevipes*. A, plant; B, leaf; C, median cells.

Plants small, slender, light green to green, leaves erect-spreading, on shaded soil in wet places, scattered in eastern North America from Minnesota, Michigan to Pennsylvania south to Missouri, Indiana and West Virginia.

5a Stem leaves erect-spreading, usually aquatic. ....
  .... *Leptodictyum vacillans* (**Sull.**) **Broth.**

Plants yellowish green, prostrate, stem leaves rather distant, Ontario to New England.

5b Stem leaves widely spreading, plants in moist habitats but not aquatic. Fig. 289.
  .... *Leptodictyum trichopodium* (**Schultz.**) **Warnst.**

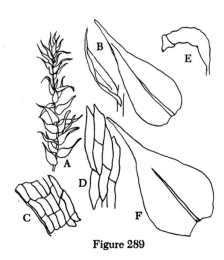

Figure 289

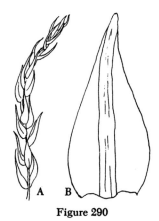

Figure 290

Figure 289 *Leptodictyum trichopodium*. A, shoot; B, leaf and stem; C, alar region; D, median cells; E, capsule; F, leaf.

Plants in loose tufts or mats, light-to-yellowish-green, stems 2-4 cm long, leaves moderately distant, on moist wet substrate, throughout North America.

Hygroamblystegium

A number of species and forms of this genus have been recognized. However, Crum (Mosses of the Great Lakes Forest, revised edition, 1976) has recognized only two species and this seems to be a wise decision for this highly variable and perplexing aquatic genus.

1a   Plants large, coarse, and generally rigid; costa very broad, terete in cross section and filling the apex of the leaf, stoutly and shortly excurrent. Fig. 290. .... *Hygroamblystegium noterophilum* (Sull. & Lesq. *ex* Sull.) Warnst.

Figure 290 *Hygroamblystegium noterophilum*. A, shoot; B, leaf.

Plants in large, rigid, dark-green or brownish mats, often encrusted with lime, attached to wet calcareous rocks, usually submerged in running water, Minnesota, Ontario, New York south to Pennsylvania and Missouri.

1b   Plants not large and coarse; costa narrower, not terete, percurrent to excurrent as a slender point. 2

2a   Leaves gradually acuminate and bluntly acute; costa percurrent. Fig. 291. .... *Hygroamblystegium tenax* (Hedw.) Jenn. var. *tenax*

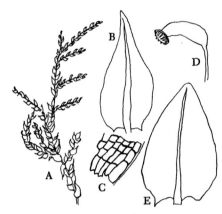

Figure 291

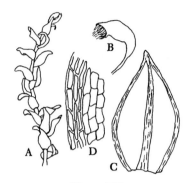

Figure 292

Figure 291 *Hygroamblystegium tenax.* A, plant; B, leaf; C, basal cells; D, capsule; E, leaf.

Plants dark-to-yellow-green, usually in rather soft mats, on rocks in or adjacent to brooks and streams, Minnesota to New Brunswick south to Arizona, Texas, Louisiana and Florida.

2b Leaves abruptly subulate; costa excurrent as a slender point. .... *Hygroamblystegium tenax* var. *spinifolium* (Schimp.) Jenn.

On rocks in flowing water, Indiana to Pennsylvania, Ohio, Ontario south to Arkansas, Missouri, New Jersey.

Sciaromium

Sciaromium lescurii (Sull.) Broth. (Fig. 292)

Figure 292 *Sciaromium lescurii.* A, shoot; B, capsule; C, leaf; D, border of leaf.

Plants in loose, dark green mats, branching irregularly, on rocks in streams, Eastern Canada south to Arkansas and the Gulf States. *S. tricostatum* (Sull.) Mitt. had been reported from Cape Arago, Oregon.

Amblystegium

1a Costa weak, ending near middle of leaf. 2

1b Costa strong, reaching apex of leaf or nearly so. Fig. 293. .... *Amblystegium varium* (Hedw.) Lindb.

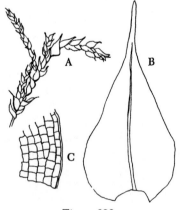

Figure 293

Figure 293 *Amblystegium varium.* A, plant; B, leaf; C, alar cells.

Plants small, in yellowish or green mats, on soil, humus, logs, rocks of moist shady habitats, widespread in North America.

2a Leaves erect-spreading; basal cells quadrate to transversely elongate. Fig. 294. .... .. *Amblystegium serpens* (Hedw.) B.S.G.

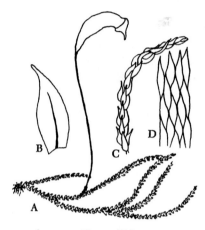

Figure 294

Figure 294 *Amblystegium serpens.* A, shoot with capsule; B, leaf; C, dry branch; D, median cells.

Plants small, in loose, soft, green or yellowish mats, on moist soil, humus, wood in wet habitats, widespread in North America.

2b Leaves widely spreading; basal cells vertically elongate. .... *Amblystegium juratzkanum* Schimp.

Similar to the previous species and often difficult to separate from it, on moist rocks, soil, wood, widespread in North America.

Platydictya

Four species of this very small moss are reported from North America.

1a Plants growing on trees. Fig. 295A-D. .... *Platydictya subtile* (Hedw.) Crum

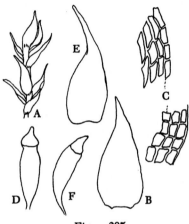

Figure 295

Figure 295 *Platydictya subtile.* A, shoot; B, leaf; C, alar and median cells; D, capsule and operculum; E, leaf of *P. confervoides*; F, capsule of same.

Small plants in dark green to brownish thin mats, on trunks of hardwoods, Minnesota to southeastern Canada south to Arkansas and North Carolina.

1b Plants growing on limestone rocks. Fig. 295E. .... *Platydictya confervoides* (Brid.) Crum

Plants small, dark green with creeping stems and erect-ascending branches, on moist, shaded limestone, southeastern Canada south to the Great Lakes and Arkansas.

Drepanocladus

Thirteen species of this genus are reported from North America.

1a Leaves plicate or finely ridged when dry. 2

1b Leaves smooth or if finely ridged, the alar cells conspicuously inflated. 3

2a Leaves plicate with a few enlarged, hyaline cells. Fig. 296A-C. .... *Drepanocladus uncinatus* (Hedw.) Warnst.

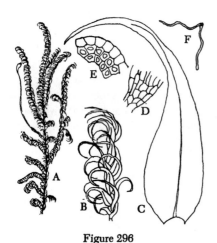

Figure 296

Figure 296 *Drepanocladus uncinatus*. A, plant; B, branch; C, leaf; D, alar cells; E, section of stem, and F, section of leaf of *D. vernicosus*.

Plants in slender to robust, loose, green, yellowish or brownish mats or tufts, irregular to subpinnately branched, on rocks, decaying wood, humus, base of trees, Alaska to Quebec, Newfoundland and Nova Scotia south to California, Utah, Colorado, Minnesota, Michigan, Pennsylvania and Ohio.

2b Leaves finely ridged without enlarged alar cells. Fig. 296E-F. .... *Drepanocladus vernicosus* (Lindb. *ex* C. Hartm.) Warnst.

Plant usually in green or yellowish tufts, stems pinnately branched, on wet soil or humus in calcareous fens and swamps. British Columbia to Nova Scotia and Newfoundland south to Oregon, Idaho, Minnesota, Indiana, New England.

3a Leaves entire. 4

3b Leaves serrulate, at least at bases. 7

4a Alar cells not or only slightly differentiated. Fig. 297. .... *Drepanocladus revolvens* (Sw.) Warnst.

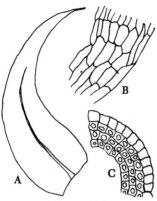

Figure 297

Figure 297 *Drepanocladus revolvens*. A, leaf; B, alar cells; C, section of stem.

Robust plants in dense, usually reddish to blackish-purple tufts, stems erect or ascending, irregularly to regularly branched, on wet soil in bogs and swamps, Alaska to Quebec and Nova Scotia south to British Columbia, Colorado, Iowa, Michigan, Ohio.

4b   Alar cells inflated, in large and conspicuous groups. 5

5a   Cells of the lower half of leaf oblong-hexagonal. Fig. 298C-D. .... *Drepanocladus aduncus* var. *polycarpus* (Bland. *ex* Voit) Roth.

Plants robust, in soft, loose to dense, dark-green to brownish mats, irregularly to pinnately branched, in water of swamps, bogs, lakes, meadows, streams, etc., Arctic North America south to California, New Mexico, Missouri, Indiana, and Pennsylvania.

5b   Cells of the lower half of leaf linear. 6

6a   Leaves of middle portion of stem falcate-secund, oblong-lanceolate, acumination long, slender, and channelled. Fig. 298A. .... *Drepanocladus aduncus* (Hedw.) Warnst. var. *aduncus*

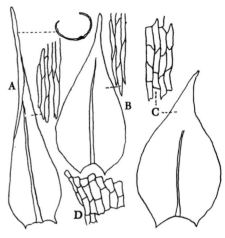

Figure 298

Figure 298 *Drepanocladus aduncus*. A, leaf and median cells of var. *aduncus*; B, same, of var. *kneiffii*; C, same, of var. *polycarpus*; D, alar cells.

Robust plants in loose, soft dark-green to brownish mats or tufts, irregularly to pinnately branched, on soil in wet, calcareous habitats, Alaska to Quebec south to California, Nevada, Colorado, Nebraska, Missouri, Indiana, Ohio, Pennsylvania, Virginia.

6b   Leaves of middle portion of stem not at all falcate or secund to only slightly so, broadly lanceolate to oblong-ovate, acumination short, flat. Fig. 298B. .... *Drepanocladus aduncus* var. *kneiffii* (B.S.G.) Moenk.

Plants in green to brownish mats, on soil in wet calcareous habitats. Range probably same as var. *aduncus*.

7a Leaves relatively flat, often finely ridged when dry; costa broad, percurrent, alar cells inflated, in large, decurrent auricles usually reaching the costa. Fig. 249. .... ***Drepanocladus exannulatus*** (B.S.G.) Warnst.

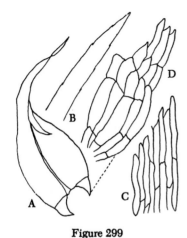

Figure 299

Figure 299 *Drepanacladus exannulatus*. A, leaf; B, acumination; C, median cells; D, alar cells.

Plants in green to yellow-brown mats, sometimes red or red-purple tinged, stems more or less pinnately branched, in wet places, often submerged, Arctic North America south to California, Colorado, Indiana and Pennsylvania.

7b Leaves concave, smooth; costa narrow ending well above middle of leaf; alar cells often somewhat enlarged but not abruptly differentiated or in decurrent auricles. .... ***Drepanocladus fluitans*** (Hedw.) Warnst.

Plants in green, yellow-green, or brown, loose mats, sparcely branched, leaves rather distant, on wet soil, Arctic North America south to Oregon, Montana, Wyoming, Nebraska, Minnesota, Indiana, Pennsylvania, West Virginia.

Hygrohypnum

Thirteen species of *Hygrohypnum* are reported from North America.

1a Outer cells of stem thin-walled, clear and colorless. Fig. 300. .... ***Hygrophypnum ochraceum*** (Turn. *ex* Wils.) Loeske

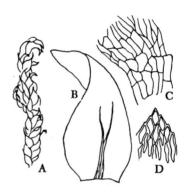

Figure 300

Figure 300 *Hygrohypnum ochraceum*. A, shoot; B, leaf; C, alar and adjacent cells; D, apex of leaf.

Plants in dense to loose, soft, green, yellow-green or brownish mats, stems 2-8 cm long, branching irregularly, on soil and rocks along cool mountains streams, Alaska to Ontario, Quebec, New England south to California, Nevada, Colorado, Minnesota, Michigan, Pennsylvania and Tennessee.

1b Outer cells of stem small and thick-walled. 2

2a Leaves very large, 1.4-3.4 mm long, rigid, harsh. Fig. 301A. ..... *Hygrohypnum bestii* (Ren. & Bryhn *ex* Ren.) Holz. *ex* Broth.

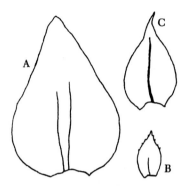

Figure 301

Figure 301 *Hygrohypnum bestii*. A, leaf; B, leaf of *H. eumontanum*; C, leaf of *H. luridum*.

Plants normally large in soft to stiff, harsh, mats or loose tufts, green, yellow-green or brownish, stems branching irregularly, 3-12 cm long, on stones in or near water, Alaska south to California, South Dakota.

2b Leaves smaller, less than 1.5 mm long. 3

3a Leaves small, 0.7 × 0.3 mm, lanceolate, serrulate to the base. Fig. 301B. .... *Hygrophypnum eumontanum* Crum, Steere & Anderson

Plants in wide, soft, light green mats, stems with numerous branches, 1-2 cm long, on shaded rocks along mountain brooks, Quebec, Newfoundland south to Tennessee.

3b Leaves larger, entire, serrate above the middle, or finely serrulate at apex. 4

4a Leaves distinctly serrulate above the middle; alar cells large, clear, with brown walls. Fig. 302. .... *Hygrohypnum novae-caesareae* (Aust.) Grout

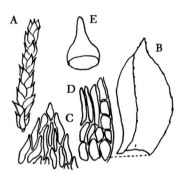

Figure 302

Figure 302 *Hygrohypnum novae-caesareae*. A, shoot; B, leaf; C, apex of leaf; D, basal cells; E, operculum.

Plants in tangled, light green mats, stems with few branches, 2-5 cm long, on wet rock along streams, Vermont to Georgia.

4b Leaves entire, or finely serrulate at apex. 5

5a Leaves entire, subtubulose near apex; alar cells not inflated. Fig. 303E. .... *Hygrohypnum luridum* (Hedw.) Jenn.

Plants small to medium size, in green to yellow-green (sometimes reddish-brown) tufts or mats, stems branching irregularly, 2-7 cm, Quebec, New Brunswick south to California, Utah, Colorado, Ohio, New Jersey, and Tennessee.

Musci—The Mosses   187

5b Leaves entire, denticulate or finely serrate near apex, flat, not subtubulose near apex. 6

6a Alar cells abruptly enlarged, clear, the inner ones with brown walls. Fig. 303A-D. .... *Hygrohypnum eugyrium* (B.S.G.) Loeske

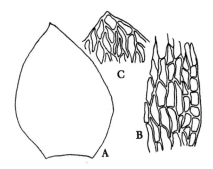

Figure 304

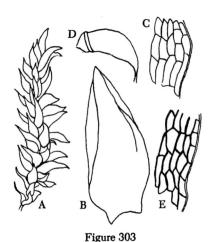

Figure 303

Figure 304 *Hygrohypnum molle.* A, leaf; B, alar and adjacent cells; C, apex of leaf.

Plants in soft, green to yellow-green mats, stems branched, 1-2 or 3 cm long, on moist stones in shaded streams, British Columbia to Great Lakes and New Brunswick, south to Oregon, Montana, Wyoming, and South Carolina.

Calliergon

Calliergon cordifolium (Hedw.) Kindb. (Fig. 305)

Figure 303 *Hygrohypnum eugyrium.* A, shoot; B, leaf; C, alar cells; D, capsule; E, alar cells of *H. luridum.*

Plants in wide, dense, bright-green patches, on moist rocks in streams, Ontario, New England, south to Georgia.

6b Alar cells colored but not inflated, leaves broadly ovate to suborbicular. Fig. 304. .. *Hygrohypnum molle* (Hedw.) Loeske

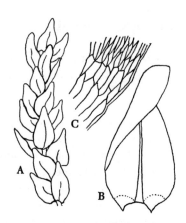

Figure 305

Figure 305 *Calliergon cordifolium*. A, shoot; B, leaf; C, alar cells.

Robust plants in soft, loose, green to yellow-green mats of tufts, stems 4-16 cm long, simple to sparingly branched, in bogs and swamps and other wet areas, Alaska to Ontario, Labrador and New England south to Oregon, Colorado, Illinois, Pennsylvania, Tennessee, New Jersey. Ten similar species are found in North America.

Calliergonella

Calliergonella cuspidata (Hedw.) Loeske (Fig. 306)

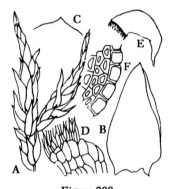

Figure 306

Figure 306 *Calliergonella cuspidata*. A, shoot; B, leaf; C, apex of leaf; D, alar region; E, capsule; F, section of stem.

Plants medium to large, in loose glossy, yellow-green mats or tufts, stems branching ascending to more or less erect, 6-12 cm long, in upland bogs, fens, and swamps, Alaska to Quebec, Nova Scotia and New Brunswick, south to California, Idaho, Wyoming, Iowa, Wisconsin, Michigan, Ohio, Pennsylvania, New Jersey, Tennessee, North Carolina. *C. conardii* Lawt. is known Yellowstone National Park and was first collected by Dr. Conard. It is an aquatic plant floating in water.

Scorpidium

Scorpidium scorpioides (Hedw.) Limpr. (Fig. 307)

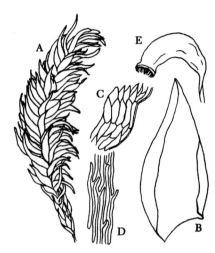

Figure 307

Figure 307 *Scorpidium scorpioides*. A, shoot; B, leaf; C, alar region; D, median cells; E, capsule.

Plants large, in yellow-golden-or-brownish-green mats or tufts, stems often branched, erect or prostrate, 5-15 cm long, in bogs and ponds, Alaska to Ontario, New England and Nova Scotia south to British Columbia to Michigan, Ohio. *S. turgescens* (T. Jens.) Loeske has loosely imbricate to spreading leaves (as opposed to more or less falcate-secund in *S. scorpioides*) and is found from Alaska to Michigan, Ontario and Newfoundland south to British Columbia, Colorado, Michigan.

BRACHYTHECIACEAE

The treatment of this family is adapted in part from "Generic Revisions of North American Brachytheciaceae" by Harold Robinson (Bryologist 65:73-146. 1962).

Tomenthypnum

Tomenthypnum nitens (Hedw.) Loeske (Fig. 308)

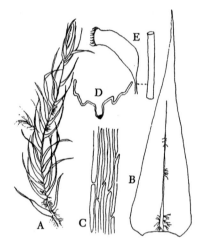

Figure 308

Figure 308 *Tomenthypnum nitens.* A, shoot; B, leaf; C, median cells; D, section of leaf; E, capsule and seta.

Plants in loose or dense, yellow-green to golden-brown tufts, stems pinnately branched, 5-15 cm long, in bogs and swamps, Alaska, to Ontario south to Colorado, Wisconsin, Michigan, Pennsylvania.

Homalothecium

Seven species of *Homalothecium* are recognized in North America.

1a Leaves dentate at base with sharp, usually recurved teeth. Fig. 309. .... *Homalothecium nuttallii* (Wils.) Jaeg. & Sauerb.

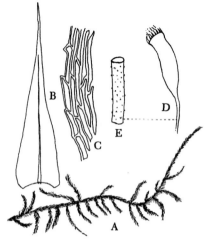

Figure 309

Figure 309 *Homalothecium nuttallii.* A, shoot; B, leaf; C, cells of leaf; D, capsule; E, seta.

Plants in yellow-green mats, stems 10 cm or more long, on tree trunks, logs, occasionally on rocks, Alaska to California.

1b Leaves entire or with only a few small teeth at base. 2

2a Plants large and coarse, regularly pinnate, alar cell walls pitted. .... *Homalothecium megaptilum* (Sull.) Robins.

Plants in mats, large, glossy, yellow-green, stems 5-16 cm long, on litter in coniferous forests, occasionally on logs or soil over rocks, British Columbia to California, Idaho, Montana.

2b Plants smaller, irregularly pinnate, alar cell walls not pitted. .... *Homalothecium nevadense* (Lesq.) Ren. & Card.

Plants medium to large, in green to yellow-green mats, stems up to 12 cm long, on rocks, rarely on logs, British Columbia to California, Arizona, Utah, Idaho, Montana, Colorado.

Homalotheciella

Homalotheciella subcapillata (Hedw.) Broth. (Fig. 310)

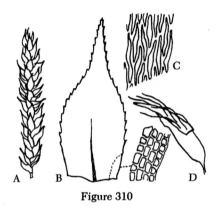

Figure 310

Figure 310 *Homalotheciella subcapillata*. A, shoot; B, leaf; C, cells of leaf; D, capsule with calyptra.

Plants in thin, glossy, yellowish mats, irregularly to subpinnately branches; calyptra hairy, on tree trunks and decaying wood, throughout the eastern United States.

Isothecium

1a Leaves loosely imbricate; apices long acuminate with cells mostly 5 or more times as long as wide; lower leaf margins serrulate. Fig. 311. .... *Isothecium spiculiferum* (Mitt.) Ren. & Card.

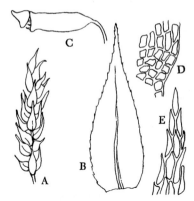

Figure 311

Figure 311 *Isothecium spiculiferum*. A, shoot; B, leaf; C, capsule; D, alar region of *I. cristatum*; E, leaf apex of same.

Plants in dark to light green, yellow-green, or brownish mats, stem branching irregularly, on trees, soil, and rocks, Alaska to California, Idaho, Montana.

1b Leaves closely imbricate when dry; apices short-acuminate with cells mostly 2-3 times as long as wide; lower leaf margins entire. .... *Isothecium cristatum* (Hampe) Robins.

Plants in dense mats or tufts, dark-to-brownish-green, on trees and rocks, British Columbia to California.

Brachythecium

There are at least thirty species of *Brachythecium* reported from North America. This is one of our most difficult genera and identification is difficult without sporophytes.

1a Leaves plicate. 2

1b Leaves not plicate. 8

2a Alar cells small, numerous, quadrate, to subquadrate, sharply differentiated. 3

2b Alar cells not as above, lax or inflated, if quadrate to subquadrate, not sharply differentiated. 5

3a Plants small, quadrate alar cells forming a distinct triangular area from margin to costa. Fig. 312A-E. .... *Brachythecium acuminatum* (Hedw.) Aust.

3b Plants medium to large, alar cells not forming a distinct triangular area from margin to costa. 4

4a Plants large, in dense mats; leaves straight (occasionally falcate); seta smooth; dioicous, eastern North America. Fig. 313A-C, E. .... *Brachythecium oxycladon* (Brid.) Jaeg. & Sauerb.

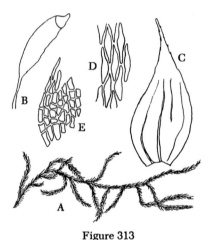

Figure 313

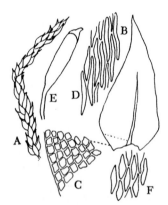

Figure 312

Figure 312 *Brachythecium acuminatum.* A, shoot; B, leaf; C, basal cells; D, median cells; E, capsule; F, median cells of var. *cyrtophylla.*

Plant slender, in green or yellowish shiny mats, stems creeping with erect, short, terete branches, seta smooth, on tree bases, soil, rocks, Saskatchewan to New Brunswick south to Florida and Texas. *B. acuminatum* var. *cyrtophyllum* (Kindb.) Redf. *ex* Crum (Fig. 312F) is recognizable by its leaves that are shorter, broader, more concave and shorter tips.

Figure 313 *Brachythecium oxycladon.* A, shoot; B, capsule; C, leaf; D, medium cells of var. *dentatum;* E, alar region.

Plants in large, yellow-green or yellow-brown mats with freely branching creeping stems, branches closely foliate and terete, very common on shaded soil, rocks, logs, tree bases, Minnesota to southeastern Canada south to the Gulf States. *B. oxycladon* var. *dentatum* (Lesq. & James) Grout (Fig. 313D) grows on moist rocks or even submerged and has median leaf cells that are shorter and broader than the linear leaf cells of *B. oxycladon. B. oxycladon* may be confused with *B. salebrosum,* however, in this latter species the alar cells are distinctly larger.

4b Plants medium size, loosely spreading; leaves falcate-secund; seta rough; monoicous; western North America. Fig. 314. .... *Brachythecium leibergii* Grout

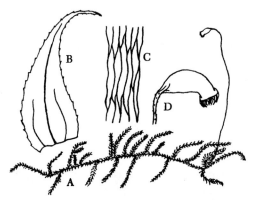

Figure 314

Figure 314 *Brachythecium leibergii*. A, shoot; B, leaf; C, medium cells; D, capsule.

Plants in extensive dark green to yellow green mats, stem prostrate, pinnately branched, on soil and logs, British Columbia and Alberta to Washington, Idaho, Montana.

5a Alar cells much larger than those above, often with a row of large clear cells across base of leaf, probably belong to stem; seta rough; western North America. Fig. 315. .... *Brachythecium frigidum* (C. Muell.) Besch.

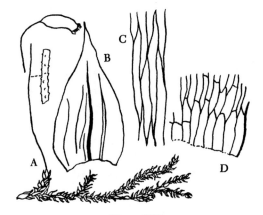

Figure 315

Figure 315 *Brachythecium frigidum*. A, shoot; B, leaf; C, median cells; D, basal cells.

Plants in large, dense tufts form deep cushions, yellow-green and usually glossy, on soil in wet places, Alaska, Alberta to California, Nevada, Utah, Montana, Wyoming.

5b Alar cells not as above. 6

6a Upper leaf margins entire or nearly so. Fig. 316A-C. .... *Brachythecium albicans* (Hedw.) B.S.G.

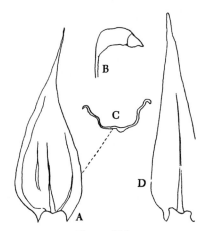

Figure 316

Figure 316 *Brachythecium albicans.* A, leaf; B, capsule; C, section of leaf; D, leaf of *B. acutum.*

Plants in medium to large light to whitish-green mats, stems prostrate, irregularly branched, branches terete to julaceous, on soil and rocks, Alaska to Ontario and Newfoundland south to California, Montana, Wyoming.

6b Upper leaf cells distinctly serrate. 7

7a Seta smooth throughout; monoicous. Fig. 317. .... *Brachythecium salebrosum* (Web. & Mohr.) B.S.G.

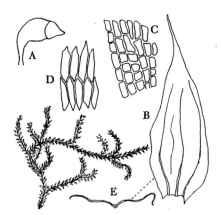

Figure 317

Figure 317 *Brachythecium salebrosum.* A, capsule; B, leaf; C, alar cells; D, median cells; E, section of leaf.

Plants medium size, in usually glossy green to yellow-green mats, stems prostrate, branching irregularly, on shaded soil, rocks, tree bases, logs, throughout most of North America.

7b Seta lightly papillose above; autoicous. .... *Brachythecium campestre* (C. Muell.) B.S.G.

Plants in wide dark green or yellow-green, glossy mats; stems irregularly branching, prostrate, on soil, rocks, roots, tree bases, Ontario to New England south to Iowa, Indiana, Tennessee.

8a Leaves straight, not secund. 9

8b Leaves more or less secund. 15

9a Leaves broadly ovate-deltoid; seta rough. 10

9b Leaves narrower, ovate-lanceolate, triangular-ovate; to long acuminate, seta smooth or rough. 11

10a Alar cells inflated walls pitted, stongly decurrent; growing in wet habitats. Figs. 67B, 318A-D. ....
.... *Brachythecium rivulare* B.S.G.

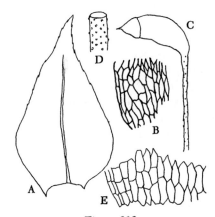

Figure 318

Figure 318 *Brachythecium rivulare*. A, leaf; B, alar region; C, capsule; D, seta; E, alar region of *B. nelsoni*.

Plants medium to large, in yellow-green to dark green mats, stems prostrate, branching irregularly, on wet soil or rock in or near streams, Alaska to Quebec and Labrador south to California, Arizona, Utah, Arizona, Colorado, Arkansas, Indiana, Tennessee, North Carolina. *B. nelsonii* Grout (Fig. 317E) is found in Colorado and the Pacific Northwest and is distinguished from *B. rivulare* by leaves that are slenderly long acuminate.

10b Alar cells enlarged, subquadrate to rectangular, not inflated, slightly decurrent, growing in mesic habitats. Fig. 319. .... *Brachythecium rutabulum* (Hedw.) B.S.G.

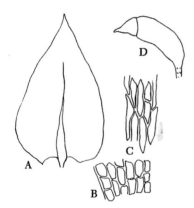

Figure 319

Figure 319 *Brachythecium rutabulum*. A, leaf; B, alar region; C, median cells; D, capsule.

Plants medium to large, in green to yellow-green, often glossy, mats, stems prostrate, irregularly branching on soil, soil over rocks, logs, roots, Montana to Ontario and New England south to Colorado, Oklahoma, Louisiana, Indiana, Kentucky, Pennsylvania, New Jersey.

11a Costa ending near apex to percurrent. 12

11b Costa not extending more than 2/3 the length of the leaf. 13

12a Plants small; branch leaves 0.6-0.9 mm long, costa ending in the acumen; upper leaf cells 3-5:1; seta rough throughout. Fig. 320. .... *Brachythecium reflexum* (Starke *ex* Web. & Mohr.) B.S.G.

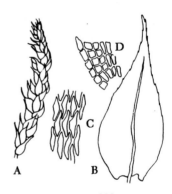

Figure 320

Figure 320 *Brachythecium reflexum*. A, plant; B, leaf; C, median cells; D, alar region.

Plants dark-to-yellow-green in loose to dense mats, stems pinnately branched, terete, on decaying wood, soil, rocks, tree bases in dry to moist woods, Alaska to Labrador south to Oregon, Michigan, Virginia.

12b Plants larger; branch leaves 1-1.6 mm long, costa percurrent or nearly so; upper leaf cells 5-8:1; seta lightly papillose above. Fig. 321E. ....

.... *Brachythecium populeum* (Hedw.) B.S.G.

Relatively slender plants in dense green to yellow-green mats, branches slender and terete, on rocks and soil, British Columbia and Manitoba to northeastern Canada south to Wisconsin, New Jersey, Virginia.

13a Branch leaves twisted at apex. .... *Brachythecium curtum* (Lindb.) Limpr.

Plants moderate in size, in loose, shiny, yellow or light green mats, stems ascending or arched, irregularly branched, on various substrate, often in bogs; Minnesota to Nova Scotia south to North Carolina and Tennessee.

13b Branch leaves not twisted at apex. 14

14a Leaves slenderly triangular-ovate or wedge-shaped, gradually tapering from base to a slender-pointed apex. Fig. 316D. ....
...... *Brachythecium acutum* (Mitt.) Sull.

Plants moderately robust, in thin green to yellow-green mats, stems prostrate, occasionally floating, distantly and irregularly branched, branches often sub-complanate, on soil and rotten wood, northeastern United States.

14b Leaves lanceolate. Fig. 321A-D. ....
.... *Brachythecium plumosum* (Hedw.) B.S.G.

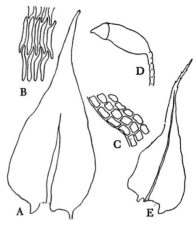

Figure 321

Figure 321 *Brachythecium plumosum.* A, leaf; B, median cells; C, alar region; D, capsule; E, leaf of *B. populeum*.

Plants medium size, in green to golden green, shiny mats, stems subpinnately branched, leaves often homomallus, concave, on moist rocks in or near streams, British Columbia to Quebec and Newfoundland south to California, Arizona, Colorado, Louisiana, Tennessee and North Carolina.

15a Seta very rough; leaf cells 12:1. Fig. 322. .... *Brachythecium velutinum* (Hedw.) B.S.G.

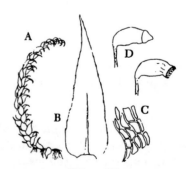

Figure 322

Figure 322 *Brachythecium velutinum*. A, shoot; B, leaf; C, alar region; D, capsule, with operculum, and open.

Plants small, in loose green to yellow-green mats, stems slender, prostrate, irregularly branched, branches often curved, on soil and rotten wood, Yukon to Quebec and New England south to California, Utah, Colorado, Idaho, Montana, Pennsylvania, New Jersey.

15b Seta smooth or only slightly roughened; leaf cells 3-6:1. Fig. 323. .... *Brachythecium collinum* (Schleich. *ex* C. Muell.) B.S.G.

Bryhnia

1a Leaves ovate or ovate-lanceolate, acute to broadly acuminate, twisted at the apex, costa smooth on back, upper leaf cells minutely papillose on back. Fig. 324D-E. .... *Bryhnia novae-angliae* (Sull. & Lesq. *ex* Sull.) Grout

Medium-sized plants in soft, dense, bright-or-yellow-green mats, on soil, humus, logs in wet shady places, especially in seepage areas, southeastern Alaska to Newfoundland south to Missouri, Indiana, Tennessee and South Carolina.

1b Leaves lanceolate, acuminate, tips flexuose but not twisted, costa tooth at back ending in a spine, upper leaf cells strongly papillose on back. Fig. 324A-C. .... *Bryhnia graminicolor* (Brid.) Grout

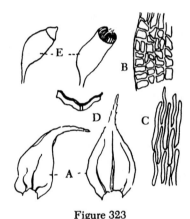

Figure 323

Figure 323 *Brachythecium collinum*. A, leaf; B, alar region; C, median cells; D, section of leaf; E, capsule with and without operculum.

Plants in green, often glossy, close mats, stems slender, prostrate, or ascending, irregularly branched, usually julaceous, on soil, often in rock crevices, Western North America east to North Dakota, Colorado, Utah.

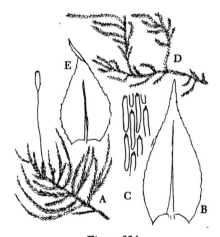

Figure 324

Figure 324 *Bryhnia graminicolor*. A, plant; B, leaf; C, cells and papillae; D, plant, and E, leaf of *B. novae-angliae*.

Plants slender, in dense, yellow-brown to green mats, on moist rocks and soil of road banks, ditches and streams, Minnesota to Ontario and New England south to Arkansas, Tennessee and Georgia.

Cirriphyllum

1a Costa of stem leaves usually short to ending near middle of leaf; Colorado and northwestern North America. Fig. 325C. .... *Cirriphyllum cirrosum* (Schwaegr. *ex* Schultes) Grout

Plants in green to yellow-green mats, stems creeping, irregularly branching, up to 7 cm long, on soil over rocks.

1b Costa of stems leaves extending 2/3 their length. Fig. 325D. .... *Cirriphyllum piliferum* (Hedw.) Grout

Medium-sized plants in soft, shiny bright-or-yellow-green mats, on humus in shady, wet places, Alaska, Victoria Island, to Newfoundland south to Washington, Michigan, South Carolina.

Bryonandersonia

Bryoandersonia illecebra (Hedw.) Robins. (Fig. 325A-C)

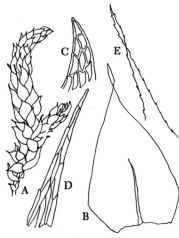

Figure 325

Figure 325 *Bryoandersonia illecebra*. A, shoot; B, leaf; C, apex of leaf; D, leaf apex of *Cirriphyllum piliferum;* E, leaf apex of *C. cirrosum.*

Plants robust, in extensive light to golden or brownish mats, branches julaceous, erect, leaves closely imbricate, on soil, rock ledges, humus, underside of ledges, Ontario, Iowa, Missouri, Texas and throughout the eastern United States.

Scleropodium

Five species of *Scleropodium* are reported from North America.

1a Leaf apices obtuse or rounded, sometimes with a very short straight apiculus, plants aquatic. Fig. 326. .... *Scleropodium obtusifolium* (Jaeq. & Sauerb.) Kindb. *ex* Macoun & Kindb.

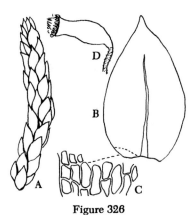

Figure 326

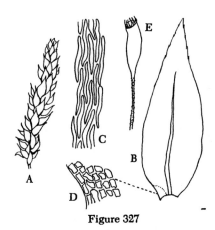

Figure 327

Figure 326 *Scleropodium obtusifolium.* A, shoot; B, leaf; C, alar region; D, capsule.

Plants robust, in dark green mats or tufts, branches strongly julaceous, on rocks in or near streams, Alaska to California, Nevada, Utah.

1b Leaf apices acuminate to abruptly narrowed to a short apiculus, plants terrestrial. 2

2a Plants small to medium sized, leaves ovate to ovate-lanceolate; basal leaf cells rectangular, in 2-6 rows; capsule erect to somewhat inclined. Fig. 327. .... *Scleropodium cespitans* (C. Muell.) L. Koch

Figure 327 *Scleropodium cespitans.* A, shoot; B, leaf; C, median cells; D, alar cells; E, capsule.

Plants in loose, low mats, stems and branches slender, more or less julaceous, on logs, tree roots and rocks, British Columbia, Alberta to California and Arizona.

2b Plants usually robust, leaves relatively broad; basal leaf cells short rectangular in 1-2 rows; capsule horizontal and somewhat zygomorphic. 3

3a Branches plainly julaceous; apiculus of leaves recurved. .... *Scleropodium tourettei* (Brid.) L. Koch var. *tourettei*

Plants in compact mats, green to yellow-green or yellow-brown, branches typically tumid or turgid, on soil, British Columbia to California.

3b Branches not julaceus; apiculus of leaves rarely recurved. .... *Scleropodium touretti* var. *colpophyllum* (Sull.) Lawt. *ex* Crum

Plants usually less compact, in straggling or loose mats, on soil, rocks logs, British Columbia to California.

Musci—The Mosses

Rhyncostegium

Rhynchostegium serrulatum (Hedw.) Jaeg. & Sauerb. (Fig. 328)

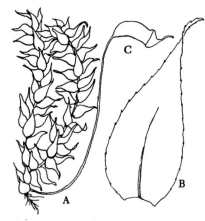

Figure 328

Figure 328 *Rhynchostegium serrulatum*. A, shoot; B, leaf; C, capsule and operculum.

Plants moderate in size, in flat, shiny, bright- or yellow-green mats, stems and branches not crowded, somewhat complanate, on soil, humus, decaying wood, rocks, widespread in eastern North America west to Kansas and Nebraska.

Rhynchostegiella

Conardia compacta (Hook.) H. Robinson Phytologia 33: 293–295 (Fig. 329)

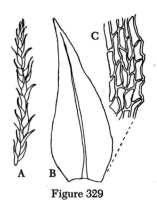

Figure 329

Figure 329 *Conardia compacta*. A, shoot; B, leaf; C, cells of leaf.

Plants in dense, shiny yellow or green mats, stems crowded and ascending, irregularly branched, costa often bearing gemmae or rhizoids at back, on wet calcareous rocks, underside of ledges, tree bases, Yukon to Quebec south to California, Arizona, Colorado, Missouri, Pennsylvania.

Eurhynchium

1a Broad-leafed aquatic; seta smooth. Fig. 330. .... *Eurhynchium riparioides* (Hedw.) Rich.

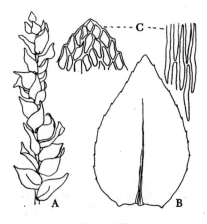

Figure 330

Figure 330 *Eurhynchium riparioides.* A, shoot; B, leaf; C, cells of middle and apex of leaf.

Robust wiry plants in dark-green, green, to yellowish-green mats, branching irregularly, leaves erect-spreading, serrulate nearly to base, on rocks along and in streams or near waterfalls, British Columbia to Newfoundland south to California, Oklahoma, Arkansas, Tennessee and Georgia.

1b   Leaves narrower, plants on soil and tree bases in woods. 2

2a   Leaves broadest 1/3 above base; seta rough. Fig. 331A-E. .... *Eurhynchium hians* (Hedw.) Sande Lac.

Quebec south to the Gulf States, also in Arizona and New Mexico.

2b   Leaves broadest near the base; seta smooth. Figs. 331F and 332. ....
.... *Eurhynchium pulchellum* (Hedw.) Jenn.

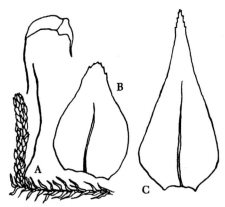

Figure 332

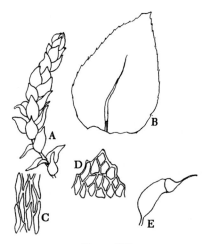

Figure 331

Figure 331 *Eurhynchium hians.* A, shoot; B, leaf; C, cells from middle, and D, from apex of leaf; E, capsule.

Plants in loose, dull or shiny green or yellowish mats, leaves ovate to triangular ovate, costa ending in a dorsal spine, on soil; widespread eastern North America from Minnesota to

Figure 332 *Eurhynchium pulchellum.* A, branch; B, leaf from branch; C, stem leaf.

Plants relatively slender, usually in shiny, green, yellowish, or brownish extensive mats, stems creeping, irregularly to pinnately branched, branches horizontal to erect-ascending, flattened to terete, on soil or humus, rotten logs, tree bases, or rocks, Alaska to Newfoundland south to Arizona, Texas, Louisiana, Georgia.

Stokesiella

1a   Stem and branch leaves not clearly differentiated, branch leaves not or only slightly decurrent; seta smooth or nearly so. ....
.... *Stokesiella brittoniae* (Grout) Robins.

Plants small, in green mats, on tree bases and rotten logs, Washington to California.

1b Stem and branch leaves clearly differentiated, branch leaves often strongly decurrent; seta rough. 2

2a Plants branching irregularly. .... *Stokesiella praelonga* (Hedw.) Robins. var. *praelonga*

Plants slender, in green to yellow-green spreading mats, stem leaves squarrose to erect-spreading, on moist soil, often in wet places, British Columbia to Washington.

2b Plants regularly pinnate. 3

3a Branch leaves usually less than 1 mm long; main stem commonly divided, with divisions often arching. .... *Stokesiella praelonga* var. *stokesii* (Turn.) Crum

Plants more common and usually larger than the preceeding variety, in more compact mats, on humus, soil, rotten wood, Alaska to California, Nevada, Idaho, Montana.

3b Branch leaves usually more than 1 mm long; main stems simple or with few divisions. Fig. 333. .... *Stokesiella oregana* (Sull.) Robins.

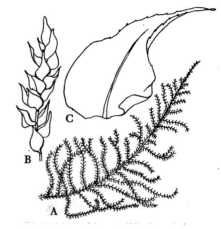

Figure 333

Figure 333 *Stokesiella oregana*. A, shoot; B, twig; C, stem leaf; stems to 25 cm long, in large cushions.

Plants in large, green to yellow-green, mats, stems 6-30 cm long or longer, prostrate with numerous brown rhizoids, on humus and rotten wood; Alaska to California and Idaho.

ENTODONTACEAE
Orthothecium

Orthothecium chryseum (Schwaegr. ex Schultes) B.S.G. (Fig. 334A-B)

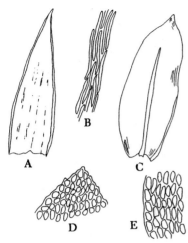

Figure 334

Figure 334 *Orthothecium chryseum*. A, leaf; B, median leaf cells. *Stereophyllum radiculosum*; C, leaf; D, leaf cells of leaf apex; E, median marginal leaf cells.

Plants in large, glossy, yellow-green to golden-green tufts, stems red, 6-10 cm long, leaves imbricate, deeply plicate, on rock or soil in wet places, Alaska, Yukon, Alberta, Manitoba south to Washington. Five other species are reported from North America.

Entodon

Seven species of *Entodon* are reported from North America.

1a Leaves gradually and narrowly acuminate; inner peristome segments adhearing to outer peristome segments. Fig. 335H. .... *Entodon brevisetus* (Hook. & Wils. *ex* Wils.) Lindb.

Plants in wide, dense, mats, light green above, dark green below, leaves closely imbricate, on tree bases and limestone rocks, Minnesota to New Brunswick south to Arkansas and North Carolina.

1b Leaves acute to apiculate. 2

2a Branches julaceous, scarcely complanate-foliate. Fig. 335A-G. ....
.... *Entodon seductrix* (Hedw.) C. Muell.

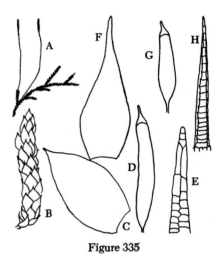

Figure 335

Figure 335 *Entodon seductrix*. A, plant; B, shoot; C, leaf; D, capsule; E, peristome tooth; F, leaf; G, capsule; H, peristome tooth of *E. brevisetus*.

Plants in green, yellowish, or brownish-tinged, extensive mats, stems creeping, subpinnately branched, stems and branches terete to slightly flattened, on moist shaded rocks, soil, tree bases and trunks, Minnesota, Ontario to New England south to the Gulf States.

2b Branched flattened. 3

3a Plants slender, branches less than 1 mm wide with complanate leaves slanting towards the substratum from middle of stem. Fig. 336. ....
.... *Entodon challengeri* (Par.) Card.

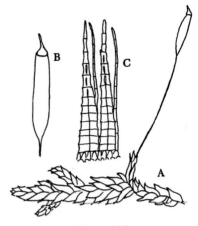

Figure 337

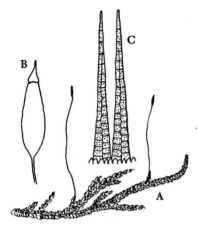

Figure 336

Figure 337 *Entodon cladorrhizans*. A, shoot; B, capsule with operculum; C, teeth of peristome.

Figure 336 *Entodon challengeri*. A, shoot; B, capsule; C, teeth of peristome.

Plants in wide, yellowish-green, sometimes glossy, mats, stems and branches flattened, on soil, logs, tree bases; South Dakota to Rhode Island south to Missouri, Indiana, North Carolina.

3b Plants moderately robust, branches more than 1 mm wide. 4

4a Seta reddish brown. Fig. 337. ....
.... *Entodon cladorrhizans* (Hedw.) C. Muell.

Plants in bright- to yellow-green mats, often tinged with golden brown, both stems and branches strongly flattened, on soil, humus, tree bases and logs in shady places; Manitoba to Quebec south to Arkansas and the Gulf States.

4b Seta yellow. .... *Entodon macropodus* (Hedw.) C. Muell.

Similar to the preceeding species, on shaded rocks, humus, tree bases, Illinois to Virginia south to Gulf States and Texas.

Pleurozium

Pleurozium schreberi (Brid.) Mitt. (Fig. 338)

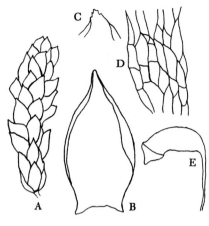

Figure 338

Figure 338 *Pleurozium schreberi*. A, shoot; B, leaf; C, apex of leaf; D, alar region; E, capsule.

Plants robust, in loose, light-green or yellowish, more or less shiny mats, stems red, erect-ascending, pinnately branched, leaves smooth, concave with turned-back points, alar cells enlarged, porose and orange, on humus, soil and rock ledges, across boreal North America south to Oregon, Arkansas, Tennessee, North Carolina.

PLAGIOTHECIACEAE
Stereophyllum

Stereophyllum radiculosum (Hook.) Mitt. (Fig. 334C-E)

Plants large, in yellow-green mats, stems and leaves flattened, sparingly branched with numerous brown radicles on underside of stem, leaves costate, on tree trunk and bases, logs, limestone, Florida and Texas. *S. leucostegum* (Brid.) Mitt. is found in Florida, has gradually acuminate leaves as opposed to the abruptly acuminate leaves in *S. radiculosum*.

Plagiothecium

Six species are reported from North America. The following key is adapted from a revision of this genus by R. R. Ireland (A taxonomic revision of the genus *Plagiothecium* for North America north of Mexico. Publication in Botany No. 1, National Museum of Natural Sciences, 1969, Ottawa).

1a   Plants whitish green, robust, leaves strongly undulate, up to 4 mm or longer; stems 3-4 mm wide; western North America. Fig. 339. .... *Phagiothecium undulatum* (Hedw.) B.S.G.

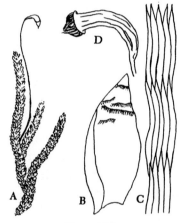

Figure 339

Figure 339 *Plagiothecium undulatum*. A, shoot; B, leaf; C, median cells; D, capsule.

Plants robust, dull to somewhat glossy, light to whitish green, stems up to 10 cm or more long, stems complanate-foliate, leaves strongly undulate, on rotten logs, stumps and tree bases, humus or soil, Alaska to California and Idaho.

1b   Plants smaller, leaves not strongly undulate, less than 4 mm long. 2

2a Leaf apex abruptly contracted to a long, filiform, flexuose acumen, sometimes 1/3 the length of the leaf; western North America. Fig. 340. ....
.... *Plagiothecium piliferum* (Sw. *ex* C. J. Hartm.) B.S.G.

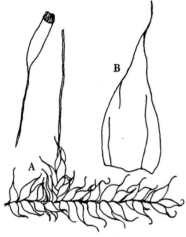

Figure 340

Figure 340 *Plagiothecium piliferum.* A, shoot; B, leaf.

Plants glossy, in light green to yellowish green mats, stems and branches prostrate, complanate-foliate to subjulaceus, on logs and noncalcareous rocks, Alaska to Oregon, Idaho, and Montana.

2b Leaf apex not abruptly contracted to a filiform apex. 3

3a Leaves symmetric or near so, typically concave, only loosely and irregularly complanate, decurrent portion of leave not auriculate. Fig. 341. ....
Plagiothecium cavifolium (Brid.) Iwats.

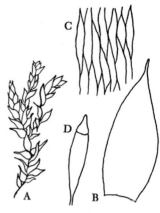

Figure 341

Figure 341 *Plagiothecium cavifolium.* A, shoot; B, leaf; C, median cells; D, capsule.

Plants in glossy (rarely dull) pale green to yellowish green mats, stems and branches usually julaceus, on soil, humus, stumps, logs, tree bases, Alaska to Labrador and Newfoundland south to Washington, Idaho, Minnesota, Arkansas, Georgia.

3b Leaves asymmetric, flat, distinctly complanate, decurrent portion of leaf auriculate. Fig. 342. .... *Plagiothecium denticulatum* (Hedw.) B.S.G.

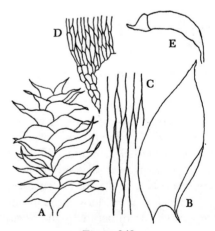

Figure 342

206 Musci—The Mosses

Figure 342 *Plagiothecium denticulatum*. A, shoot; B, leaf; C, median cells; D, alar region; E, capsule.

Plants in dull or glossy, dark green to yellow green mats, on logs, stumps, tree bases, soil and humus, Alaska to Labrador south, Oregon, Alberta, Idaho, Montana, Utah, Colorado, New Mexico, Michigan, Tennessee and North Carolina.

SEMATOPHYLLACEAE
Heterophyllium

Heterophyllium affine (Hook.) Fleisch. (Fig. 343)

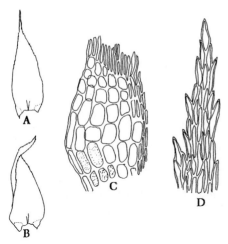

Figure 343

Figure 343 *Heterophyllium affine*. A-B, leaves; C, inflated alar cells; D, cells at apex of leaf.

Plants robust, in bright to pale green, wide, loose, mats, stems more or less regularly pinnate, leaves, concave, ovate to oblong-ovate, margins serrate, costa short and double or lacking, on logs and tree bases, Tennessee, Virginia south to Georgia.

Brotherella

1a Plants occurring in the Pacific Northwest. .... *Brotherella roellii* (Ren. & Card. *ex* Roell) Fleisch

Plants small, in thin silky mats, leaves often secund, ovate-lanceolate, acuminate, costa absent or short and double, alar cells differentiated, 1-2 inflated cells on margin and usually a row of enlarged cells extending across the leaf base, on rotten logs, British Columbia and Washington.

1b Plants occurring in eastern North America. 2

2a Plants slender, stems and branches not complanate. ....
.... *Brotherella tenuirostris* (Bruch & Schimp. *ex* Sull.) Fleisch.

Plants in thin, glossy, green- to yellow-green mats, branches slender, leaves oblong-lanceolate to slenderly acuminate, slightly curved to secund, costa absent or short and double, alar cells differentiated with angular cells inflated and hyaline, on decaying logs, tree bases and rocks; Wisconsin to New York south to Arkansas, Tennessee and South Carolina.

2b Plants medium, distinctly complanate-foliate with falcate leaves. Fig. 344. .... *Brotherella recurvans* (Michx.) Fleisch.

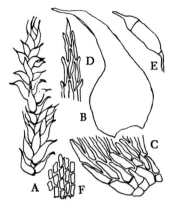

Figure 344

Figure 344 *Brotherella recurvans.* A, shoot; B, leaf; C, alar cells; D, apex of leaf; E, capsule with operculum; F, outer cells of capsule.

Plants in golden sheen mats, on soil, humus, bases of trees, Great Lakes to Nova Scotia south to Iowa, Indiana, Tennessee, North Carolina and Louisiana.

Sematophyllum

1a Plants growing on rocks; capsules more or less inclined and unsymmetric. 2

1b Plants growing on bark of trees and decaying wood; capsules erect and symmetric or nearly so. 3

2a Plants robust; leaves 1.2-1.8 mm long; median leaf cells 8-12:1. .... *Sematophyllum marylandicum* (C. Muell.) Britt.

Plants in dark green, loosely interwoven mats, on rocks near streams and seepage areas near waterfalls, West Virginia to Georgia and in Arkansas.

2b Plants slender; leaves averaging less than 1.2 mm long; median leaf cells 6-8:1. Fig. 345. .... *Sematophyllum demissum* (Wils.) Mitt.

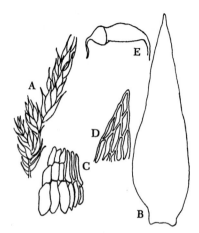

Figure 345

Figure 345 *Sematophyllum demissum.* A, shoot; B, leaf; C, alar cells; D, apex of leaf; E, capsule with operculum.

Plants in yellow-green mats closely attached to moist rocks, Michigan to southeastern Canada south to Gulf States.

3a Median leaf cells linear-fusiform, 8-12:1. Fig. 346. .... *Sematophyllum adnatum* (Michx.) Britt.

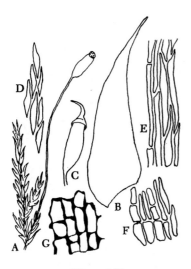

Figure 346

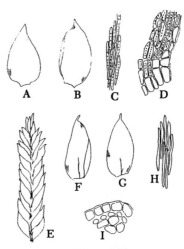

Figure 347

Figure 346 *Sematophyllum adnatum.* A, shoot; B, leaf; C, capsule with operculum; D, apex of leaf; E, median cells; F, alar cells; G, cells of capsule wall.

Plants green to golden green, in thin mats, on trees and logs, Iowa, Kentucky, Pennsylvania south to the Gulf States.

3b Median leaf cells fusiform-rhomboidal, 3-7:1. .... *Sematophyllum caespitosum* (Hedw.) Mitt.

Plants in dark to yellowish green, loose patches, on trees and decaying logs, Florida.

Taxithelium

Taxithelium planum (Brid.) Mitt. (Fig. 347A-D)

Figure 347 *Taxithelium planum.* A-B, leaves; C, median leaf cells; D, alar cells. *Isopterygiopsis muelleriana;* E, branch; F-G, leaves; H, median leaf cells; I, section of stem.

Plants in delicate, thin, light green mats, leaf cells with a single row of papillae down the middle, on decaying wood, bases of trees and roots; Florida.

HYPNACEAE
Platygyrium

Platygyrium repens (Brid.) B.S.G. (Fig. 348)

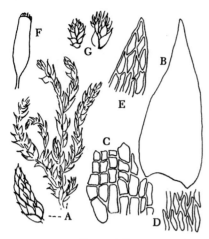

Figure 348

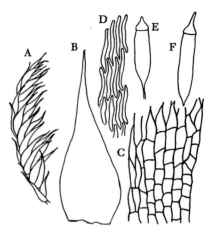

Figure 349

Figure 348 *Platygyrium repens.* A, shoot; B, leaf; C, alar region; D, median cells; E, apex of leaf; F, capsule; G, gemmae.

Plants in flat, dark-, yellowish-, golden-, or brownish-green, glossy mats, branches ascending, usually bearing clusters of minute gemmae, capsules erect and symmetric, on logs, stumps, tree trunks, rocks, British Columbia to New Brunswick and throughout eastern North America.

Pylaisiella

1a Inner peristome attached to outer. 2

1b Inner peristome free from outer. Fig. 349. ....
.... *Pylaisiella polyantha* (Hedw.) Grout

Figure 349 *Pylaisiella polyantha.* A, shoot; B, leaf; C, alar region; D, median cells; E, capsule and operculum; F, capsule and operculum.

Plants in glossy, green to yellowish dense mats, on trees, circumpolar south to Arizona, New Mexico, Colorado and Alabama.

2a Inner peristome partially adherent to outer; spores 18-24 μ in diameter; alar cells numerous (15-20 along margin of leaf). Fig. 350. .... *Pylaisiella selwynii* (Kindb.) Crum, Steere & Anderson

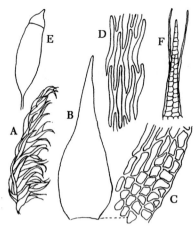

Figure 350

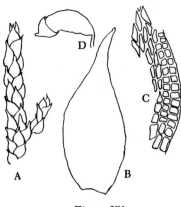

Figure 351

Figure 350 *Pylaisiella selwynii*. A, shoot; B, leaf; C, alar region; D, median cells; E, capsule; F, teeth and segments of peristome.

Plants in glossy, green to yellow-green mats, on trees, Minnesota to Quebec south to the Gulf States and Arizona.

2b Inner peristome wholly adherent to outer; spores 25-30 μ, alar cells less numerous (8-10 along margin of leaf). .... *Pylaisiella intricata* (Hedw.) Grout

Plants in wide, glossy, yellow-green mats, on tree trunks, Minnesota to New Brunswick south to the Gulf States and Arizona.

Homomallium

Homomallium adnatum (Hedw.) Broth. (Fig. 351)

Figure 351 *Homomallium adnatum*. A, shoot; B, leaf; C, alar and median cells; D, capsule and operculum.

Plants small, in dull, dark-green, blackish, or brownish mats, on rocks infrequently on tree bases (especially limestone), southeastern Canada south to Georgia, Arkansas, Texas. *H. mexicanum* Card. is found in West Texas, Arizona and New Mexico. It is distinguished from *H. adnatum* by its terminal secund leaves.

Hypnum

Nineteen species of *Hypnum* are known from North America.

1a Cortical cells of stems (make a cross-section) hyaline, larger than cells within. 2

1b Cortical cells small, thick-walled, not larger than cells within. 7

2a Alar cells usually plainly inflated. 3

2b Alar cells not plainly inflated, often quadrate. 6

3a Leaves strongly circinate, slenderly acuminate; Pacific Northwest. ....
...... **Hypnum callichroum** Funck *ex* Brid.

Plants medium-sized in green to yellow-green, glossy, extensive mats, on soil and rock, Alaska to Alberta and Washington.

3b Leaves falcate-secund, sometimes nearly straight, acute to broadly acuminate. 4

4a Stems prostrate, closely and regularly pinnate, leaves serrulate to well below the apex. .... **Hypnum fertile** Sendtn.

Plants slender, in soft, yellow-green to yellow-brown, flat mats, on rotten logs, Manitoba to Labrador south to Michigan and Georgia.

4b Stems usually ascending, loosely and irregularly branched; leaves entire or serrulate only near apex. 5

5a Leaves somewhat complanate, usually strongly concave and distinctly falcate-secund; alar cells abruptly inflated in large decurrencies delimited above by small cells. Fig. 352. .... **Hypnum lindbergii** Mitt.

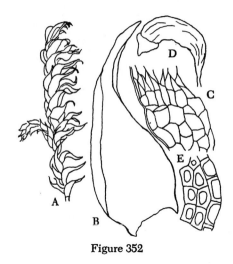

Figure 352

Figure 352 *Hypnum lindbergii*. A, shoot; B, leaf; C, alar cells; D, capsule; E, section of stem.

Plants large, in loose light-green to yellowish on brownish, shiny mats, stems red, on wet soil, humus, logs, along streams and near lakes and ponds, circumpolar south to Washington, Arizona, Colorado, to the Gulf Coast in the east.

5b Leaves loosely complanate, moderately concave, curved-secund at tips, alar cells enlarged but not particularly inflated, in gradually differentiated groups. .... **Hypnum pratense** Koch *ex* Spruce

Plants medium in size, in yellowish to golden, shiny tufts, on calcareous wet soil or humus, circumpolar south to Oregon, Colorado, Michigan and to North Carolina in the east.

6a Leaves rounded to the insertion; alar cells plainly differentiated, hyaline, rectangular to more or less inflated. .... **Hypnum pratense** Koch *ex* Spruce

6b Leaves not rounded to insertion; differentiated alar cells few, quadrate. Fig. 353. .... *Hypnum subimponens* Lesq.

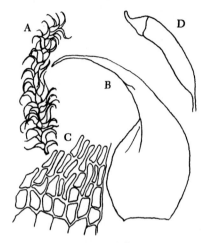

Figure 353

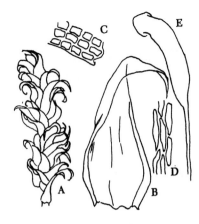

Figure 354

Figure 353 *Hypnum subimponens*. A, shoot; B, leaf; C, alar cells with thin-walled cortical cells of stem; D, capsule.

Plants medium-sized, in green to yellow-green extensive mats, on calcareous rocks and soil, Alaska to Quebec, south to Washington, Arizona, Utah, Colorado, Minnesota, Ontario.

7a Margins of leaves revolute nearly to apex, entire, western North America. Fig. 354. ....
.... *Hypnum revolutum* (Mitt.) Lindb.

Figure 354 *Hypnum revolutum*. A, shoot; B, leaf; C, alar cells; D, median cells; E, capsule.

Plants medium sized, in green to brownish-green or brown mats, irregularly to regularly pinnate, on rocks and soil over rocks, Alaska and Yukon to California, Nevada, Utah, New Mexico, Colorado, South Dakota.

7b Margins plain or more or less reflexed. ...
.... 8

8a Leaves cordately contracted to the stem.
.... 9

8b Leaves not cordate at base, tapering to insertion. 10

9a Leaves entire, or serrulate near apex; decurrent cells at basal angles inflated, colorless; eastern North America. Fig. 355. ....
.... *Hypnum curvifolium* Hedw.

Musci—The Mosses   213

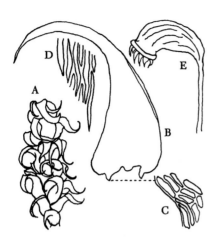

Figure 355

Figure 355 *Hypnum curvifolium*. A, shoot; B, leaf; C, alar region; D, median cells; E, capsule.

Plants relatively large, in yellow-green to brownish, shiny extensive flat mats, capsules furrowed, on rocks, soil, logs and tree bases, throughout eastern North America (except Florida). Often used by florists as "sheet moss".

9b   Leaves serrulate; acumination very long and slender; circinately coiled; western North America. Fig. 356. .... *Hypnum circinale* Hook.

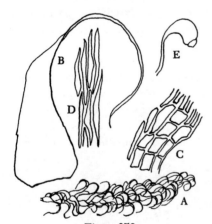

Figure 356

Figure 356 *Hypnum circinale*. A, shoot; B, leaf; C, alar region; D, median cells; E, capsule.

Plants small, in green to yellow-green, extensive mats, capsules usually horizontal and zygomorphic, on logs, tree trunks, rock, Alaska to California, Idaho and Montana.

10a   Leaves entire or nearly so; alar cells very numerous, opaque. 11

10b   Leaves serrulate to base. 13

11a   Median leaf cells 10-15:1; alar cells 6-10 (rarely 15) in the marginal row; leaves usually gradually narrowed from base to apex. 12

11b   Median leaf cells 4-6:1; alar cells 12-25 in the marginal row; leaves abruptly acuminate. Fig. 357F-G. .... *Hypnum vaucheri* Lesq.

Plants medium-sized, usually in yellow-green to brownish-green, thick cushions, on calcareous rocks and soil, Alaska to Ontario and Quebec south to Arizona, Nevada, Utah, Colorado, Nebraska, Minnesota.

12a   Branches often hooked at the ends, not extremely slender, and filiform. Fig. 357A-E. .... *Hypnum cupressiforme* Hedw. var. *cupressiforme*

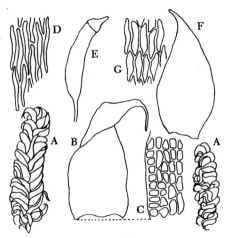

Figure 357

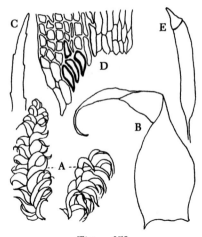

Figure 358

Figure 357 *Hypnum cupressiforme* var. *cupressiforme*. A, shoot from top (left) and from side (right); B, leaf; C, alar cells; D, median cells; E, capsule; F, leaf, and G, median cells of *H. vaucheri*.

Plants medium-sized to large, in yellow-green to brownish, glossy, rather large mats, on calcareous soil and rocks, trees and decaying wood, Alaska to Newfoundland and Labrador south to Washington, Arizona, Utah, Colorado, Nebraska, Arkansas, Tennessee and North Carolina.

12b Branches long, straight, filiform and extremely slender. .... *Hypnum cupressiforme* var. *filiforme* Brid.

Plants hanging in sheets of green threads on bark and rocks, ranges with the preceeding species.

13a Leaves 2 mm long; alar cells thick-walled, orange brown when old. Fig. 358. .... *Hypnum imponens* Hedw.

Figure 358 *Hypnum imponens*. A, shoot from top (left) and from side (right); B, leaf; C, apex of leaf; D, alar cells; E, capsule.

Plants large, in yellow-brown to yellow-green mats, capsules nearly erect, on rotten logs, soil, humus, tree bases in moist woods, Yukon to Labrador to Newfoundland south to Arkansas, Alabama and Georgia.

13b Leaves about 1 mm long; quadrate alar cells very numerous, none inflated. Fig. 359. .... *Hypnum pallescens* (Hedw.) P.-Beauv.

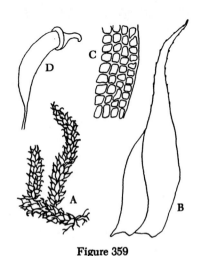

Figure 359

Musci—The Mosses   215

Figure 359 *Hypnum pallescens.* A, shoot; B, leaf; C, alar cells; D, capsule.

Small plants in dark-or yellow-green, dense mats, on logs, humus, tree bases, Alaska to Newfoundland south to British Columbia, Arizona, Utah, Wyoming, Montana, Missouri, Indiana, Tennessee, North Carolina.

Callicladium

Callicladium haldanianum (Grev.) Crum (Fig. 360)

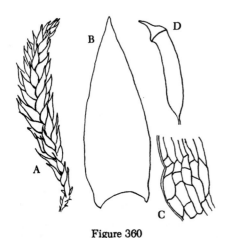

Figure 360

Figure 360 *Callicladium Haldanianum.* A, shoot; B, leaf; C, alar cells; D, capsule.

Plants in loose, dark to brownish green, wide mats, leaves entire, median leaf cells 12-18:1, costa short, double or absent, not decurrent, on soil, logs, tree trunk and base, rocks, Manitoba, Montana, California, Arizona in the far west, Minnesota to Quebec and Ontario south to Missouri, Tennessee, and South Carolina.

Isopterygiopsis

Isopterygiopsis muelleriana (Schimp.) Iwats. (Fig. 347E-I)

Plants small to median-sized, in thin, yellowish-green flat mats, on moist shaded rocks, soil, Alaska, Michigan, Minnesota, Nova Scotia to New Brunswick south to Arkansas, Tennessee, and South Carolina.

Isopterygium

Five species of *Isopterygium* are reported from North America.

1a Pseudoparaphyllia present, filimentous, of 1-2 rows of cells; alar cells in marginal row usually quadrate or transversely elongate; asexual reproductive bodies with papillose cells. Fig. 361A-D. .... *Isopterygium tenerum* (Sw.) Mitt.

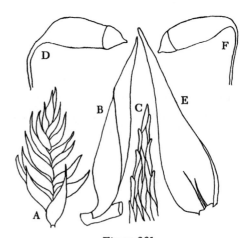

Figure 361

Figure 361 *Isopterygium tenerum*. A, shoot; B, leaf; C, apex of leaf; D, capsule with operculum; E, leaf and F, capsule of *I. elegans*.

Plants in thin, light green to yellowish green mats, on rotten logs, bases of trees, sandy soil, southern New York to Massachusetts south to Gulf Coast, Texas, Missouri, Arkansas.

1b Pseudoparaphylla lacking; alar cells in marginal row rectangular to quadrate; asexual reproductive bodies with smooth cells. Fig. 361E-F. ....
.... *Isopterygium elegans* (Brid.) Lindb.

Plants in thin to dense, dark or light green to yellowish green mats, on shaded soil, humus, tree bases, and rotten logs, Alaska to Labrador and Newfoundland, south to California in the west, in the east south to Missouri, Tennessee, and Georgia.

Taxiphyllum

Four species of *Taxiphyllum* are reported from North America.

1a Leaves obtuse at apex, distant on stem. Fig. 362. .... *Taxiphyllum taxirameum* (Mitt.) Fleisch.

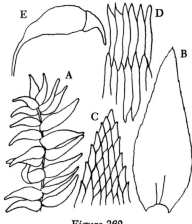

Figure 362

Figure 362 *Taxiphyllum taxirameum*. A, shoot; B, leaf; C, apex of leaf; D, median cells; E, capsule.

Plants glossy, in dark green to yellowish green, dense mats, Iowa to New York south to Florida, Louisiana, Texas, Arizona.

1b Leaves acuminate, often abruptly narrowed to an acute or filiform apex, closely imbricated. Fig. 363. ....
.... *Taxiphyllum deplanatum* (Bruch & Schimp. ex Sull.) Fleisch.

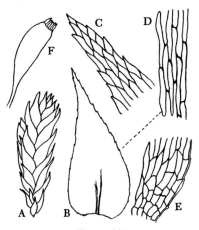

Figure 363

Musci—The Mosses

Figure 363 *Taxiphyllum deplanatum*. A, shoot; B, leaf; C, apex of leaf; D, median cells; E, alar region; F, capsule.

Plants glossy, in this to dense, light green to golden green mats, Saskatchewan to Quebec south to Louisiana, Tennessee, and North Carolina.

Vesicularia

Vesicularia vesicularis (Schwaegr.) Broth. (Fig. 364)

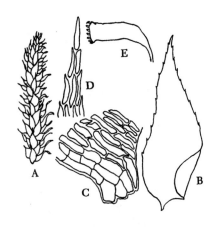

Figure 365

Figure 365 *Herzogiella striatella*. A, shoot; B, leaf; C, alar cells; D, apex of leaf; E, capsule.

Plants in thin to dense mats, yellowish-green to dark green, brownish green with age, on shaded soil, humus, rotten logs, stumps tree bases and roots and non-calcareous cliffs and rocks, Alaska to Washington in the west, Wisconsin to Newfoundland and Labrador south to Michigan, Ohio, Kentucky, Tennessee, and North Carolina.

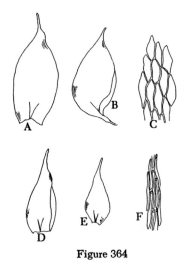

Figure 364

Figure 364 *Vesicularia vesicularis*. A-B, leaves; C, median leaf cells. *Mittenothamnium diminutivum*. D-E, leaves; F, median leaf cells.

Plants in thin, spreading, yellowish-green mats, on decaying wood and moist limestone, central Florida. Three other species are reported from southern Georgia and Florida.

Herzogiella

Herzogiella striatella (Brid.) Iwats. (Fig. 365)

Ctenidium

Ctenidium molluscum (Hedw.) Mitt. (Fig. 366)

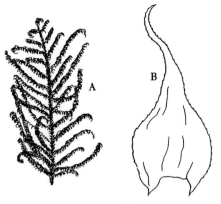

Figure 366

Figure 366 *Ctnedium molluscum.* A, shoot branching; B, leaf.

Plants small to medium-sized, in soft, dense, green, yellowish or golden-brown shiny mats, on soil, humus, rocks, decaying wood, Minnesota to Newfoundland south to Louisiana, Alabama and Georgia.

Ptilium

Ptilium crista-castrensis (Hedw.) De Not. (Fig. 367)

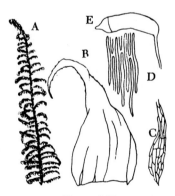

Figure 367

Figure 367 *Ptilium crista-castrensis.* A, shoot; B, leaf; C, alar cells; D, median cells; E, capsule.

Plants plumose, in extensive bright-green, yellow, or golden, shiny mats, on humus and old logs, Alaska to Newfoundland south to Oregon, Idaho, Montana, Iowa, Michigan, Georgia. The most beautiful of feathery or frondose mosses, said to be on the Coat of Arms of the House of Lancaster.

Mittenothamnium

Mittenothamnium diminutivum (Hampe) Britt. (Fig. 364D-F)

Plants in slender, delicate, loose, yellow-green mats, leaf cells minutely papillose at upper ends on dorsal side, on decaying logs, limestone, Florida, Georgia.

RHYTIDIACEAE
Rhytidium

Rhytidium rugosum (Hedw.) .Lindb. (Fig. 368)

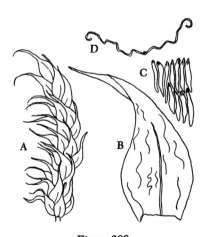

Figure 368

Figure 368 *Rhytidium rugosum.* A, branch with leaves; B, leaf; C, papillae; D, cross section of leaf.

Large, robust plants forming extensive dull green to yellow-green mats, on soil over rocks, humus, Alaska to Ontario south to Washington, Montana, Arizona, Colorado, South Dakota, Missouri, Tennessee, and North Carolina.

Rhytidiopsis

Rhytidiopsis robusta (Hedw.) Broth. (Fig. 369)

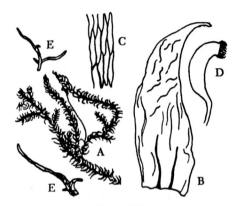

Figure 369

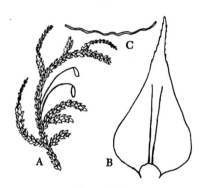

Figure 370

Figure 370 *Rhytidiadelphus triquetrus.* A, plant; B, leaf; C, cross-section of leaf.

Plants robust, in green to yellow-green mats, on soil and humus, Alaska to Newfoundland and Nova Scotia south to California, Nevada, Montana, Arkansas, Ohio, Tennessee and North Carolina.

1b Leaf cells smooth, costa absent or short and double. 2

2a Leaves plicate, falcate-secund. Fig. 371. .... *Rhytidiadelphus loreus* (Hedw.) Warnst.

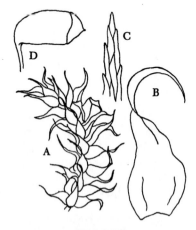

Figure 371

Figure 369 *Rhytidiopsis robusta.* A, shoot; B, leaf with 2 unequal ribs; C, median cells; D, capsule; E, *paraphyllium.*

Plants in large, loose, yellow-green to brownish mats, on soil, Pacific Northwest.

Rhytidiadelphus

1a Leaf cells spinose-papillose by projecting cell ends; costa double and strong, often reaching the middle of the leaf. Fig. 370. .... *Rhytidiadelphus triquetrus* (Hedw.) Warnst.

Figure 371 *Rhytidiadelphus loreus.* A, shoot; B, leaf; C, apex of leaf; D, capsule.

Plants in large, loose masses, light-green to yellow-green, on soil and humus in woods, Alaska to New Brunswick and Nova Scotia south to California, Ohio, North Carolina.

2b Leaves not plicate, plainly squarrose. .... *Rhytidadelphus squarrosus* (Hedw.) Warnst.

Plants in light-green to yellow-green mats, on logs, soil or in lawns among grass, Alaska to Labrador south to Oregon, Nevada, Idaho, Montana, Ohio, Tennessee and North Carolina.

HYLOCOMIACEAE
Hylocomium

1a Costa single, reaching middle of leaf, rarely forking. .... *Hylocomnium pyrenaicum* (Spruce) Lindb.

Plants medium-sized to large, in green to yellow-green mats, on logs and wet rocks, Yukon to Labrador south to British Columbia, Colorado, Minnesota, Michigan, New York.

1b Costa double, usually reaching middle of leaf. 2

2a Leaf cells papillose. Fig. 372. .... *Hylocomnium splendens* (Hedw.) B.S.G.

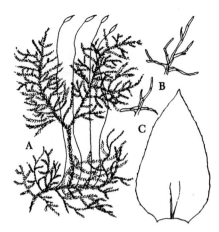

Figure 372

Figure 372 *Hylocomium splendens.* A, plant; B, paraphyllia; C, leaf.

Plants usually large, in wide, loose, green, yellow-green to brown mats, often carpeting rocks, humus, rotten logs in coniferous woods, Alaska to Quebec and Labrador south to California, Colorado, South Dakota, Iowa, Michigan, Georgia.

2b Leaf cells smooth. 3

3a Stem leaves decurrent and coarsely tooth at base. Fig. 373A-D. .... *Hylocomnium umbratum* (Hedw.) B.S.G.

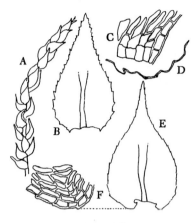

Figure 373

Figure 373 *Hylocomium umbratum*. A, shoot; B, leaf; C, alar cells; D, cross-section of leaf; E, leaf of *H. brevirostre*; F, alar region of same.

Similar to the preceeding species, on logs, humus, rocks, in moist shaded places of mountain woods and cool ravines, Michigan to Ontario and Labrador, south to Tennessee and North Carolina.

3b Stem leaves rounded-cordate at base and finely toothed. Fig. 373E-F. ....
Hylocomnium brevirostre (Brid.) B.S.G.

Plants robust, in loose, dark green tufts, on boulders, humus, and soil at higher elevations, Ontario to Newfoundland and Nova Scotia south to Georgia.

Order Buxbaumiales

BUXBAUMIACEAE
Buxbaumia

Buxbaumia aphylla Hedw. (Fig. 374A)

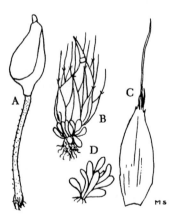

Figure 374

Figure 374 *Buxbaumia aphylla*. A, plant complete; B, plant of *Diphyscium foliosum*; C, perichaetial leaf; D, foliage leaves.

Small, annual plants growing from persistent protonema, capsules oblique to horizontal on a stout, elongate seta, on sandy or clayey soil, logs, stumps, along trails and roads, rare, Yukon to Colorado, Iowa, Michigan to Nova Scotia south to Illinois, Delaware. Three other species of *Buxbaumia* are known from North America.

Diphyscium

Diphyscium foliosum (Hedw.) Mohr. (Fig. 374B-D)

Plants small, in rigid, dark green, brown, to blackish extensive tufts, capsules nearly sessile in forests on soil, humus, shaded bank, cliff, rock walls of ravines and gorges, widespread in eastern North America. *D. cumberlandianum* Harvill is endemic to the Southern Appalachians (North Carolina to Alabama and Kentucky) and differs from *D. foliosum* is that it has smooth leaf cells.

Order Tetraphidales

TETRAPHIDACEAE
Tetraphis

Tetraphis pellucida Hedw. (Figs. 43, 375)

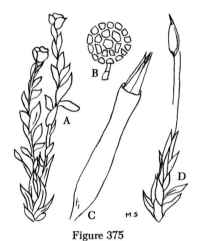

Figure 375

Figure 375 *Tetraphis pellucida*. A, plant with gemma cups; B, gemma; C, capsule with peristome; D, capsule with seta and perichaetial leaves.

Plants small, 8-15 mm high, tufted, gregarious, stems erect, capsules cylindric, peristome teeth four, on rotten stumps, humus, soil, sandstone rocks, Alaska to Labrador south to California, Arizona, Colorado, Arkansas, Alabama and Georgia. *T. geniculata* Girg. *ex* Milde is more northern in distribution and has a crooked seta.

Order Polytrichales

POLYTRICHACEAE
Atrichum

Eight species of *Atrichum* are reported from North America. The following key follows R. R. Ireland (Taxonomic studies on the genus *Atrichum* in North America, Can. J. Bot. 47: 353-368. 1969. Reproduced by permission of the National Research Council of Canada from the *Canadian Journal of Botany*, Volume 47, pp. 353-368, 1969.)

1a Upper leaf cells averaging 12-17 μ in longest dimension; leaves narrow, those near the middle of the stem usually less than 1 mm wide, often with 6 or more lamellae covering 1/3 of leaf near middle. Fig. 376. ....
...... *Atrichum angustatum* (Brid.) B.S.G.

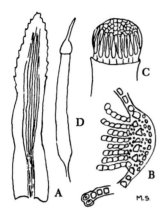

Figure 376

Figure 376 *Atrichum angustatum*. A, leaf with lamellae; B, cross section of leaf; C, peristome; D, capsule with operculum.

Plants medium-sized, 10-20 mm high, in loose, dull, dark green to brownish-green tufts, on light, often sandy soil in woods, ditches, roadbanks, disturbed areas, Manitoba to Newfoundland south the Gulf States.

1b Upper leaf cells averaging 17 μ or more in longest dimension; leaves broad, those near middle of stem usually more than 1 mm wide with 0-6 (rarely up to 8) lamellae commonly covering less than 1/3 of leaf. 2

2a Median leaf cells thin-wall, not collenchymatous to only slightly so, leaves frequently obtuse, broad, usually 1.5 m wide or more. 3

2b Median leaf cells often thick-walled and strongly collenchymatous, leaves narrow, seldom over 1.5 mm wide. 4

3a Lamellae 0-3 (rarely 4), 1-4 cells high; cuticle of marginal leaf cells usually covered with verrucose or striate papillae. .. .. *Atrichum crispum* (James) Sull.

Plants small to moderately robust, in dull or yellowish-green tufts, on soil in mountains, Ontario to Quebec and Nova Scotia south to Iowa, Tennessee and Florida.

3b Lamellae 2-6, 2-13 cells high; marginal leaf cells smooth. .... *Atrichum selwynii* Aust.

Plants in loose, light to dark green tufts, stems erect, 1-4 cm high, on soil, Alaska to Manitoba south to California, Idaho, Montana, Utah, and Colorado.

4a Plants dioicous, stems commonly over 3 cm high, densely clothed with whitish rhizoids, leaves dense. .... *Atrichum oerstedianum* (C. Muell.) Mitt.

Plants in loose, light to dark green tufts, stems erect, 2-6 cm high, on humus and soil in shaded habitats, Wisconsin and Ontario to Labrador south to Louisiana and Mississippi.

4b Plants monoicous (at least some plants), stems seldom over 3 cm high and not noticeably clothed with whitish rhizoids, leaves distant. Fig. 377. .... *Atrichum undulatum* (Hedw.) P.-Beauv.

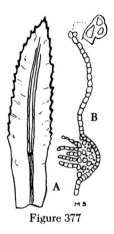

Figure 377

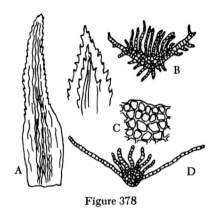

Figure 378

Figure 377 *Atrichum undulatum.* A, leaf with lamellae; B, cross section of leaf.

Plants relatively robust, 1.5-7 cm high, in light to dark-green tufts, leaves undulate, on rich humus, soil, clay, mud, in moist woods and ravines along streams, Alaska to Nova Scotia south to British Columbia and the Gulf States.

Oligotrichum

1a Lamellae on ventral surface of leaf wavy or undulate. 2

1b Lamellae straight and entire or nearly so. Fig. 378D. ....
Oligotrichum parallelum (Mitt.) Kindb.

Plants 2-5 cm high with leaves crisped and contorted when dry, on soil, arctic, alpine, Alaska to Washington.

2a Lamellae conspicuous on costa and upper leaf surface. Fig. 278A, B, C. ....
.... *Oligotrichum aligerum* Mitt.

Figure 378 *Oligotrichum aligerum.* A, leaf; B, cross section of leaf; C, margin of lamella; D, leaf of *O. parallelum.*

Plants in loose or scattered, dark green tufts, 1-2 cm high, leaves crisped and contorted when dry, on soil and clay banks. Alaska to Oregon and Idaho.

2b Lamellae few and small; arctic-alpine. ....
Oligotrichum hercynicum (Hedw.) DC.

Similar to the preceeding species, on soil, Alaska south to Oregon and Montana.

Bartramiopsis

Bartramiopsis lescurii (James) Kindb. (Fig. 379)

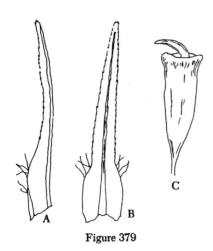

Figure 379

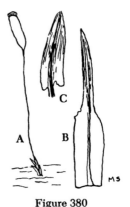

Figure 380

Figure 379 *Bartramiopsis lescurii*. A-B, leaves; C, capsule.

Plants in dark green, reddish brown to nearly black tufts, stems erect, 2-5 cm high, on soil, tree roots, Alaska to Washington.

Pogonatum

1a Stems short or obsolete, seta 1-2.5 cm tall. 2

1b Stems 1.0-18 (avg. 5-10) cm tall, sometimes branched. 3

2a Leaves with 10-15 lamellae, serrate in upper half; plants usually scattered. Fig. 380A-B. .... *Pogonatum pensilvanicum* (Hedw.) P.-Beauv.

Figure 380 *Pogonatum pensilvanicum*. A, plant natural size; B, leaf enlarged; C, leaf of *P. brachyphyllum*.

Plants arising from perennial protonema, in small, erect, scattered or open tufts, on clayey banks, Iowa to Ontario, Newfoundland south to Arkansas, Georgia. Only with care does one find leaves at the base of the seta.

2b Leaves with 20-70 lamellae, entire; plants forming dense tufts. Fig. 380C. .... *Pogonatum brachyphyllum* (Michx.) P.-Beauv.

Plants in dense tufts, leaves short, thick, on sandy soil, Missouri to New Jersey south to the Gulf States.

3a Leaves thin, curley when dry, toothed halfway down on the sheathing base; teeth of peristome 32. .... *Pogonatum contortum* (Brid.) Lesq.

Stems erect, 3-12 cm long, on soil, usually clay banks, Alaska to California.

3b Leaves thick; sheaths entire; end cells of lamellae as seen in cross section papillose; peristome teeth. 4

4a End cells of lamellae as seen in cross-section rounded to acute; cavity pear-shaped or ventrically elliptical. 5

4b End cells of lamellae as seen in cross section flat on top; papillae large; cavity transversely rectangular. Fig. 381E. ....
.... *Pogonatum dentatum* (Brid.) Brid.

Stems erect, 1-3 cm high, usually not branched, on bare soil, Alaska to New England and Quebec south British Columbia, Alberta, and New York.

5a Cavity of end cell of lamellae pear-shaped; as high as wide or higher than wide; teeth of peristome 40-64. Fig. 381D. ....
.... *Pogonatum alpinum* (Hedw.) Roehl.

Plants 2-10 cm tall, on soil, Alaska to Labrador, south to California, Colorado, Wisconsin, Michigan, North Carolina.

5b Cavity of end cell vertically elliptical; teeth of peristome 32. Fig. 381A-C. ....
.... *Pogonatum urnigerum* (Hedw.) P.-Beauv.

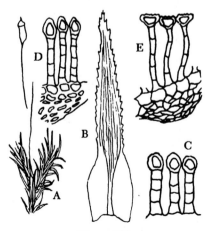

Figure 381

Figure 381 *Pogonatum urnigerum*. A, plant; B, leaf; C, cross section of lamellae; D, lamellae of *P. alpinum*; E, lamellae of *P. dentatum*.

Plants 1-5 cm tall, Alaska to Quebec and Nova Scotia south to California, Idaho, Montana, New England and New York.

Polytrichum

Ten species of *Polytrichum* are found in North America.

1a End calls of lamellae as seen in cross section, roughened with shallow, elongate pits; calyptra only slightly hairy. Fig. 382A-F. ....
.... *Polytrichum lyallii* (Mitt.) Kindb.

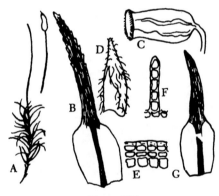

Figure 382

Figure 382 *Polytrichum lyallii*. A, plant; B, leaf; C, capsule; D, calyptra; E, side view of lamella; F, section of lamella; G, leaf of *Polytrichum sexangulare*.

Plants 1-5 cm high in dark green scattered tufts, on soil in mountains, Yukon south to California, Arizona, New Mexico.

1b End cells of lamellae smooth; calyptra densely hairy. 2

2a Leaves blunt at tip, entire; capsule 6-angled. Fig. 382G. ....
.... *Polytrichum sexangulare* Brid.

Plants 1-2 cm high, on soil in mountains, Alaska south to Washington, Montana, and Wyoming.

2b Leaves ending in an awn. 3

3a Margins of leaf translucent, rolled up over upper surface of leaf completely covering the lamellae. 4

3b Margins of leaf scarcely taller than lamellae, toothed from apex to at least middle of leaf. 5

4a Awn of leaf reddish. Fig. 383A-F. ....
.... *Polytrichum juniperinum* Hedw.

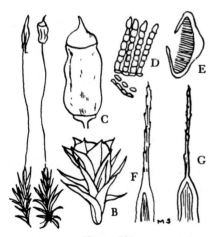

Figure 383

Figure 383 *Polytrichum juniperinum*. A, fruiting plant; B, antheridial head; C, capsule; D, cross section of lamellae; E, cross section of leaf; F, apex of leaf; G, apex of leaf of *P. piliferum* with its abrupt narrowing to a long white hair; the leaf is otherwise similar to that of *P. juniperinum*.

Plants rather small to robust, in loose, green, bluish-green, or red-brown tufts, 1-13 cm high, on soil and rocks, usually in dry, exposed to partially shaded places, throughout North America.

4b Awn of leaf white or hyaline. Fig. 383G.
.... *Polytrichum piliferum* Hedw.

Short, stout plants (1-4 cm high) in loose brown to glaucous-green tufts, on dry, often sandy or gravelly soil in exposed places, circumpolar south to California, Colorado, Missouri, Tennessee, North Carolina.

5a End cells of lamellae in cross section similar to the cells below; margins 4-9 cells wide. 6

5b End cells peculiar in shape and thickening of walls. 7

6a Terminal cells of lamellae narrowly elliptic; cells of leaf sheath oblong-linear; margin of upper portion of leaf very narrow (usually less than 4 cells wide), the cells about 10-12 μ wide. Fig. 384E-F. ....
.... *Polytrichum formosum* Hedw. var. *formosum*

Plants in dark green to brownish loose tufts, 4-7 cm high, on soil and rocks, Alaska to Labrador south to Oregon, Colorado and the Gulf of Mexico in the east.

6b Terminal cells rounded or shortly oval; cells of sheath shorter (3-5:1), oblong; margin of upper part of leaf some what wider, the cells about 15-18μ wide. Fig. 384A-C. .... *Polytrichum formosum* var. *aurantiacum* (Brid.) C. J. Hartm.

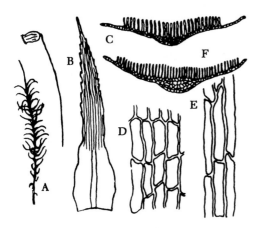

Figure 384

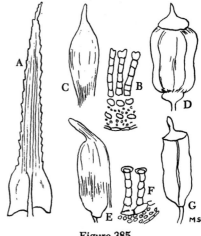

Figure 385

Figure 385 *Polytrichum commune*. A, leaf; B, section of lamellae; C, calyptra enclosing a capsule; D, capsule alone; E, capsule and calyptra of *P. ohioense*; F, section of lamellae; G, capsule alone, with tapering base.

Figure 384 *Polytrichum formosum* var. *aurantiacum*. A, plant; B, leaf; C, section of leaf; D, cells of sheath; E, cells of sheath of *P. formosum*; F, section of leaf of same.

Tufted plants, 2-5, rarely 15 cm high, on soil and humus, Alaska to Labrador, south to Colorado, Michigan and North Carolina.

Robust, dark-green or brownish plants in loose to dense tufts, 4-45 cm tall, on soil in moist areas, widespread in North America.

7a End cells notched with thickened walls. Fig. 385A-D. .... *Polytrichum commune* Hedw.

7b End cells wider than high, the upper wall thickened. Fig. 385E-G. .... *Polytrichum ohioense* Ren. & Card.

Robust plants in dark-green to brown loose tufts, 1.5-6 cm high, on soil and humus in woods, widespread in eastern North America west to Kansas, Oklahoma, and Texas.

Class I. Anthocerotae— The Hornworts

ANTHOCEROTACEAE

KEY TO THE GENERA

1a Capsules erect, 1-3 cm long, becoming black after splitting. 2

1b Capsules horizontal, short, slightly projecting from the margins of the thallus. .... (p. 231) *Notothylas*

2a Spores black. .... (p. 230) *Anthoceros*

2b Spores yellow .... (p. 231) *Phaeoceros*

Anthoceros

Six species of *Anthoceros* are reported from North America.

1a Sporophytes filiform, 10-30 mm high. Fig. 386A, 387D. .... *Anthoceros punctatus* L.

Figure 386

Figure 386 *Anthoceros punctatus*. A, thallus with sporophyte; *A. macounii*. B, thallus with sporophyte.

Thallus yellowish-green to dark green with deeply divided lobes, on damp soil, widespread in North America.

1b Sporophytes short cylindrical, 3-10 mm high. Fig. 386B. .... *Anthoceros macounii* M. A. Howe

Thallus deeply lobed, dark green, on moist soil, Minnesota and Quebec east to New England.

Phaeoceros

Phaeoceros laevis (L.) Prosk. (Fig. 387A-C)

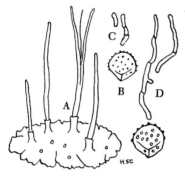

Figure 387

Figure 387 *Phaeoceros laevis*. A, plant; B, spore; C, elaters; D, spore and elaters of *Anthoceros punctatus*.

Thallus dark green, 0.5-3 cm in diameter, deeply lobed, on moist soil and rocks, widespread in North America. Two other species are recognized from North America.

Notothylas

Notothylas orbicularis (Schwein.) Sull. (Fig. 388)

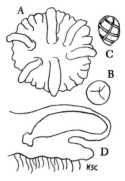

Figure 388

Figure 388 *Notothylas orbicularis*. A, plant; B, spore; C, elater; D, section of sporophyte in perianth.

Plants in small green rosettes up to 2.5 cm in diameter with deeply cut lobes, on moist soil, Minnesota to New England south to Texas, Louisiana, Tennessee, and North Carolina. *N. breutelii* Gott. is reported from Louisiana.

Class II. Hepaticae—
The Liverworts

Order 1. Takakiales

TAKAKIACEAE
Takakia

Takakia lepidozioides Hatt. & H. Inoue (Fig. 35)

Small yellow-green plants in loose to dense tufts, leaves typically divided into 2-3 segments to the base, sheltered rocks and rock crevices, Queen Charlotte Islands of British Columbia and in Sitka and Ketchikan areas of Alaska. A primitive bryophyte with a low chromosome number: $n=4$.

Order 2. Calobryales

HAPLOMITRIACEAE
Haplomitrium

Haplomitrium hookeri (Sm.) Nees. (Fig. 9)

Plants green, in loose tufts, 2-9 mm high, leaves transversely inserted, on moist sites, subarctic-subalpine, Maine, New Hampshire. This plant might be mistaken for a juvenile species of *Bryum*, however, it will lack rhizoids.

Order 3. Jungermanniales

KEY TO GENERA

(Adapted in part from R. M. Schuster. *The Hepaticae and Anthocerotae of North America*, Vol. I. New York: Columbia University Press, 1966, pp. 677-691 by permission of the publisher.)

1a Leaves deeply divided for 0.6-0.9 their length into many filiform segments 1-2 cells broad or into cilia. Fig. 389 A-C. ... 2

1b Leaves entire, or toothed, or divided at tip into 2, 3, or 4 lobes. Fig. 389D-H. 6

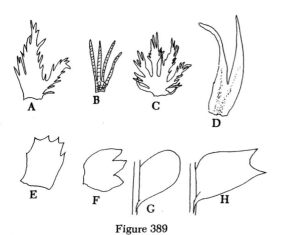

Figure 389

Figure 389 Leaves deeply divided into threads or rows of cells. A, *Ptilidium pulcherrimum*; B, *Blepharostoma trichophyllum*; C, *Trichocolea tomentella*. Leaves entire, toothed, or divided into lobes; D, *Herbertus aduncus*; E, *Plagiochila sullivantii*; F, *Barbilophozia floerki*; G, *Odontoschisma prostratum*; H, *Lophocolea heterophylla*.

2a Plants robust. 3

2b Plants minute, filiform and slender. 4

3a Plants whitish-green; cells without trigones. Fig. 24A. .... (p. 242) *Trichocolea*

3b Plants green to fulvous or reddish-brown; cells with distinct trigones. Fig. 24D. .... (p. 241) *Ptilidium*

4a (2b) Leaf segments formed of delicate, elongated cells (50-100 μ); shiny when dry. .... (p. 243) *Telaranea*

4b Leaf segments formed of short, rigid cells (under 50 μ long); dull when dry. 5

5a Basal part of divisions 2 cells wide. .... (p. 242) *Kurzia*

5b Basal part of divisions only 1 cell wide. .. .. (p. 241) *Blepharostoma*

6a (1b) Plants robust; leaves bilobed, more or less falcate-secund, with bifurcate vitta of elongated cells extending to or beyond the middle of the lobes. Fig. 389D. .... (p. 241) *Herbertus*

6b Plants not as above; if leaves bilobed, bifurcate vitta absent. 7

7a Leaves transversely attached, or succubous; attached obliquely so that the edge of the leaf on upper surface of stem is attached nearer the base of the stem than the lower edge; thus the leaf slopes toward the apex of the stem. Fig. 21. 8

7b Leaves incubous; attached obliquely to the stem so that the edge of leaf on upper surface of stem is attached nearer the apex of the stem than the lower edge; thus the leaf slopes toward the base of the stem. Fig. 20. 39

8a Leaves entire and unlobed; underleaves absent or considerably smaller than leaves. 9

8b   Leaves lobed or their margins with distinct teeth. 16

9a   Plants bearing leafless stolons; leaves often margined with more or less thick walled cells. .... (p. 248) *Odontoschisma*

9b   Plants without leafless stolons. 10

10a Cells of leaf middle large, 45 μ or more wide, with large coarse, nodose trigones. Fig. 24E. .... (p. 257) *Mylia*

10b Cells of leaf middle smaller, 36 μ or less wide, trigones small or absent. 11

11a Rhizoids restricted to small areas at the bases of large lanceolate or bifid underleaves. 12

11b Rhizoids scattered over the stem; underleaves absent or small and inconspicuous. 13

12a Perianth triangular, split in three at the top, overtopped by the larger, obovate calyptra (archegonium) when sporophyte emerges, borne on a short branch; underleaves present, conspicuous, 2-lobed; antheridia in axils of leaves in groups along the stem. ....
.... (p. 251) *Chiloscyphus*

12b Similar to the above, but calyptra remaining deep within the perianth, which terminates a main shoot; antheridia just below the perianth. ....
.... (p. 250) *Lophocolea*

13a (11b) Underleaves present, small, undivided, lanceolate; leaves orbicular. ....
.... (p. 258) *Nardia*

13b Underleaves lacking or divided into several cilia. 14

14a Leaves succubous, long-decurrent on upper side of stem with margins turned back; underleaves minute but distinct, formed of 2- several short cilia tipped by slime papillae. .... (p. 251) *Plagiochila*

14b Leaves not as above; underleaves lacking or present only as an unlobed vestige.
.... 15

15a Mouth of perianth contracted, and fringed with many-celled hairs. ....
.... (p. 258) *Jamesoniella*

15b Mouth of perianth plaited but not fringed. .... (p. 259) *Jungermannia*

16a (8b) Leaves complicated-bilobed, the smaller antical (upper) lobe bent over on the dorsal side of the larger postical (lower) lobe; keel between lobes sharp; underleaves absent. Fig. 22. 17

16b Leaves not complicate-bilobed. 18

17a Postical (larger) lobe oblong-lanceolate; perianth cylindric and furrowed. ....
.... (p. 261) *Diplophyllum*

17b Postical (larger) lobe rounded-ovate; perianth with a wide flattened mouth. ....
.... (p. 262) *Scapania*

18a (16b) Leaves transversely inserted on the stem. 19

18b Leaves succubously inserted and thus more or less obliquely oriented. 27

19a Leaves 3-4 lobed. 20

19b Leaves mostly bilobed. 23

20a Most leaves 4-lobed, underleaves very large, usually 2-lobed. 21

20b Most leaves 3-lobed, underleaves small or absent. 22

21a Underleaves large, almost as long as leaves. .... (p. 253) *Chandonanthus*

21b Underleaves smaller, less than half the length of leaves. .... (p. 255) *Lophozia*

22a (20b) Underleaves small; antical lobe of leaves clearly the larger; antical end of leaf unsertion oblique, running down the stem. .... (p. 255) *Lophozia*

22b Underleaves absent; antical lobe of leaves smaller or equal to others; antical half of leaf insertion transverse or running up towards stem apex. .... (p. 256) *Tritomaria*

23a (19b) Lobes of leaves ciliate. .... (p. 248) *Nowellia*

23b Lobes of leaves obtuse, acute, or rounded. 24

24a Leaf lobes acute or subacute, somewhat unequal, the antical smaller; leaves loosely complicate; vinaceous to reddish or purplish gemmae present on leaves; plants unusually robust, 0.75-2.5 mm wide. .... (p. 257) *Anastrophyllum*

24b Leaf lobes equal or nearly so, not at all complicate; gemmae absent or smoothly ovoid or ellipsoidal: plants not robust, usually less than 1 mm wide. 25

25a Leaf cells collenchymatous, trigones often buldging; underleaves and gemmae absent. .... (p. 261) *Marsupella*

25b Leaf cells not collenchymatous, trigones absent; leaves bilobed; underleaves present; gemmae often present. 26

26a Cells very small, 6-15 μ wide at base of lobes; outer cells of stem without large, pellucid cells. .... (p. 249) *Cephaloziella*

26b Cells larger, 12-40 μ wide at base of lobes; outer cells of stem large and pellucid. .... (p. 246) *Cephalozia*

27a (18b) Underleaves of sterile shoots present, well developed, often large. 28

27b Underleaves of sterile shoots absent or poorly developed. 34

28a Rhizoids restricted to bases of underleaves; underleaves bifid, usually with one or two teeth on outer margins. .... (p. 250) *Lophocolea*

28b Rhizoids scattered over the lower surface of shoots. 29

Class II. Hepaticae

29a Leaf margins dentate but not lobed, rarely entire. .... (p. 251) *Plagiochila*

29b Leaf margins entire except at apex that is clearly bilobed. 30

30a Underleaves unlobed, ovate to lingulate to ovate-lanceolate. 31

30b Underleaves bilobed, ciliate or dentate. 33

31a Cells equally thick-walled. .... (p. 246) *Cladopdiella*

31b Cells thin-walled with conspicuous, often bulging trigones. 32

32a Branches mainly ventral in origin. .... (p. 249) *Harpanthus*

32b Branches all lateral. .... (p. 258) *Nardia*

33a (30b) Underleaves bifid almost to base into 2 parallel, lanceolate segments. .... (p. 249) *Geocalyx*

33b Underleaves not as above, more or less ciliate or dentate. .... (p. 255) *Lophozia*

34a (27b) Leaves (at least some) 3 or more lobed, or irregularly lobed or bearing several to many teeth. 35

34b Leaves uniformly and simply bilobed. 36

35a Leaves with filiform lobes, or spinose dentate on margins. .... (p. 251) *Plagiochila*

35b Leaves bearing 3-4 subequal, lobes without accessory teeth. .... (p. 253) *Barbilophozia*

36a (34b) Most branches, or at least the sexual branches, ventral in origin; leaf cells noncollenchymatous. 37

36b Branches uniformly lateral in origin, plants usually blackish. .... (p. 257) *Gymnocolea*

37a Outer cells of stem large, pellucid; underleaves absent. ..... (p. 246) *Cephalozia*

37b Outer cells of stem small, opaque. 38

38a Plants over 0.5 mm wide; leaves with rounded or obtuse lobes. .... (p. 246) *Cladopodiella*

38b Plants very small, usually less than 0.5 mm wide, leaves with acute to subacute lobes. .... (p. 249) *Cephaloziella*

39a (7b) Leaves not complicate-bilobed. 40

39b Leaves complicate-bilobed. Fig. 23. .. 42

40a Leaves divided to middle into 3-4 lobes. .... (p. 243) *Lepidozia*

40b Leaves 2-3 dentate at apex or entire. .. 41

41a Leaves 2-3 dentate at apex; plants opaque, green to brownish. .... (p. 243) *Bazzania*

41b Leaves entire to minutely bidendate at apex; plants generally pellucid, pale to bluish-or grayish-green. .... (p. 244) *Calypogeja*

42a (39b) Underleaves absent. 43

42b Underleaves present. 44

43a Rhizoids originating on ventral lobes; plants opaque, greenish. .... (p. 264) *Radula*

43b Rhizoids on stem; plants pellucid or yellow-green. .... (p. 275) *Cololejeunea*

44a (42b) Underleaves unlobed. 45

44b Underleaves bilobed. 49

45a Lobule lingulate or oblong, lying parallel with stem; large plants, 2.5-5 mm wide. Fig. 23. .... (p. 265) *Porella*

45b Lobule inflated, lying parallel with posterior leaf margin, attached to it by a long keel; minute to small plants, 0.25-2.5 mm wide. Fig. 449A. 46

46a Margin of lobule with 4-6 minute teeth. .... (p. 270) *Ptychocoleus*

46b Margin of lobule with 1-2 teeth at or near apex, or entire. 47

47a Plants grayish-green. .... (p. 271) *Leucolejeunea*

47b Plants fuscous to blackish-green in color. 48

48a Mature plants closely appressed to substrate, shiny and flat when dry. .... (p. 270) *Lopholejeunea*

48b Mature plants ascending or suberect, dull; leaves convolute when dry. .... (p. 270) *Mastigolejeunea*

49a (44b) Lobule conspicuous, attached by a very narrow, stalk-like base, appearing free from stem; plants opaque, deep green to reddish-brown; underleaves inserted on over 4 rows of stem cells. Fig. 23. 50

49b Lobules usually broadly attached to stem (or obsolete); plants usually pale green to yellow-green; underleaves attached to only 2 rows of large ventral stem cells. Fig. 453C. 51

50a Plants dark green; dorsal lobe pointed; cell walls thin; leaf subtending a branch partly attached to the branch. .... (p. 270) *Jubula*

50b Plants black to red-brown or green; dorsal lobe usually not pointed; cell walls usually thick, with conspicuous trigones and often with bead-like thickenings along the walls (intermediate thickenings); leaf subtending a branch attached only to subtending a branch attached only to stem. .... (p. 266) *Frullania*

51a (49b) Underleaves widest near middle with lobes nearly erect or connivent. .. 52

51b Underleaves with lobes uniformly and distinctly divergent, thus widest at apex. 55

52a Plants brownish-tinged, shiny when dry; apical tooth of lobule prominent; with 1-several large paracysts at base. .... (p. 271) *Ceratolejeunea*

52b Plants green not brownish-tinged; apical tooth of lobe usually small; paracysts usually absent. 53

53a Leaves clearly caducous (falling off stem); perianths strongly dorsiventrally compressed. .... (p. 272) *Rectolejeunea*

53b Leaves persistent; perianths not dorsiventrally compressed. 54

54a Leaves always broadly rounded at apex; cells with vestigal to small trigones and without or few intermediate thickenings. .... (p. 273) *Lejeunea*

54b Leaves apiculate or mucronate: cells with large trigones and numerous intermediate thickenings. .... (p. 272) *Crossotolejeunea*

55a (51b) Underleaves as many as lateral leaves; lobule with a long, filiform tooth. .... (p. 274) *Diplasiolejeunea*

55b Underleaves half as many as lateral leaves; lobules with a small 1-celled apical tooth. .... (p. 272) *Drepanolejeunea*

Order 4. Metzgeriales

KEY TO GENERA

1a Plants divided into a stem and into leaves, rhizoids purplish. .... (p. 275) *Fossombronia*

1b Plants thallose or shallowly lobed. 2

2a Plants shallowly lobed, with scattered lumps of blue-green algae imbedded in thallus; gemmae present in bottle-shaped containers or with tiny star-shaped gemmae. .... (p. 278) *Blasia*

2b Margins of thallus wavy or even, but not shallowly lobed. 3

3a Midrib well defined, bulging like a cord along lower side of plant, rest of thallus only 1 cell thick. 4

3b Midrib ill-defined, merely the gradually thickened central part of the plant, thallus is only 1 cell thick at the margin if at all. 6

4a Plant 1-2 mm wide, much longer than wide, of very even width, forking; sex organs underneath. .. (p. 277) *Metzgeria*

4b Plant 3-4 mm wide, often irregularly wavy; sex organs on upper side along midrib. 5

5a (b) Midrib with a central strand of small, thick-walled cells. .... (p. 276) *Pallavicinia*

5b Midrib without a central strand. .... (p. 276) *Moerckia*

6a (3b) Plant 4-16 mm wide, usually crowded in wide (10-50 mm) patches; elaters attached at base of capsule, on moist shaded ground. ...... (p. 277) *Pellia*

6b Plant 1-5 mm wide, variously lobed or branched, in very wet places or in shallow water. .... (p. 279) *Riccardia*

Order 5. Sphaerocarpales

KEY TO GENERA

1a Plants aquatic submerged, with a long erect stem and one longitudinal wing-like expansion on the vein. .... (p. 280) *Riella*

1b Plants terrestrial, prostrate or nearly so, with no wing-like expansion of vein. .... (p. 280) *Sphaerocarpus*

Order 6. Marchantiales

KEY TO GENERA

1a Air pores visible without lens, each in a polygonal area; capsules borne on the underside of an umbrella-shaped receptacle, with spirally banned elaters among spores; walls of capsules with ring-shaped thickenings. 8

1b Air pores absent, or if present visible only with a strong lens. 2

2a Plants submerged or floating, or in circular rosettes on very wet ground; air pores, if any, not visible with a hand lens; capsules imbedded in the plant, with no elaters among the rough spores. 3

2b Plants on moist or dry rocks or banks, rarely, if ever, in neat rosettes; capsules borne on the underside of an umbrella-shaped receptacle. 4

3a Lobes of the thallus 5-10 mm wide, with air spaces in 3 or 4 irregular layers; in rosettes 2-3 cm across on muddy shores or floating in triangular pieces bearing many scales beneath; plants with a distinct sharp median groove throughout the length of the thalli. .... (p. 285) *Ricciocarpus*

3b Lobes 1-3 mm wide, with air chambers in 1 or 2 layers or with mere chinks between the chains of upper cells; plants without a sharp median groove, or with groove confined to tips of thalli. .... (p. 286) *Riccia*

4a (2b) Thallus without or with only vestigial pores. .... (p. 285) *Dumortiera*

4b Thallus with pores. 5

5a (2b) Cell walls radiating from pores strongly thickened, giving the pore a star-like (stellate) appearance. Fig. 471B. .... (p. 283) *Athalamia*

5b Pores not stellate in appearance, or absent. 6

6a Cells of the epidermis with thin walls and large, bulging trigones; pores inconspicuous. 11

6b Trigones of epiderminal cells small, not bulging. 7

7a Cells around pores in 2-3 concentric rows, thick-walled; cells of epidermis collenchymatous (walls thickened all around and at corners); white scales around the base and summit of stalk. Fig. 467. .... (p. 281) *Mannia*

7b Cells around pores irregular, thin-walled; cells of epidermis not collenchymatous; stalk of umbrella naked at both ends, capsule surrounded by several scales, remnants of a tubular pseudoperianth. .... (p. 282) *Asterella*

8a (1a) Thalli with open or half-cups of dish-shaped gemmae on the thallus; archegonia (and sporophytes) on the underside of long-fingered umbrellas with 4-9 fingers. 9

8b Thalli without gemmae, and without marginal scales on underside. 10

9a Gemmae cups round, fringed; female umbrellas 9-lobed; thalli with thin scales along the margin beneath; air pores elliptic. .... (p. 284) *Marchantia*

9b Gemmae in half-cups; female umbrellas 4-lobed; thalli without marginal scales beneath; found only in greenhouses and sterile, except in southern California. .... (p. 283) *Lunularia*

10a (8b) Air pore on a low mound of colorless cells; antheridia in a warty spot on the thallus; sporophytes beneath a cone-shaped umbrella. .... (p. 283) *Conocephalum*

10b Air pores circular, surrounded by a low cylinder of cells; antheridia and sporophytes on upraised scalloped umbrellas. .... (p. 284) *Preissia*

11a (6a) Pores surrounded by 4-5 concentric rows of thick walled cells. .... (p. 281) *Reboulia*

11b Pores not surrounded by concentric rows of thick walled cells, western North America. .... (p. 281) *Targionia*

Order Jungermanniales

PSEUDOLEPICOLEACEAE
Blepharostoma

Blepharostoma trichophyllum (L.) Dum. (Fig. 390)

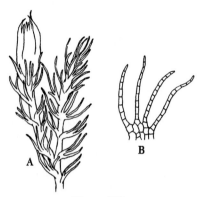

Figure 390 *Blepharostoma trichophyllum*. A, plant; B, leaf.

Plants in light-green to yellow-green loose patches, on moist decaying logs, vertical rocks, predominantly in forests, Alaska to Labrador south to California, Idaho, Montana, Colorado, Minnesota, Illinois, Tennessee, Georgia. *B. arachnoideum* M. A. Howe, has 2 or 3 divisions with cells nearly twice as long as those of *B. trichophyllum* and occurs from British Columbia to California and New Mexico.

HERBETACEAE
Herbertus

Herbertus aduncus (Dicks.) S. Gray (Fig. 391)

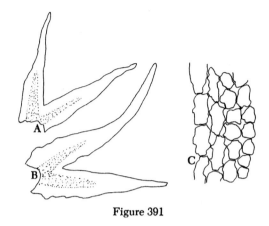

Figure 391 *Herbertus aduncus*. A-B, bifurcate leaves; C, median leaf cells (larger cells part of vitta.

Plants forming extensive polsters or pendulous from rock faces or tree trunks, yellowish to brownish to reddish-brown or short tips rose-tinged, stems rigid, 2-20 cm long, Alaska to Washington and in the Appalachians from New York to Tennessee and South Carolina.

PTILIDIACEAE
Ptilidium

1a Leaves with many cilia at margin. 2

1b Leaves without cilia or only few cilia at margin; western North America. .... *Ptilidium californicum* (Aust.) Underw.

Plants in dense, appressed mats, bright-green, yellowish-green, or reddish-brown, stema 1-5 cm long, on trees, Alaska south to Montana, Idaho, and California.

2a Leaves with largest lobe 15-25 cells wide at base, cilia relatively short. .... *Ptilidium ciliare* (L.) Hampe

Plants usually copper colored, in deep, luxuriant, lax tufts, robust, stems 2.5-6 mm long, on soil over rocks, widespread in North America, Alaska to Quebec, New Brunswick, Labrador south to British Columbia, Montana, Minnesota, Wisconsin, Indiana, Pennsylvania, Virginia.

2b Leaves with largest lobe 6-10 or 12 cells wide at base, cilia abundant and long. Fig. 392. .... *Ptilidium pulcherrimum* (G. Web.) Hampe

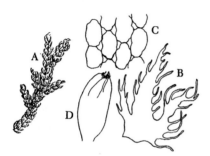

Figure 392

Figure 392 *Ptilidium pulcherrimum*. A, plant; B, leaf; C, cells of leaf; D, perianth.

Plants in low, flat, dense, green to tawny brown tufts, on trees, decaying logs, clifts and ledges, Alaska to New Brunswick, Quebec, Newfoundland, south to Washington, Montana, Idaho, Missouri, Indiana, Pennsylvania, Kentucky, North Carolina and Tennessee.

TRICHOCOLEACEAE
Trichocolea

Trichocolea tomentella (Ehrh.) Dum. (Fig. 393)

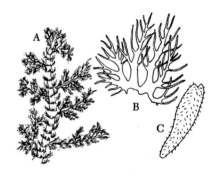

Figure 393

Figure 393 *Trichocolea tomentella*. A, plant; B, leaf; C, perianth.

Plants in robust light yellowish-green to pale green loose mats, stems 5-15 cm long, 1.5-3 cm wide, on moist soil, edges of streams and brooks and in swamps, Wisconsin to Ontario, Nova Scotia and Newfoundland south to Arkansas, Tennessee, Georgia, Florida.

LEPIDOZIACEAE
Kurzia

1a Underleaves normally 3-lobed; never in peat bogs. Fig. 394. .... *Kurzia sylvatica* (Evans) Crolle

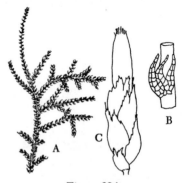

Figure 394

242 Class II. Hepaticae

Figure 394 *Kurzia sylvatica*. A, plant; B, leaf; C, perianth.

Stems thread-like, 5-20 mm long, in dull to deep green or brownish-green dense tufts or patches, on moist acid, often peaty shaded soils, eastern United States from Quebec to Nova Scotia south to the Gulf States and Texas.

1b Underleaves normally 4-lobed, in peat bogs. .... *Kurzia setacea* (Web.) Grolle

Small, delicate plants scattered among and creeping over peat mosses in bogs, green to dull brown or fuscous, Alaska to Newfoundland and Nova Scotia south in eastern North America to Michigan, Indiana, Ohio, New Jersey.

Lepidozia

Lepidozia reptans (L.) Dum. (Fig. 395)

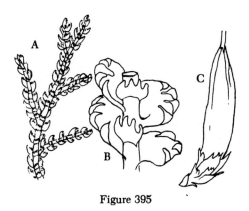

Figure 395

Figure 395 *Lepidozia reptans*. A, plant; B, leaves and underleaves; C, perianth.

Plants in loose, flat, light green mats, shoots 1.5-3 cm long, 1-pinnately branched, on humus, decaying wood or peaty soil, occasionally on rocks, in shaded moist sites, Alaska to Newfoundland and Nova Scotia south to California, Montana, Idaho, New Mexico, Wyoming, Minnesota, Iowa, Indiana, Tennessee and North Carolina. Two other species are reported from North America.

Telaranea

Telaranea nematodes (Gott. *ex* Aust.) M. A. Howe (Fig. 396)

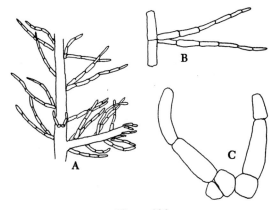

Figure 396

Figure 396 *Telaranea nematodes*. A, shoot, ventral view; B, leaf; C, underleaf.

Very minute delicate, filmy plants, pale to whitish-green, in wet sites on peaty to sandy-peaty substrates, New York south to Florida, Arkansas and Mississippi.

Bazzania

Six species of *Bazzania* are recognized in North America.

1a Plants large, normally 3-6 mm wide. Fig. 397A-D. ....
.... *Bazzania trilobata* (L.) S. Gray

Class II. Hepaticae   243

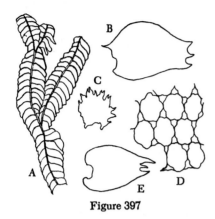

Figure 397

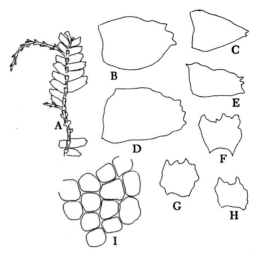

Figure 398

Figure 397 *Bazzania trilobata*. A, plant; B, leaf; C, underleaf; D, cells of leaf; E, leaf of *B. tricrenata*.

Robust plants in dense tufts or polsters, on shaded moist humus, logs and vertical rock faces, Alaska to Labrador and Newfoundland south to Arkansas, Illinois, Alabama, and North Carolina.

1b   Plants less than 2.5 mm wide. 2

2a   Many leaves broken off leaving parts of stem bare (cauducous). Fig. 398. ....
.... *Bazzania denudata* (Torrey *ex* Gott. et al.) Trev.

Figure 398 *Bazzania denudata*. A, shoot with cauducous leaves; B-E, leaves; F-H, underleaves; I, median leaf cells.

Plants in mats, light-green to green to yellowish-green, shoots 1-2.5 cm long, tree bases or rocks in moist sheltered areas, Alaska to Quebec and Nova Scotia, south to Washington, Alberta, Tennessee, Georgia.

2b   Leaves not cauducous. Fig. 397E. ....
Bazzania tricrenata (Wahlenb.) Lindb.

Plants in loose patches or tufts, on moist shaded soil over rocks, occasionally on decaying wood, Alaska to Newfoundland and Nova Scotia south to California, Idaho, Pennsylvania, North Carolina and Tennessee.

CALYPOGEJACEAE
Calypogeja

Nine species of *Calypogeja* are known from North America.

1a Leaves all 2-toothed or 2-lobed at apex; lobes of underleaves narrowly acute, with a tooth on each side. .... *Calypogeja sullivantii* Aust.

Plants deep green (in shade) to rarely yellowish-green (in sun), shoots 1.2-2.0 mm wide, prostrate, stems with large-celled hyaline cortex, usually in moist shaded sites on soil or rock faces, especially along streams, Ohio to Maine and Nova Scotia south to Arkansas, Alabama, and Florida.

1b Leaves entire or very shallowly 2-pointed; lobes of underleaves never narrowly acute. 2

2a Marginal cells of leaves about twice as long as wide; sinus of underleaves shallow or lacking. .... *Calypogeja neesiana* (Mass. & Carest.) K. Müll

Plants dull, whitish-green to gray-green, 1.5-2.1 mm wide, leaves usually slightly longer than wide, on acid organic soils, rarely on bases of trees, eastern North America from Wyoming to Northwest Territories, Quebec and Newfoundland south to South Carolina and Georgia.

2b Marginal cells hardly longer than wide. 3

3a Plants when living appearing turquoise blue at apices, when dead often with cell walls or the cytoplasmic layer adjacent to them bluish-pigmented. Fig. 399A-C. .... *Calypogeja trichomanis* (L.) Corda

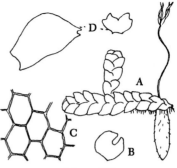

Figure 399

Figure 399 *Calypogeja trichomanis*. A, shoot from dorsal surface, with subterranean "perigynium" and sporophyte; B, underleaf; C, cells of leaf; D, leaf and underleaf of *C. fissa* subsp. *neogaea*.

Plants bluish to grayish-green in dark patches, shoots 2-3 mm wide and 1-2.5 cm long, the apices of living plants strongly turquoise blue, rare, on humus and decaying logs, Quebec, Nova Scotia, Maine, Vermont, New York, South Carolina and Wyoming. According to Schuster (The Hepaticae and Anthocerotae of North America, Vol II, pp. 152-153. 1969) this species is not widely distributed and many reports are more likely the next species.

3b Plants when living not as above. 4

4a Underleaves typically ovate to suborbicular, up to 1.4× as wide as long, sinus descending to 1/3 rarely more the underleaf length. Fig. 400A-C. .... *Calypogeja muelleriana* (Schiftn.) K. Müll. subsp. *muelleriana*

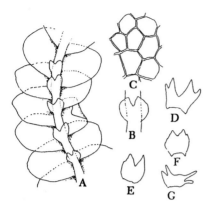

Figure 400

Figure 400 *Calypogeja muelleriana* subsp. *muelleriana*. A, section of shoot, ventral view; B, underleaf; C, median leaf cells. *C. muelleriana* subsp. *blomquistii;* D-G, underleaves.

Plants moderate to robust, branches, 1.5-3.5 mm wide and up to 5 cm long, light yellowish-green to whitish-green, in thin patches on soil, humus, and rocks of moist sites, widespread in eastern North America from Ontario, Quebec and Newfoundland south to Georgia and west to Kansas. In the Southern Appalachians and the Ouachitas *C. muelleriana* subsp. *blomquistii* Schust. (Fig. 400D-G) may be encountered and is distinguished from subsp. *muellerian* by more or less crenulate leaf margins and at least some underleaves that have erect, parallel lobes that are greatly elongated.

4b Underleaves strongly transverse, typically 1.3-1.8× as wide as long, divided almost to base. Fig. 399D. .... *Calypogeja fissa* (L.) Raddi subsp. *neogaea* Schust.

Plants in flat patches, whitish green, shiny when dry, shoots 2-3 mm wide, on moist shaded soil, New York and Vermont south to Texas, Arkansas, Mississippi and Florida.

CEPHALOZIACEAE
Cladopodiella

Cladopodiella fluitans (Nees) Joerg. (Fig. 401)

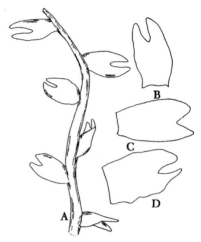

Figure 401

Figure 401 *Cladopodiella fluitans*. A, portion of sterile shoot; B-D, leaves.

Shoots 1-2 mm wide and up to 5 cm long, green to usually clear chestnut brownish to blackish, creeping over or among *Sphagnum*, rarely forming tufts, British Columbia and Washington east to Quebec, Newfoundland, Nova Scotia south to Minnesota, Michigan, West Virginia and New Jersey. One other species, *C. francisci* (Hook.) Joerg. occurs in the same range as *C. fluitans*. The leaves of *C. francisci* are bilobed 0.15-0.2 their length whereas the leaves of *C. fluitans* are bilobed 0.2-0.4 their length.

Cephalozia

Eleven species of *Cephalozia* are reported from North America.

1a Leaves divided half or more into slender, nearly parallel lobes, not decurrent. 2

1b Leaves divided about 1/3-1/2 into two short converging lobes, decurrent. 3

2a Stolons abundantly developed; median leaf cells at bases of lobes 23-30 μ wide and 35-50 μ long or less. Fig. 402. ....
.... *Cephalozia bicuspidata* (L.) Dum. subsp. *bicuspidata*

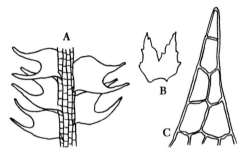

Figure 402

Figure 402 *Cephalozia bicuspidata*. A, shoot; B, bract; C, apex of lobe.

Plants in patches, green to whitish-green, shoots 0.6 to 1 mm wide, 5-25 mm long, on moist, acid rocks or sandy soil at edges of streams and brooks, Alaska to Quebec, Labrador and Newfoundland south to Oregon, Idaho, Colorado, Kansas, Iowa, Illinois, Kentucky, Georgia. The long parallel lobes of the leaves easily characterize this species. It often has erect, small leafed "flagella."

2b Stolons absent or rare; median leaf cells at bases of lobes large, 40-50 μ wide × 45-70 μ long. ....
.... *Cephalozia bicuspidata* subsp. *lammersiana* (Hüb.) Schust.

Similar to preceeding subspecies, on moist to wet mineral soils, Minnesota to Quebec and south to Tennessee.

3a Median leaf cells large, 42-48 × 45-60 μ, lobes more or less strongly conivent. Fig. 403D. ....
.. *Cephalozia connivens* (Dicks.) Lindb.

Plants in thin, flat patches, rather shiny, hyaline, pale yellowish to whitish-green, usually on acid sand, peat, or decaying logs, Alaska to Quebec, Labrador and Newfoundland south to British Columbia, Minnesota, Kansas, Missouri, Texas and the Gulf States.

3b Median leaf cells small, 16-35 μ wide × 18-36 μ long at lobe bases. 4

4a Leaves long decurrent with strongly connivent lobes. Fig. 403A-C. ....
.... *Cephalozia lunulifolia* (Dum.) Dum.

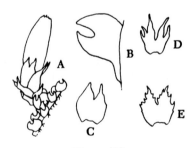

Figure 403

Figure 403 *Cephalozia lunulifolia*. A, shoot; B, leaf; C, bract; D, bract of *C. connivens*; E, bract of *C. catenulata*.

Plants small, shoots 0.4-0.6 mm wide, in flat, pale to yellowish-green patches or creeping over and among other mosses, on moist rocks, decaying logs soil banks, Alaska to Newfoundland and Nova Scotia south to California,

Nevada, Wyoming, Kansas, Missouri, Illinois, Indiana, Georgia, Florida.

4b Leaves moderately to short-decurrent with lobes not or hardly conivent. Fig. 403E. ....
.... *Cephalozia catenulata* (Hüb.) Lindb.

Plants in thin patches, deep green to pale brown or fulvous or brownish, shoots 0.4-0.6 mm wide to 15 mm long, on moist decaying logs or occasionally sandstone, British Columbia to Quebec and Newfoundland south to California, Minnesota, Kansas, and the Gulf States.

Nowellia

Nowellia curvifolia (Dicks.) Mitt. (Fig. 404)

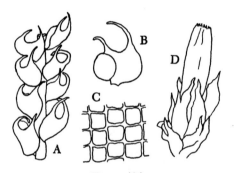

Figure 404

Figure 404 *Nowellia curvifolia*. A, shoot; B. leaf; C, cells of leaf; D, perianth with bracts.

Plants in loose, prostrate patches, usually green though sometimes rosy or reddish tinged, with unique leaves, on moist decaying logs, eastern North America from Ontario to Quebec and Newfoundland south to Arkansas, Tennessee and Georgia.

ADELANTHACEAE
Odontoschisma

Six species are reported from North America.

1a Leaves with 2-4 rows of marginal cells, more or less elongated at right angles to margin and more or less equally thick-walled. Fig. 405A-B. ....
Odontoschisma prostratum (Sw.) Trev.

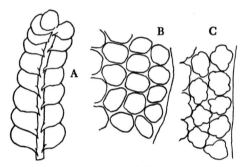

Figure 405

Figure 405 *Odontoschisma prostratum*. A, plant; B, cells of leaf; C, cells of *O. denudatum*.

Often in extensive mats or creeping among other bryophytes, pale to dull green, leaf apices often brownish or reddish-brown, shoots 1.6-2.1 mm wide to 3 cm long, on wet to moist banks and shaded vertical rocks, Illinois and Indiana to New York and Massachusetts and Maine south to Arkansas, Texas, Mississippi, Alabama, and Florida.

1b Leaves without a marginal border of elongated cells. Fig. 405C. ....
.... *Odontoschisma denudatum* (Nees *ex* Mart.) Dum.

Plants in loose patches, yellow-green with tips of leaves brownish to reddish-brown or even purplish black, shoots 1.0-1.9 mm wide to 4 cm long, on moist decaying logs and shaded acid rocks, eastern half of North America, Minnesota to Quebec, Newfoundland and Nova Scotia south to the Gulf States.

CEPHALOZIELLACEAE
Cephaloziella

This genus has at least 21 species in North America and it may be considered, because of the small size of its species, to be one of our most difficult groups of liverworts.

1a   Plants usually fertile, monoecious; underleaves of sterile shoots lacking. Fig. 406F. .... *Cephaloziella hampeana* (Nees) Schiffn.

Plants minute, in green to brownish-green patches, on sandy or peaty soil, British Columbia to Nova Scotia south to California, Wyoming, Minnesota, Missouri, Michigan, New York and Massachusetts.

1b   Plants usually sterile, dioicous; underleaves present on robust sterile shoots. Fig. 406A-E. .... *Cephaloziella divaricata* (Sm.) Schiffn.

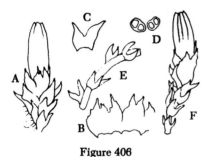

Figure 406

Figure 406 *Cephaloziella divaricata*. A, shoot with perianth; B, bract; C, leaf; D, gemmae; E, shot; F, perianth of *C. hampeana*.

Similar to the preceeding species, on rotten wood and sandy soil, Alaska to Quebec, Labrador and Nova Scotia south to California, Wyoming, Missouri, Illinois, Tennessee and North Carolina.

GEOCALYACEAE
Harpanthus

Harpanthus scutatus (Web. & Mohr.) Spruce (Fig. 407)

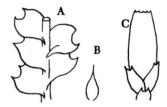

Figure 407

Figure 407 *Harpanthus scutatus*. A, shoot; B, underleaf; C, perianth.

Plants pale green, in patches, shoots about 1 mm wide and up to 1.5 mm long, underleaves large and lanceolate and often united base to the base of a leaf, on decaying logs, damp soil, on sandstone, British Columbia, Minnesota to Quebec and New Brunswick, south to the Ozarks, Tennessee and North Carolina. Two other species occur in North America.

Geocalyx

Geocalyx graveolans (Schrad.) Nees (Fig. 408)

Class II. Hepaticae

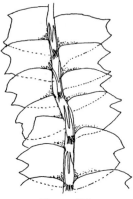

Figure 408

Figure 408 *Geocalyx graveolans,* portion of shoot, ventral view.

Plants opaque, yellow to bright green, in patches or scattered over mosses, shoots 0.2-0.3 mm wide and up to 2 cm long, underleaves large and bifid nearly to base, Alaska to Quebec and New Brunswick south to California, Idaho, Ozarks, Tennessee and North Carolina.

LOPHOCOLEACEAE
Lophocolea

Six species are reported from North America.

1a Plants 0.6-1 mm wide; gemmae in yellow green masses on margins of leaves and bracts; leaf margins irregular due to presence of gemmae. Fig. 409E-F. .... *Lophocolea minor* Nees

Plants yellowish-green, in mats, on shaded rocks, stream banks, rotten logs, tree bases, Yukon to Ontario south to British Columbia, Colorado, Arkansas, Illinois, Pennsylvania and Virginia.

1b Plants 0.8 to 3 mm wide; gemmae absent; leaf margins regular. 2

2a Leaves bilobed to entire on same plant; rhizoids numerous in tufts at bases of underleaves. Fig. 409A-D. .... *Lophocolea heterophylla* (Schrad.) Dum.

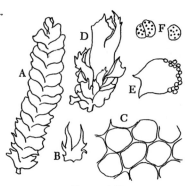

Figure 409

Figure 409 *Lophocolea heterophylla.* A, plant; B, underleaf; C, cells of leaf; D, perianth; E, leaf of *L. minor* with gemmae; F, gemmae.

Plants in patches or mats, pale yellowish-green, leafy shoots 1-2 mm wide, up to 2 cm long, on rotten wood, trunks of trees, soil and rocks, Southern Canada and the United States. The variability of this species has been extensively investigated by R. H. Hatcher (Brittonia 19:178-201. 1967).

2b Leaves mostly uniform in shape; divided into two long-acuminate lobes. 3

3a Perianths rarely found; plants unisexual, cells in middle of leaf 25-30 μ. Fig. 410A-D. .... *Lophocolea bidentata* (L.) Dum.

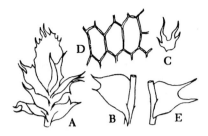

Figure 410

Figure 410 *Lophocolea bidentata*. A, shoot; B, leaf; C, underleaf; D, cells of leaf; E, leaf of *L. cuspidata*.

Plants in loose patches, pale to whitish-green, shoots 2.5-3 mm wide, up to 5 cm long, on wet or submerged rocks, soil, or rotten wood, Washington to Ontario and New England south to California, Idaho, Arkansas, Tennessee and North Carolina.

3b Perianths common; plants bisexual; cells in middle of leaf 30-50 μ. Fig. 40E. .... *Lophocolea cuspidata* (Nees) Limpr.

Plants pale yellowish-green, in patches; shoots about 3 mm wide, up to 2 cm long, on moist rocks, soil, decaying wood, Alaska, Montana, Idaho to Pnnsylvania south to California, the Ozarks, and Tennessee, probably also in northeastern North America.

Chiloscyphus

1a Cells of leaf-middle 35-60 μ, more or less strongly hyaline. Fig. 411D. .... *Chiloscyphus pallescens* (Ehrh. ex Hoffm.) Dum. var. *pallescens*

Plants in thin patches, shoots 2-3.5 mm wide, up to 3 cm long, on moist shaded soil and logs, Alaska to Quebec south to California, New Mexico, Colorado, Ozarks, Tennessee, North Carolina. *C. pallescens* var. *fragilis* (Roth.) K. Müll. is darker, leaf cells hardly hyaline and occurs throughout the range of var. *pallescens*.

1b Cells of leaf-middle 18-28 × 20-35 μ or less, little or not hyaline and pellucid. Fig. 411A-C. .... *Chiloscyphus polyanthus* (L.) Corda var. *polyanthus*

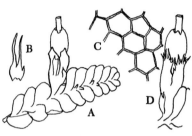

Figure 411

Figure 411 *Chiloscyphus polyanthus* var. *polyanthus*. A, plant; B, underleaf; C, cells of leaf; D, perianth and calyptra of *C. pallescens*.

Plants in green to brownish-green patches, shoots 2-4 mm wide and up to 6 cm long, on moist soil and logs, across North America south to California, Wyoming, Ozarks, Tennessee and North Carolina. *C. polyanthus* var. *rivularis* (Schrad.) Nees is an aquatic form that occurs on stones in cold streams across North America south to California and Texas, to North Carolina.

PLAGIOCHILACEAE
Plagiochila

There are at least 25 species of *Plagiochila* reported from North America. The following key is based upon a monographic treatment of family and genus by R. M. Schuster (Amer. Mid. Nat. 62(1):1-166; 62(2):257-395; 63(1): 1-130. 1959-60).

1a Leaves caducous (falling off) the stem. Fig. 412). .... ***Plagiochila sullivantii*** Gott. *ex* Evans

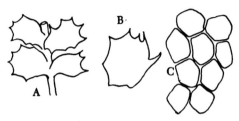

Figure 412

Figure 412 *Plagiochila sullivantii*. A, shoot; B, leaf; C, cells of leaf.

Plants differentiated into a leafless caudex and an erect to suberect leaf aerial shoots, deep green to brownish, leafy shoots 1.4-1.8 mm wide and up to 18 mm long, on moist shaded rock walls and ledges, often around water falls, Virginia, West Virginia, Tennessee, North Carolina, South Carolina, Georgia.

1b Leaves not caducous. 2

2a Median leaf cells 25-33 μ wide; leaf margins with the teeth ranging from fine to vestigal to absent. Fig. 413. .... ***Plagiochila asplenioides*** (L.) Dum.

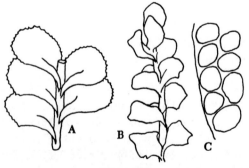

Figure 413

Figure 413 *Plagiochila asplenioides*. A, wet shoot; D, dry shoot; C, cells of leaf.

Plants in large patches or mats, green, olive green, to yellowish-green, shoots 1.8-6 mm wide and up to 10 cm long, in wet to dry sites on rocks, soil, bases of trees, Alaska across North America to Labrador and Newfoundland south to California, Arizona, New Mexico, Colorado, Arkansas, Tennessee, Georgia.

2b Median leaf cells 18-25 μ wide; leaf margins with a limited number of coarse teeth or entire or nearly so and strongly crispate-undulate on postical margin. .. 3

3a Leaves closely imbricate and postical margin strongly crispate-undulate. Fig. 414A. .... ***Plagiochila undata*** Sull.

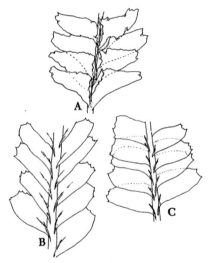

Figure 414

Figure 414 A, *Plagiochila undata*, ventral view; B, *P. floridana*, dorsal view; C, *P. virginica*, dorsal view.

Plants robust, shoots 2.5-3.5 mm wide, up to 2.5 cm long, in dense tufts or masses, green, olive-green to olive-brown, on rocks, humus, bases of trees, Arkansas, Southern Illinois, Tennessee, North Carolina south to the Gulf States.

3b Leaves laxly imbricate; leaf margins with a limited number of coarse teeth. 4

4a Mature shoots with leaves spreading at 65°-75° from stem apex; not found on Coastal Plain. Fig. 414C. .... *Plagiochila virginica* Evans

Plants in patches, green to deep green, leaf shoots, 2.75-3.2 mm wide, up to 2 cm long, damp to dry limestone and sandstone ledges, West Virginia to Virginia south to Mississippi and Georgia.

4b Mature shoots with leaves spreading at an angle of 40°-45° from the stem apex; Coastal Plain. Fig. 414B. .... *Plagiochia floridana* Evans

Plants in green to brownish-green patches, leafy shoots 2.4-2.8 mm wide and up to 25 mm long, propagula often present on leaves, mostly growing on tree roots and bases in evergreen swamp forests, North Carolina to Florida and Mississippi.

JUNGERMANNIACEAE
Chandonanthus

Chandonanthus setiformis (Ehrh.) Lindb. (Fig. 415)

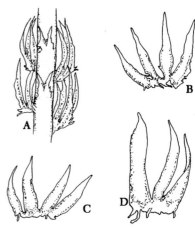

Figure 415

Figure 415 *Chandonanthus setiformis*. A, section of shoot, ventral view; B-D, leaves.

Plants in deep, loose, tufts, deep olive-green to yellowish-brown, robust, shoots 0.7-1.1 mm wide and up to 6-12 cm long, arctic-alpine, on granite rocks and soil between boulders, Alaska to Labrador south to British Columbia, Ontario, and New England, absent in the Rocky Mountains.

Barbilophozia

Eleven species are reported from North America. Many authors include this genus in *Lophozia*.

1a Leaves flat when wet; lobes usually obtuse; lower margin of leaf with cilia (hairlike appendages). Fig. 416. .... *Barbilophozia barbata* (Schmid. *ex* Schreb.) Loeske

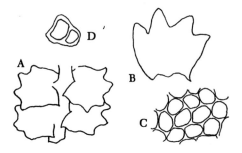

Figure 416

Figure 416 *Barbilophozia barbata*. A, shoot; B, leaf; C, cells of leaf; D, gemma.

Plants in loose, depressed mats, dark green to brownish-green, shoots 2.0-5.0 mm wide, up to 8 cm long, at higher elevations in dry to moist sites, on loamy soil and rocks, Alaska to Newfoundland south to Washington, Idaho, New Mexico, Minnesota, Wisconsin, Indiana, Pennsylvania and North Carolina.

1b   Lower margins of leaf with 1-4 cilia; lobes acuminate; underleaves present, cleft in two, ciliate margined. 2

2a   Cells of cilia much longer than broad. .. 3

2b   Cells of cilia about as broad as long. 4

3a   Leaves mostly 3-lobed. Fig. 417. .... *Barbilophozia hatcheri* (Evans) Loeske

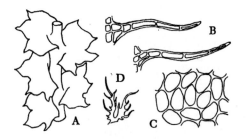

Figure 417

Figure 417 *Barbilophozia hatcheri*. A, shoot; B, cilia; C, cells of leaf; D, underleaf.

Plants in loose, green to brown patches; shoots 1-2.7 mm wide, up to 5 cm long, arctic-alpine, on vertical rocks, across northern North America south to California, Colorado, Minnesota, Michigan and New England, disjunct in North Carolina.

3b   Leaves mostly 4-lobed. .... *Barbilophozia lycopodioides* (Wallr.) Loeske

Plants in depressed, loose and flocculant mats, often mixed with *Dicranum* and other mosses, very robust, leafy stems 3-5 mm wide and up to 8 cm long, on acid soil or soil-covered rocks, high subarctic-subalpine, across northern North America south to Washington, New Mexico and New England.

4a   Leaves 2- or 3-lobed about 1/3 of length. 5

4b   Leaves 4-lobed about 1/2 of length. Fig. 418. .... *Barbilophozia quadriloba* (Lindb.) Loeske

Figure 418

Figure 418 *Barbilophozia quadriloba*. A, leaf; B, cilium; C, underleaf.

Plants slender, in loose tufts to dense patches, dull olive-green to brownish, shoots 0.8-1.3 mm wide, appearing terete, up to 4 cm high, arctic-alpine, on soil, humus, and rocks, across northern North America south to British Columbia, Michigan, Newfoundland.

5a Leaves predominantly 2-lobed. Fig. 419F. .... *Barbilophozia kunzeana* (Hüb.) Gams

Plants scattered, creeping to erect, among other spahgna or mosses, or forming loose tufts, yellow-brown to chestnut brown, shoots 0.6-0.8 mm wide, up to 5 cm long, arctic-alpine, peaty ledges and shallow soils, across northern North America south to British Columbia, Colorado, Minnesota, Michigan and New England.

5b Leaves predominantly 3-lobed. Fig. 419A-C. .... *Barbilophozia floerkei* (Web. & Mohr.) Loeske

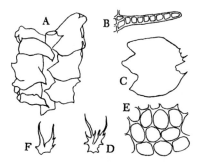

Figure 419

Figure 419 *Barbilophozia floerkei*. A, shoot; B, cilium; C, leaf; D, underleaf; E, cells of leaf; F, underleaf of *B. kunzeana*.

Plants in loose to compact tufts, pale or yellowish-green to dark green, often forming extensive masses, shoots 4-5 mm wide, up to 5 cm long, on humus and boulders in deep shade at high elevations, across northern North America south to Washington, New Mexico and upper New England.

Lophozia

There are at least 32 species of incredibly and complex genus recorded for North America.

1a Plants pale, opaque, bluish green in color. Fig. 420. .... *Lophozia incisa* (Schrad.) Dum.

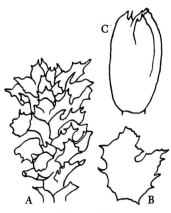

Figure 420

Figure 420 *Lophozia incisa*. A, shoot; B, leaf; C, perianth.

Plants in compact, often extensive patches, shoots small, 1-2.2 mm wide, up to 10 mm long, on moist acid rocks, logs, and humus, higher elevations and latitudes, across northern North America south to California, Nevada, New Mexico, Minnesota, Tennessee, and North Carolina.

1b Plants not bluish green in color. 2

2a Gemmae vinaceous to purplish or brownish purple. Fig. 421F. .... *Lophozia excisa* (Dicks.) Dum.

Class II. Hepaticae   255

Plants in small, compact tufts or patches, shoots 1.2-2.2 mm wide and up to 3 cm long or high, on acid soils, rocks, and humus, across northern North America south to California, Wyoming, Minnesota, Wisconsin, Tennessee, and North Carolina.

2b Gemmae with walls colorless walls, thus appearing greenish. 3

3a Trigones of leaf cells strongly bulging; on logs. Fig. 421E. .... *Lophozia guttulata* (Lindb. & H. Arnell) Evans

Plants usually green or pale green, shoots 0.8-1.5 mm wide, up to 2 cm long, almost always on decaying wood, across northern North America and south to California, Utah, Colorado, Minnesota, Wisconsin, Michigan, New York, and Connecticut.

3b Trigone of leaf cells not strongly bulging. Fig. 421A-C. .... *Lophozia ventricosa* (Dicks.) Dum.

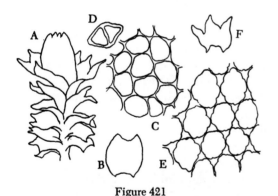

Figure 421

Figure 421 *Lophozia ventricosa*. A, shoot with perianth; B, leaf; C, cells of leaf; D, gemma; E, cells of *L. guttulata*; F, bract of *L. excisa*.

Tritomaria

Six species of *Tritomaria* are reported from North America.

1a Leaves as wide or wider than long; gemmae usually absent; shoots 2-3 mm wide. Fig. 422E. .... *Tritomaria quinquedentata* (Huds.) Buch

Plants prostrate, in patches or mats, green to yellowish-brown or pale brown, across northern North America south to British Columbia, Manitoba, Minnesota, Michigan, New York, and Connecticut.

1b Leaves distinctly longer than wide; red-brown gemmae constantly present; shoots 0.7-2 mm wide. 2

2a Gemmae angular. Fig. 422D. ....*Tritomaria exsectiformis* (Breidl.) Loeske

Growing in green to brownish patches, on decaying logs, peaty soil, and vertical rock faces, across northern North America south to British Columbia, Colorado, Iowa, Ohio, Pennsylvania, and North Carolina.

2b Gemmae ovoid, smooth. Fig. 422A-C. .... *Tritomaria exsecta* (Schrad.) Loeske

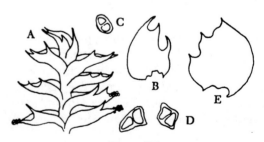

Figure 422

256 Class II. Hepaticae

Figure 422 *Tritomaria exsecta*. A, shoot; B, leaf; C, gemma; D, gemmae of *T. exsectiformis*; E, leaf of *T. quinquedentata*.

In dense, apple green to brownish-tinged patches, shoots 1.2-1.8 mm wide, up to 20 mm long, confined mainly to Spruce-Fir forests, on soil, rocks, peat, and decaying logs, across northern North America south to Washington, Colorado, Iowa, Tennessee, and Georgia.

Anastrophyllum

Anastrophyllum michauxii (Web.) Buch & Evans (Fig. 423)

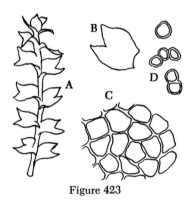

Figure 423

Figure 423 *Anastrophyllum michauxii*. A, shoot; B, leaf; C, cells of leaf; D, gemmae.

Plants robust, in extensive olive-green to brown mats or patches, shoots 1.0-2.5 mm wide, up to 5 cm long, gemmae often present, on vertical rock faces and decaying logs, across northern North America south to Washington, Wyoming, California, Minnesota, Michigan, Tennessee, and North Carolina. Eight other species are reported from North America.

Gymnocolea

Gymnocolea inflata (Huds.) Dum. (Fig. 424)

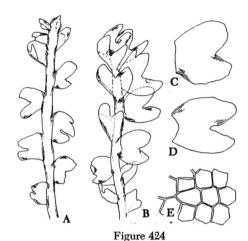

Figure 424

Figure 424 *Gymnocolea inflata*. A, shoot, ventral view; B, shoot, dorsal view; C-D, leaves; E, median leaf cells.

Plants green to brownish or purplish-black, in patches or growing over *Sphagnum*, shoots 0.75-1.5 mm wide, up to 25 mm long, on acidic rocks, other mosses in bogs, or decaying logs, across northern North America south to Oregon, Wyoming, South Dakota, Arkansas, Tennessee, and North Carolina. One other rare species, *G. acutiloba* (Schiffn.) K. Müll., occurs in Maine and Tennessee.

Mylia

1a Cuticle smooth; growing over Sphagnum. Fig. 425. ....
.... *Mylia anomala* (Hook.) S. Gray

Class II. Hepaticae

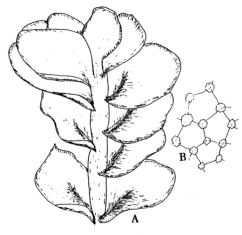

Figure 425

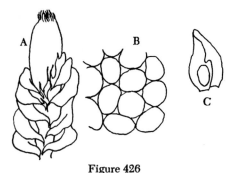

Figure 426

Figure 425 *Mylia anomala*. A, shoot, dorsal view; B, median leaf cells.

Plants prostrate, creeping or forming dense patches in peat, green to yellowish brown to fulvous, shoots, 2.4-3 mm wide and up to 3 cm long, in bogs, across northern North America south to Washington, Alberta, Wisconsin, Michigan, Pennsylvania, and West Virginia.

1b Cuticle sculptured into polygonal, coarse plates; on shaded rocks and decaying logs. .... *Mylia taylori* (Hook.) S. Gray

Robust, in carmine-red to purplish-brown, thick tufts or sods, shoots up to 5 mm wide and 12 cm long, northeastern North America south to Tennessee and North Carolina.

Jamesoniella

Jamesoniella autumnalis (DC.) Steph. (Fig. 426)

Figure 426 *Jamesoniella autumnalis*. A, plant with perianth and bracts; B, cells of leaf; C, leaf that bears an antheridium.

Plants dioicous; antheridia at the end of a special shoot, or in patches along the stem, in 4-6 pairs of bracts, each bract with 1 or 2 teeth on the upper margin; perianth on the end of a main shoot (these characters will distinguish this species from *Odontoschisma* or *Chiloscyphus*); sporophytes produced in September, common on moist sandstone or earth, usually with other bryophytes, British Columbia to Newfoundland south to Oregon, Montana, Wyoming, Arkansas, Mississippi, Alabama, and Florida.

Nardia

Nardia lescurii (Aust.) Underw. (Fig. 427)

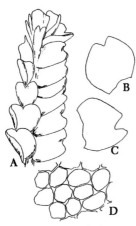

Figure 427

Figure 427 *Nardia lescurii.* A, shoot; B-C, leaves; D, median leaf cells.

Plants pale, pellucid green to purplish or rose-red, stems prostrate, 0.8-1.8 mm wide, up to 4 cm long, on damp loamy soil, humus, rocks, a "gorge" species, West Virginia and Virginia south to Tennessee and South Carolina. Five other species of *Nardia* are known from North America.

Jungermannia

Nineteen species of *Jungermannia* are reported from North America.

1a   Leaves of well developed sterile shoots circular or nearly so, widest at or near their middle. 2

1b   Leaves of well developed sterile shoots not circular, varying from rectangular to quadrate to ovate, distinctly longer than wide. 4

2a   Marginal cells of leaves 2-3 × the inner cell size, forming an obvious border, visible with a hand lens; cells thin-walled or equally thick-walled; rhizoids colorless. Fig. 428E. .... *Jungermannia gracillima* Sm.

Plants prostrate, in pellucid or whitish-green patches or sods, shoots laterally compressed, 0.9-1.3 mm wide, up to 15 mm long, on sandy, loamy, or clayey soil, or on rocks, Minnesota to Newfoundland south to Arkansas, Mississippi, Alabama, and Florida. In the northwest, *J. rubra* Gott. *ex* Underw. may key out to this species.

2b   Marginal cells of leaves 1-1.3 × the inner cell size; cells with distinct trigones; rhizoids purplish. 3

3a   Marginal leaf cells flat, not swollen; leaves below perianth and bracts usually undulate. Fig. 428A-C. .... *Jungermannia hyalina* Lyell

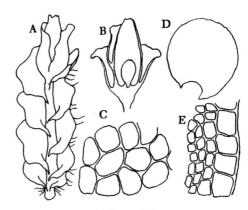

Figure 428

Class II. Hepaticae   259

Figure 428 *Jungermannia hyalina*. A, shoot with perianth; B, sction of perianth; C, cells of leaf; D, leaf of *J. gracillima;* E, marginal cells of same.

Plants in prostrate to ascending tufts, pale to yellowish to green, often carmine or reddish-pigmented on leaf bases, shoots 1.5-3.0 mm wide, up to 15 mm long, on shaded soil and soil covered rocks and on soil banks along streams, British Columbia to Newfoundland south to California, Arizona, New Mexico, Arkansas, Alabama, and Georgia.

3b Marginal leaf cells swollen, forming a turgid border that appears elevated above the rest of the leaf surface; leaves below perianth never undulate. ....
.... *Jungermannia crenuliformis* Aust.

Plants in compact prostrate patches or tufts, pellucid, pale to yellowish-green, leaf bases purplish-red, shoots 1.4-2.2 mm wide, up to 25 mm long, usually on acidic rocks along streams, Minnesota, Michigan, southern Ontario to New York and Massachusetts south to Kansas, Arkansas, Tennessee,, and Georgia.

4a Leaves rectangular to oblong-quadrate, cells usually with bulging trigones; perianth with a small apical beak set in a depressed-truncate mouth. Fig. 429. ....
.... *Jungermannia leiantha* Grolle

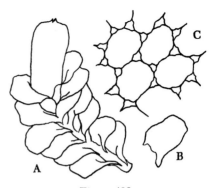

Figure 429

Figure 429 *Jungermannia leiantha*. A, plant; B, leaf; C, cells of leaf. In thin mats among mosses.

Plants in prostrate mats or patches, light green to slightly brownish, shoots 2.0-2.5 mm wide, up to 3 cm long, on damp, shaded sandstone rocks, sandy banks and soil covered rocks near streams, Alaska to Newfoundland south to California, Utah, Kansas, Arkansas, Alabama, and Georgia.

4b Leaves elliptical to ovate or cordate, usually without or with only minute trigones; perianth not beaked. 5

5a Leaves of sterile stems oval to only slightly cordate; plants small, up to 0.5-1.0 cm long; paroecious. Fig. 430E. ....
.... *Jungermannia pumila* With.

Plants small, in prostrate, flat, thin patches, olive-green to blackish, shoots 1.0-2.0 mm wide, up to 1 cm long, on damp shaded rocks, British Columbia to Newfoundland south to California, Colorado, Iowa, Tennessee, and North Carolina.

5b Leaves of sterile stems broadly cordate-ovate; plants larger, 2.5-12 cm long; dioecious. Fig. 430A-D. ....
.... *Jungermannia exsertifolia* Steph.

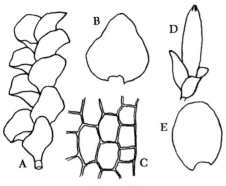

Figure 430

Figure 430 *Jungermannia exsertifolia*. A, plant; B, leaf; C, cells of leaf; D, perianth; E, leaf of *J. pumila*.

Plants in brownish-green to purplish-brown (or black) spongy tufts, shoots 2-4 mm wide, up to 12 cm long, on wet sloping rocks near streams, Alaska to Labrador and Newfoundland south to California, Colorado, Wisconsin, Michigan, New York and Connecticut.

GYMNOMITRIACEAE
Marsupella

Fourteen species of *Marsupella* are reported from North America.

1a   Leaves with sinus descending 1/4 to 1/2 the leaf length; dorsal leaf margin not reflexed. .... *Marsupella sphacelata* (Gieseke) Dum.

Plants erect to suberect, in clear to dull green, spongy patches, up to 7 cm high in aquatic phases, on rocks along and in small streams and brooks, Alaska to Newfoundland south to California, Montana, Ontario, Pennsylvania, Arkansas, Tennessee, and Georgia.

1b   Leaves with sinus descending to only 1/4 the leaf length; dorsal leaf margin reflexed. Fig. 431. .... *Marsupella emarginata* (Ehrh.) Dum.

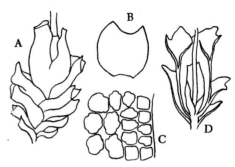

Figure 431

Figure 431 *Marsupella emarginata*. A, shoot; B, leaf; C, cells of leaf; D, section of perianth and bracts.

Plants in large patches on damp wet rocks, green to olive-green in shade, reddish-brown in sun, shoots 1.6-2.4 mm wide, up to 5 cm long, Alaska to Newfoundland south to California, Montana, Colorado, Minnesota, Michigan, Tennessee, and Georgia.

SCAPANIACEAE
Diplophyllum

Nine species of *Diplophyllum* are reported from North America.

1a   Leaf lobes with a sharply defined vitta of very long linear cells. Fig. 432A-C. .... *Diplophyllum albicans* (L.) Dum.

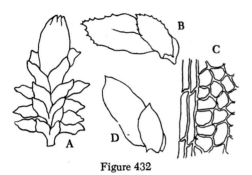

Figure 432

Figure 432 *Diplophyllum albicans*. A, shoot with perianth; B, leaf; C, leaf-cells; D, leaf of *D. apiculatum*.

Plants in compact to loose patches, 2-5 cm high, green to golden-yellow to deep brown, shoots 1.5-3.5 mm wide, on damp rocks of coastal areas, Alaska to Oregon, Newfoundland to Maine.

1b Leaf lobes without a sharply defined vitta of very long cells. 2

2a Ventral leaf lobes distinctly apiculate. Fig. 432D. .... *Diplophyllum apiculatum* (Evans) Steph.

Shoots in green to brownish or reddish-brown patches, 1.5-2.4 mm wide, up to 10 mm long, on moist banks and non-calcareous rock ledges, throughout the deciduous forests of eastern North America west to Minnesota, Kansas, Oklahoma.

2b Ventral leaf lobes rounded at apex. .... *Diplophyllum taxifolium* (Wahlenb.) Dum.

Plants in thin mats, green to yellowish-green, shoots 1-2.4 mm wide, up to 3-5 cm long, on moist non-calcareous rocks, boreal-subarctic, Alaska to Newfoundland south to California, Montana, Minnesota, Tennessee, and North Carolina.

Scapania

At least 41 species of *Scapania* have been reported from North America.

1a Basal margin of smaller (upper) leaf-lobe with long, branched cilia. Fig. 433. .... *Scapania bolanderi* Aust.

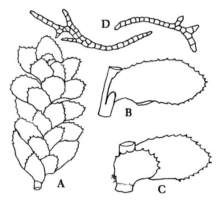

Figure 433

Figure 433 *Scapania bolanderi*. A, shoot; B, leaf and under side of stem; C, leaf and upper side of stem; D, cilia from margin of leaf.

Shoots to 8 cm long, leaves to 1 mm, common, on logs and stumps, Washington and Oregon. The teeth of the leaves are like those of other species; the cilia are unique.

1b Basal margin entire or finely toothed, not ciliate. 2

2a Lower (larger) lobe of leaf with sharply toothed margin; both upper and lower lobes decurrent. Fig. 434. .... *Scapania nemorosa* (L.) Dum.

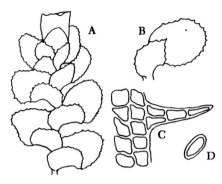

Figure 434

Figure 434 *Scapania nemorosa*. A, shoot; B, leaf and upper surface of stem; C, margin of leaf; D, gemma.

Plants polymorphic, in green to yellow-green mats, 1.5-5.6 mm wide and up to 10 cm long, teeth of leaf mostly 2-3 celled, on soil and rocks along streams and on tree bases, Alberta to Newfoundland south to Wyoming and Texas to Florida.

2b Lower lobes entire or wavy or with a few minute teeth at apex; upper lobe not decurrent. 3

3a Leaves with lower lobe narrow, longer than wide, not decurrent; plants small, mostly 0.8-1.5 cm long. .... *Scapnia curta* (Mart.) Dum.

Plants usually light green, shoots 1.3-2.5 mm wide, up to 15 mm long, on bare or soil-covered rocks, humus, Alaska to Newfoundland south to California, Colorado, Minnesota, Wisconsin, Michigan, and New York.

3b Leaves with lower lobe broad, ovate to sub-orbicular; plants larger, mostly 1.2-10 cm long. 4

4a Leaves with lower half not long-decurrent at base; marginal leaf-cells with small to distinct trigones. Fig. 435. .... *Scapania irrigua* (Nees) Gott. et al.

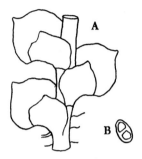

Figure 435

Figure 435 *Scapania irrigua*. A, shoot; B, gemma.

Plants in yellowish-green to brown mats, shoots 2-4 mm wide, up to 5 cm long, often in standing water, on humus, logs in bogs, Alaska to Newfoundland south to California, Minnesota, Pennsylvania, and New Jersey.

4b Leaves with lower half long-decurrent at base; marginal leaf cells without trigones, but with walls equally thickened; leaf lobes with wavy margins. Fig. 436. .... *Scapania undulata* (L.) Dum.

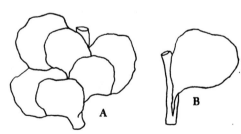

Figure 436

Class II. Hepaticae   263

Figure 436 *Scapania undulata.* A, shoot; B, leaf and lower side of stem.

An exceedlingly polymorphic species, shoots ranging from 2-4.5 mm wide and up to 20 cm long, in green to whitish-green mats, from aquatic to dry sites, on soil, rocks, Alaska to Nwfoundland south to California, Arizona, Colorado, the Ozarks, Tennessee, and Georgia. Easily confused with *S. nemorosa* from which it may be separated by its transverse, non-decurrent insertion of the dorsal lobe.

RADULACEAE
Radula

Twelve species of Radula are reported from North America.

1a Dorsal lobe of leaf only 1.5-1.7 × the length of the ventral lobe, i.e., the ventral lobe large; gemmae absent; Pacific Northwest. .... *Radula bolanderi* Gott.

Plants in dark green to yellowish patches, shoots 0.9-1.1 mm wide, up to 2 cm long, on moist rocks, bark of trees, Alaska to California.

1b Dorsal lobe 1.4-2.5 × as long as the ventral lobe. 2

2a Many leaves with dorsal lobes broken off; gemmae absent. Fig. 437B. .... *Radula obconica* Sull.

Plants in olive-green to yellow-green patches, shoots 0.5-1 mm wide, up to 1 cm long, on moist rock and tree trunks, Minnesota to Maine south to the Gulf States.

2b Leaves not broken; gemmae usually present. 3

3a Ventral lobe arching entirely across the stem, the tip broadly rounded, perianths absent. Fig. 437A. .... *Radula mollis* Lindenb. & Gott.

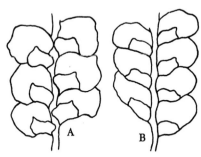

Figure 437

Figure 437 *Radula mollis.* A, under side of shoot; B, under side of shoot of *R. obconica.*

Plants in pale to dark green mats, shoots 1.3-2 mm wide, up to 2.5 cm long, on shaded moist rocks, the Ozarks to West Virginia south to Florida.

3b Ventral lobes arching about half way across stem, the tip acute to obtuse; perianths common. Fig. 438. .... *Redula complanata* (L.) Dum.

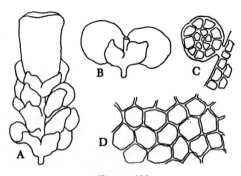

Figure 438

Figure 438 *Radula complanata*. A, shoot with perianth; B, leaves from beneath; C, gemma; D, cells of leaf.

Plants in pale to bright yellow patches, shoots 2.25-2.5 mm wide, up to 5 cm long, gemmae common on edges of leaves, on tree trunks and rocks in moist areas, widespread in North America.

PORELLACEAE
Porella

Nine species of Porella are recognized in North America.

1a   Trigones large (in old leaves), bulging into the cells; plants glossy; western North America. 2

1b   Trigones small; surface of plants dull green, not glossy. 4

2a   Underleaf greatly ruffled along its decurrent portion. Fig. 439D. .... *Porella cordaeana* (Hüb.) Moore

Plants in glossy, yellow-green to dark green patches, shoots 3-4 mm wide, up to 10 cm long, in moist sites, on rocks and trees, Alaska to Idaho and California.

2b   Underleaf not or only slightly ruffled at base. 3

3a   Most underleaves 1-1 1/2 × as wide as adjacent ventral lobes. Fig. 439A-C. .... *Porella navicularis* (Lehm. & Lindenb.) Lindb.

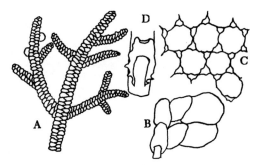

Figure 439

Figure 439 *Porella navicularis*. A, shoot from above; B, underleaves and underlobes; C, cells of leaf; D, underleaf of *P. cordaeana*.

Plants in yellowish-green to brown dense, coarse mats, shoot 1.5-3 mm wide, up to 12 cm long, in moist shaded sites on trees, logs, and rocks, Alaska south to Idaho, Montana, and California.

3b   Most underleaves twice as wide as adjacent ventral lobes. .... *Porella roellii* Steph.

Plants in shining green to yellowish-brown patches, shoots distinctly complanate, 1.5-2.1 mm wide, up to 8 cm long, on moist rocks, rarely soil and trees, Alaska to Idaho and California.

4a   Underleaves narrow, no wider than stem, flat, not decurrent; ventral lobes small, tongue shaped; aquatic. Fig. 440. .... *Porella pinnata* L.

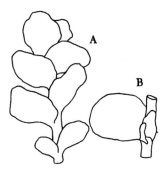

Figure 440

Figure 440 *Porlla pinnata.* A, shoot; B. underleaf and underlobe.

Plants aquatic, in dark green to almost black patches, shoots 2-4 mm wide, up to 8 cm long, on rocks and wood along streams, widespread throughout the eastern United States west to Minnesota, Oklahoma, and Texas.

4b   Underleaves decidedly wider than stem, decurrent; ventral lobes ovate, large, wider than stem. 5

5a   Ventral lobes about as wide as underleaves, broadly rounded at apex. Fig. 441A-B. .... *Porella platyphylloidea* (Schwein.) Lindb.

Figure 441 *Porella platyphylloidea.* A, shoot; B, underlobes and underleaves; C, same of *P. platyphylla;* D, perianth.

Leafy shoots 2.5-4 mm wide, in dull yellow-green to brownish-green mats, 1-2 pinnate, on rocks and trees, widespread in North America.

5b   Ventral lobes distinctively narrower than underleaves, somewhat tapering to apex. Fig. 441C-D. ....
.... *Porella platyphylla* (L.) Pfeiff.

Plants in dull yellow green to brownish patches, shoots 1-2.5 mm wide, up to 8 cm long, 2-3 pinnate, on rocks, trees, and soil, widespread in North America.

JUBULACEAE
Frullania

There are at least 22 species of *Frullania* reported from North America.

1a   Plants with erect leafless shoots ("flagella"). Fig. 442A-B. ....
.... *Frullania bolanderi* Aust.

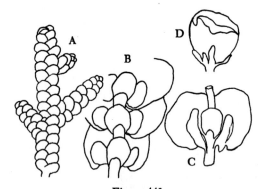

Figure 441

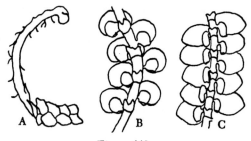

Figure 442

266   Class II. Hepaticae

Figure 442 *Frullania bolanderi*. A, plant with "flagellum"; B, under side of shoot; C, under side of shoot of *F. kunzei*.

Plants in green to reddish-brown patches, shoots about 0.5 mm wide, up to 3 cm long, on trees and vertical rocks, British Columbia to Quebec and south to California, Minnesota, and Iowa.

1b   Plants without leafless shoots. 2

2a   Apex of dorsal (upper) lobe of stem leaf acute; paracysts sometimes present. Fig. 443E. .... *Frullania tamarisci* (L.) Dum. var. *nisquallensis* (Sull.) Hatt.

Plants in yellowish-green to reddish-brown patches, shoots 1-1.3 mm wide, up to 4 cm long, on trees, rocks, and shrubs, Alaska to California.

2b   Apex of dorsal lobe of stem leaf rounded. 3

3a   Paracysts present on at least some leaves. 4

3b   Paracysts absent. 8

4a   Mature or well developed underleaves of stem cordate to auriculate at base. 5

4b   Mature or well developed underleaves cuneate to rounded at base. 6

5a   Paracysts scattered, absent in many leaves. ....
...... *Frullania californica* (Aust.) Evans

Plants yellowish-green to reddish-brown, shoots about 0.5 mm wide, up to 5 cm long, on logs, trees, and rocks, British Columbia to California.

5b   Paracysts in a row, resembling a midrib. Fig. 443A-D. ....
.... *Frullania tamarisci* (L.) Dum. var. *asagrayana* (Mont.) Hatt.

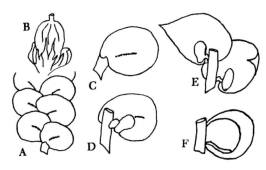

Figure 443

Figure 443 *Frullania tamarisci* var. *asagrayana*. A, shoot seen from above; B, perianth seen from beneath; C, leaf seen from above; D, leaf and underlobe; E, same of *F. tamarisci* var. *nisquallensis*; E, leaf and underlobe of *F. oakesiana*.

Plants in patches or pendulous, green to reddish-brown, shoots 0.7-1.4 mm wide, up to 5 cm long, on tree trunks, rocks, decaying wood, Minnesota to Newfoundland south to the Gulf States.

6a   Plants of the west coast of North America. .... *Frullania franciscana* M. A. Howe

Plants in greenish to reddish-brown patches, shoots 0.9-1.2 mm wide, up to 5 cm long, on trees, fences, and rocks, Alaska to California.

6b   Plants of eastern North America. 7

7a Plants of the southeastern United States; paracysts rare and scattered. Fig. 442C. .... ***Frullania kunzei*** **Lehm. & Lindenb.**

Plants in reddish-brown to blackish-red patches, 0.4-0.6 mm wide, up to 2 cm long, on trees, shrubs, rocks, sometimes creeping over lichens, Arkansas to North Carolina south to the Gulf States.

7b Plants of the northeastern United States; paracysts in rows. .... ***Frullania selwyniana*** **Pears.**

Plants in reddish-brown to brown patches, shoots about 1 mm wide, up to 2 cm long, nearly always fertile and with female bracts dentate, on trees, Minnesota to Quebec and New England south to Ohio and Michigan.

8a Lobule (underlobe) averaging 1-1/2-3 × as long as wide. 9

8b Lobule averaging less than 1-1/2 × as long as wide. 10

9a Lobule usually more than their width from the stem, diverging at a wide angle from it. .... ***Frullania obcordata*** **Lehm. & Lindenb.**

Plants in yellowish-green to reddish-brown patches, shoots 0.8-1.2 mm wide, up to 4 cm long, on trees, North Carolina to the Gulf States.

9b Lobules usually less than their width from the stem, nearly parallel with it. .... (p. 268) ***Frullania kunzei*** **Lehm. & Lindenb.**

10a Underleaves cordate at base. .... ***Frullania plana*** **Sull.**

Plants in green to brown-green patches, shoots 1-1.3 mm wide, up to 3 cm long, on shaded rocks, Michigan to Nova Scotia south to Florida.

10b Underleaves narrowed or rounded at base, the margins not decending lower than point of insertion. 11

11a Lobule large, obscuring more than half of upper lobe. Fig. 443F. .... ***Frullania oakesiana*** **Aust.**

Plants in greenish to reddish-brown patches, shoots about 0.75 mm wide, up to 2 cm long, on trees, Minnesota to Newfoundland south to Michigan, Vermont, and New Hampshire.

11b Lobule smaller, not obscuring more than half of the upper lobe. 12

12a Dorsal lobes of mature leaves truncate at base, leaf cells with or without trigones, intermediate thickenings absent. Fig. 444A-D. .... ***Frullania inflata*** **Gott.**

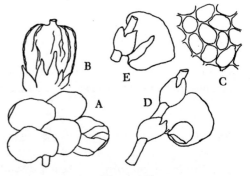

Figure 444

Figure 444 *Frullania inflata*. A, shoot seen from above, with antheridial branch; B, perianth seen from beneath; C, cells of leaf; D, underlobe and underleaf; E, underlobe and underleaf of *F. riparia*.

Plants dull green to tinged with red or brown, in loose tufts, shoots 1-1.3 mm wide, up to 3 cm long, on rocks and trees, often intermingled with other bryophytes, widespread in Notrh America.

12b Dorsal lobes of mature leaves cordate or auriculate at base; leaf cells normally with trigones and often with intermediate thickenings. 13

13a Underleaves small to moderate size, 1-2 × as wide as stem, about equal in size to lobule; lobules normally all inflated. .. 14

13b Underleaves very large, 2.5-4 × as wide as the stem and 2-3 × as large as the lobule; lobules frequently flat, not sac-like. 15

14a Lobules inflated throughout; cells with numerous intermediate thickenings. Fig. 445A-D. .... *Frulliana eboracensis* Gott.

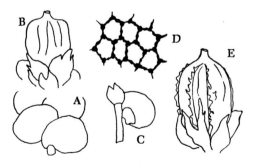

Figure 445

Figure 445 *Frullania eboracensis*. A, shoot seen from above; B, perianth seen from below; C, leaf and underleaf seen from beneath; D, cells of leaf; E, perianth of *F. squarrosa*.

Plants in patches, green, brown, to reddish or blackish, shoots 0.4-0.8 mm wide, up to 3 cm long, on trees, stumps, and logs widespread in the eastern United States.

14b Lobules collapsed near their stalks; intermediate thickenings few. .... *Frullania brittoniae* Evans

Plants in greenish to reddish-brown patches, shoots 0.75-1.1 mm wide, up to 2.5 cm long, on rocks and trees, Minnesota to New England south to New Mexico and the Gulf States.

15a Leaves nearly flat to slightly ascending when wet; lobules generally flat; plants usually green. Fig. 444E. .... *Frullania riparia* Hampe *ex* Lehm.

Shoots 0.9-1.2 mm wide, up to 1.5 cm long, in patches, on trees, rocks, Minnesota to England south to New Mexico, Oklahoma, Alabama, Tennessee, and North Carolina.

15b Leaves strongly squarrose and standing erect when set; curved down and clasping them when dry; lobules generally inflated; plants usually reddish-brown in sun. Fig. 445E. .... *Frullania squarrosa* (Reinw. et al.) Dum.

Plants in wide mats, shoots 0.8-1.3 mm wide, up to 5 cm long, on rocks, trees, and logs, widespread in the mid-west and eastern North America.

Jubula

Jubula pennsylvanica (Steph.) Evans (Fig. 446)

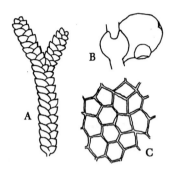

Figure 446

Figure 446 *Jubula pennsylvanica*. A, shoot; B, underleaf and underlobe; C, cells of leaf.

Plants in patches, dark green, shoots 1-2 mm wide, up to 2 cm long, dorsal lobes pointed, on wet shaded rocks or soil, eastern North America from Ohio to Nova Scotia south to Oklahoma, Arkansas, Alabama, and Georgia.

LEJEUNEACEAE
Ptychocoleus

Ptychocoleus heterophyllus Evans (Fig. 447A-B)

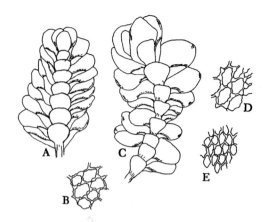

Figure 447

Figure 447 *Ptychocoleus heterophyllus*. A, shoot, ventral view; B, median leaf cells. *Mastigolejeunea auriculata*. C, shoot, ventral view; D-E, variation in median leaf cells.

Plants in patches or mats, yellowish-green to brownish green, leafy shoots complanate, 1-1.3 mm wide, up to 15 mm long, underleaves unlobed, on tree trunks, central Florida.

Mastigolejeunea

Mastigolejeunea auriculata (Wils. & Hook.) Schiffn. (Fig. 447C-E)

Plants in patches, glaucous green to brownish or purplish, stems prostrate to pendent, shoots 0.7-2 mm wide, up to 5 cm long, leaves closely imbricate, underleaves narrowed to insertion, rounded to emarginate, on trees, logs, or rocks, Florida and Louisiana.

Lopholejeunea

Lopholejeunea muelleriana (Gott.) Schiffn. (Fig. 448A-B)

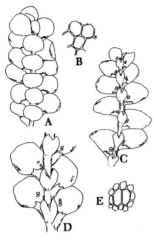

Figure 448

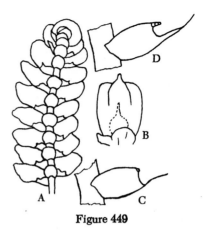

Figure 449

Figure 448 *Lopholeeunea muelleriana*. A, shoot, ventral view; B, median leaf cells. *Ceratolejeunea laetefusca;* C, small branch shoot, ventral view; D, portion of shoot, ventral view; E, ocelli.

Plants in blackish (with almost bluish cast) patches or mats, shoots 0.9-1.1 mm wide, leaves contiguous to loosely imbricate, undersleeves rounded to insertion, on trees, North Carolina and Florida. *L. subfusca* (Nees) Schiffn. is also found in these same localities.

Leucolejeunea

Four species of *Leucolejeunea* are found in North America.

1a Margin of ventral lobe straight or slightly curved from tip to keel, apical tooth 1-2 cells long. Fig. 449A-C. ....
.... *Leucolejeunea clypeata* (Schwein.) Evans

Figure 449 *Leucolejeunea clypeata*. A, lower side of shoot; B, perianth; C, margin and tooth of underlobe; D, margin and tooth of *L. unciloba*.

Plants in pale to whitish-green mats, shoots 0.5-1.2 mm wide, up to 2 cm long, leaves imbricate, underleaves distant, unlobed, orbicular, on rocks and trees, Michigan to New England south to Oklahoma and the Gulf States.

1b Margin of ventral lobe deeply curved from tip to keel, the apical tooth 3-6 cells long, curved. Fig. 449D. ....
.... *Leucolojeunea unciloba* (Lindenb.) Evans

Plants in pale glaucous green to whitish mats, shoots 1.25-2.5 mm wide, up to 2 cm long, leaves closely imbricate, underleaves contiguous to imbricate, unlobed, broadly orbicular or reniform, on trees and occasionally rocks, Missouri to Rhode Island and Delaware south to Texas and the Gulf States.

Ceratolejeunea

Ceratolejeunea laetefusca (Aust.) Schust. (Fig. 448C-E)

Plants in depressed mats, yellowish-brown to olive-brown, translucent when moist, shoots 0.8-1.1 mm wide, up to 2.5 cm long, leaves contiguous to weakly imbricate, paracysts usually 2-3, lobule strongly inflated, underleaves distant, broadly ovate to orbicular, bilobed half their length, on trees, South Carolina, Florida, Mississippi, and Louisiana. Two other species are known from Florida.

Rectolejeunea

Rectolejeunea maxonii Evans (Fig. 450A-C)

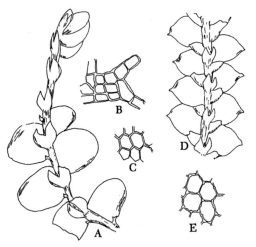

Figure 450

Figure 450 *Rectolejeunea maxonii*. A, shoot showing cadaucous leaves; ventral view; B, underlobe of leaf; C, median leaf cells. *Crossotolejeunea bermudiana;* D, shoot, ventral view; F, median leaf cells.

Plants in pale green, whitish, or yellowish mats, shoots 0.5-1.1 mm wide, leaves loosely imbricate, underleaves distant, 2-lobed about 1/2 their length, leaf cells thin walled, trigones minute, paracysts absent, on trees, Arkansas, Tennessee, and North Carolina south to Florida and Alabama. Three other species are known from Florida.

Crossotolejeunea

Crossotolejeunea bermudiana Evans (Fig. 450D-E)

Plants in yellowish-green to pale green tufts, shoots 0.5-0.9 mm wide, leaves distant to loosely imbricate, without paracysts, leaf cells thin-walled with small trigones and numerous intermediate thickenings, underleaves distant, 2-lobed about 1/2 their length, on soil, rocks, and tree trunks, Tennessee and Florida.

Drepanolejeunea

Drepanolejeunea appalachiana Schust. (Fig. 451A)

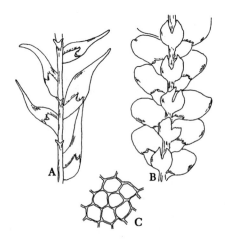

Figure 451

272   Class II. Hepaticae

Figure 451 *Drepanolejeunea appalachiana.* A, shoot, ventral view (after Frye & Clark, Hepaticae of North America, Part V, 1947, Univ. of Washington Publication in Biology). *Lejeunea flava;* B, shoot, ventral view; C, median leaf cells.

Plants in yellowish-green patches, shoots small, less than 0.5 mm wide, up to 1 cm long, leaves distant to imbricate, dorsal lobes pointed, 1-2 paracysts present, underleaves distant, 2-lobed 1/3-1/2 their length, tips widely spreading, on trees and logs, North Carolina and Tennessee.

Lejeunea

At least 17 species of *Lejeunea* are reported from North America.

1a Plants small, often filiform, less than 0.5 mm wide, leaves less than 300 μ long. .. 2

1b Plants larger, 0.6-1.6 mm wide, leaves longer than 300 μ. 3

2a Lobules well developed; dorsal lobe moderately to strongly convex dorsally, oval to orbicular to oblong-ovate. Fig. 452A & C. ....
.... *Lejeunea ulicina* (Tayl.) Gott.

Figure 452 *Lejeunea ulicina.* A, under side of leaf; eastern United States and Europe; B, leaf and underlobe of *L. laetivirens;* C, underleaf of *L. ulicina.*

Plants scattered, or in patches, green, shoots 350-550 μ wide, 4-8 mm long, leaves distant to contiguous, on trees, rocks, or on other bryophytes, Ohio to Nova Scotia south to Florida.

2b Lobules usually poorly developed; dorsal lobes plane, ovate. Fig. 452B. ....
.... *Lejeunea laetivirens* Nees & Mont.

Plants scattered or in tufts, pale green, shoots 280-460 μ wide, 2-5 mm long, leaves distant to imbricate, on trees, logs, and sandy soil, Kentucky and West Virginia south to the Gulf States.

3a Plants of the Appalachian Mountains and the Northeastern States. 4

3b Plants restricted by and large to the Coastal Plain of the Southeastern United States. 5

4a Leaves not notched at juncture of keel and dorsal lobe; underleaves clearly larger than lobules. Fig. 453A-C. ....
.... *Lejeunea cavifolia* (Ehrh.) Lindb. emend. Buch

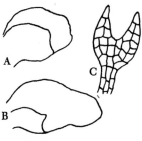

Figure 452

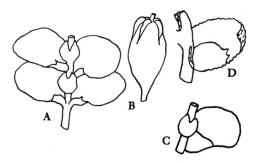

Figure 453

Figure 453 *Lejeunea cavifolia.* A, under side of shoot; B, perianths; C, under side of shoot of *L. lamacerina* subsp. *geminata;* D, leaf and underlobe of *Cololejeunea biddlecomiae.*

Plants in flat patches or mats or creeping over other bryophytes, clear to pale green, shoots 0.9-1.2 mm wide, up to 2 cm long, leaves moderately imbricate, on tree trunks and rocks, Minnesota and Ontario south to Tennessee and North Carolina.

4b Leaves more or less notched at juncture of keel and dorsal lobe; underleaves not or barely as large as lobules. Fig. 453C. .... ***Lejeunea lamacerina*** **Nees & Mont. subsp. *gemminata* Schust.**

Plants pale to whitish green, in loose patches, shoots 850-1200 μ wide, up to 15 mm long, leaves contiguous to weakly imbricate, on dry to damp shaded rocks and bark of trees, Newfoundland and south to Arkansas, Tennessee, and Georgia.

5a Lobules always well developed; underleaves contiguous to imbricate. Fig. 451A-C. .... ***Lejeunea flava*** **(Sw.) Nees subsp. *flava***

Plants in pale or whitish-green patches, more or less shiny when dry, shoots 0.9-1.3 mm wide, up to 5 cm long, leaves slightly to moderately imbricate, lobules well developed, on bark, more rarely on rocks or as an epiphyte on living leaves, Coastal Plain from Virginia to Texas.

5b Lobules, at least in part, obsolete; underleaves distant. .... ***Lejeunea cladogyna*** Evans

Plants in loose mats, green to yellowish or brownish with age, strongly shiny when dry, shoots 0.5-0.8 mm wide, leaves slightly distant to weakly imbricate, underleaves distant, tree trunks and exposed roots in shaded gullies, ravines, and in dense hammocks, Florida and Mississippi.

Diplasiolejeunea

Diplasiolejeunea rudolphiana Steph. (Fig. 454A)

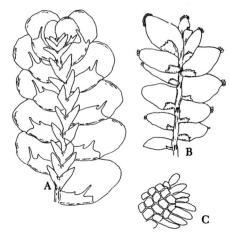

Figure 454

Figure 454 *Diplasiolejeunea rudolphiana.* A, shoot, ventral view. *Cololejeunea cardiocarpa;* B, shoot, ventral view; C, apex of leaf showing hyaline cells.

Plants in flat, pale yellow-green patches, shoots 1.1-2.0 mm wide, leaves clearly imbricate, underleaves large, distant to subimbricate, one per lateral leaf, bilobed to middle with lobes obliquely spreading, on tree bark and living leaves of evergreen trees, central and southern Florida.

Cololejeunea

Twelve species of *Cololejeunea* are reported from eastern North America with most restricted to the southeastern United States.

1a Mature leaves tipped by a cluster of finger-like hyaline cells. Fig. 454B-C. .... *Colojeleunea cardiocarpa* (Mont.) Schust.

Plants delicate, in closely appressed, flat, yellow-green patches, shoots 0.8-1.2 mm wide, underleaves absent, leaves widely spreading, somewhat imbricate, on bark and living leaves of broad-leaved evergreens, Coastal Plain, North Carolina to Florida and Louisiana.

1b Mature leaves not tipped by a cluster of hyaline cells. Fig. 453D. .... *Cololejeunea biddlecomiae* (Aust.) Evans

Plants scattered, creeping, in yellow-green patches, shoots 0.5-0.7 mm wide, up to 9 mm long, cells of dorsal lobe tuberculate, leaves distant to barely imbricate, underleaves absent, on bark of trees, rocks, and creeping over other bryophytes, Minnesota and Ontario to Quebec and Nova Scotia south to Oklahoma, and Arkansas, Mississippi and Georgia.

Order Metzgeriales

CODONIACEAE
Fossombronia

Ten species of *Fossombronia* are reported from North America.

1a Elaters few and imperfect or none; spores 36-46 μ, pale, reticulate, with 6-7 meshes across the spore. .... *Fossombronia cristula* Aust.

Plants gregarious, small, delicate green, shoots up to 5 mm long, leaves contiguous, broadly ligulate, on moist sand, Michigan to southern New England south to Ohio, West Virginia and New Jersey, also in Texas and California.

1b Elater perfect, with spiral bands. 2

2a Spores with ridges, not forming a network. Fig. 455A-B. .... *Fossombronia wondraczekii* (Corda) Dum.

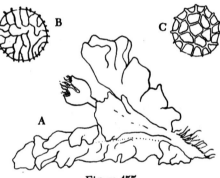

Figure 455

Figure 455 *Fossombronia wondraczekii.* A, plant; B, spore; C, spore of *F. foveolata*.

Plants large, shoots up to 10 mm long, in small pale green patches, upper leaves crowded, oblong-quadrate, sinuate-crispate, on bare moist soil of cultivated fields and paths or margins of ditches and ponds, Minnesota and Ontario to New Brunswick south to Indiana, Tennessee, and Florida.

2b Spores covered with a distinct network with 5-6 meshes across the spore. Fig. 455C. .... *Fossombronia foveolata* Lindb.

Plants in pale green patches, strongly odorous, stems up to 2 cm long, leaves obliquely obcuneate, sinuate-lobed, on moist sandy soil and in rocks crevices, British Columbia to Nova Scotia south to Washington, Texas, Louisiana, Alabama, and North Carolina.

Petalophyllum

Petalophyllum ralfsii (Wils.) Nees & Gott. (Fig. 456)

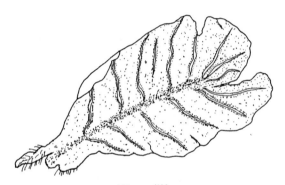

Figure 456

Figure 456 *Petallophyllum ralfsii*. Thallus showing dorsal lamellae (after Frye & Clark, *Hepaticae of North America*, Part I, 1937, Univ. of Washington Publications in Biology).

Plants in small pale green patches, up to 1 cm long, erect dorsal lamellae on the lamina, on wet to dry sand, Texas.

PALLAVICINIACEAE
Pallavicinia

Pallavicinia lyellii (Hook.) Carruth. (Fig. 457)

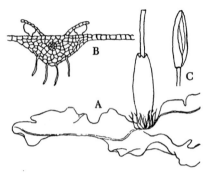

Figure 457

Figure 457 *Pallavicinia lyellii*. A, plant; B, section of antheridial shoot; C, capsule.

Thalli prostrate in thin patches, pale green, thalli 2-6 cm long, 4-5 mm wide, with a central strand, on wet banks along streams and in swamps, Minnesota to Newfoundland south to Florida, Alabama, Mississippi and Louisiana.

Moerckia

Moerckia hibernica (Hook.) Gott. (Fig. 458)

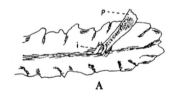

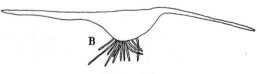

Figure 458

Figure 458 *Moerckia hibernica*. A, portion of female plant with involucre (i) and perianth (p); B, cross section of thallus (note absence of central strand).

Plants in pale green patches, thalli about 3.5 cm long, 2-4 mm wide, forked, without a central strand, on moist sand and wet soil in boggy places, Alaska to Newfoundland, south to Washington, Minnesota, Nebraska, Wisconsin, Michigan, and Connecticut.

PELLIACEAE
Pellia

1a Thallus with a flap-like, low involucres; thallus in longitudinal section with obvious bands of thickenings. 2

1b Thallus with erect tubular involucres, ciliate or lacinate at mouth; thallus lacking banks of thickenings. ....
.... *Pellia endiviifolia* (Dicks.) Dum.

Thalli dark green, 4-8 mm wide, up to 2.3 cm long, often repeatedly forking, on soil and rocks in wet places, Alaska to Nova Scotia south to California, Colorado, Michigan, North Carolina.

2a Plant paroecious; antheridia a short distance behind the female involucre. Fig. 459A-B. .... *Pellia epiphylla* (L.) Corda

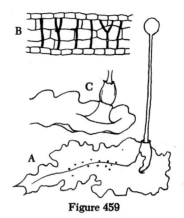

Figure 459

Figure 459 *Pellia epiphylla*. A, plant; B, section of thallus with thickened bands; C, fruiting tip of *P. neesiana* showing cylindric involucre.

Thalli dark green, often tinged with red, 10-15 mm wide, 1-7 cm long, branches few, on wet ground and rocks, in springy places, widespread in North America.

2b Dioecious; antheridia on separate plants. Fig. 459C. ....
.... *Pellia neesiana* (Gott.) Limpr.

Thalli dark green and usually tinged with red, translucent near border, dichotomously branching, 3-7 mm wide, on wet springy ground, widespread in North America.

METZGERIACEAE
Metzgeria

Eight species of *Metzgeria* are reported from North America.

1a Thallus with hairs on both upper and lower surfaces. ....
.. *Metzgeria pubescens* (Schrank) Raddi

Thalli grayish or yellowish-green, 2-3 cm long, 1-2 mm wide, on shaded rocks and tree trunks, Alaska to Oregon and Montana, New England.

1b Thallus without hairs on upper surface. 2

2a Underside of midrib 2 cells wide; marginal hairs in pairs, curved and hooked; hairs confined to margins and midrib. .... *Metzgeria leptoneura* Spruce

Thalli pale green to yellowish-green to light green, 1.5-2.5 mm wide, up to 10 cm long, irregularly dichotomously branched, on trees and rocks, Alaska, Tennessee, and North Carolina.

2b Underside of midrib 3-7 cells wide. 3

3a Marginal hairs in pairs; thallus 2 mm wide. Fig. 460D. .... *Metzgeria conjugata* Lindb.

Thalli green to yellowish-green, dichotomously branched, 2-3 cm long, on shaded rocks and tree trunks, widespread in North America.

3b Marginal hairs single; thallus 1 mm or less wide. Fig. 460A-C. .... *Metzgeria furcata* (L.) Dum.

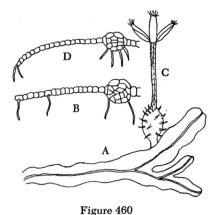

Figure 460

Figure 460 *Metzgeria furcata*. A, plant; B, cross section of thallus; C, involucre, seta, open capsule with elaters; D, section of thallus of *M. conjugata*.

Thalli green to yellowish-green, dichotomously branched, 5-25 mm long, on wet rocks and bark of trees, British Columbia to Oregon, Michigan to Nova Scotia south to Arkansas, Tennessee, and North Carolina.

BLASIACEAE
Blasia

Blasia pusilla L. (Fig. 461)

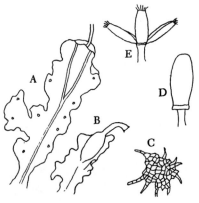

Figure 461

Figure 461 *Blasia pusilla*. A, plant; B, gemma-bottle; C, star-shaped gemma; D, capsule; E, opened capsule.

Thalli 1.5-2.5 cm long, 3-5 mm wide, green to yellowish-green, in patches, vein wide and flat on underside, margins lobed, on wet clayey or gravelly ground, Alaska to Nova Scotia south to California, New Mexico, Iowa, Michigan, Pennsylvania, and Virginia.

ANEURACEAE
Riccardia

Five species of *Riccardia* are known from North America.

1a Plants large, 4-10 mm wide and 10-15 cells thick; plants green and greasy in appearance. .. *Riccardia pinguis* (L.) Gray

Thalli with irregular lateral branches, up to 6 cm long, on wet banks and humus along streams and in swamps, widespread in North America.

1b Plants smaller, delicate, thallus less than 2 mm wide, 4-9 cells thick. 2

2a Thallus regularly 3-pinnately branched, the ultimate branches with a hyaline border one cell thick and 2-3 cells wide. Fig. 462A-B. ....
.... *Riccardia multifida* (L.) S. Gray

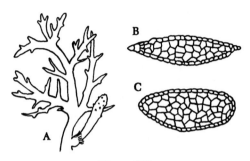

Figure 462

Figure 462 *Riccardia multifida*. A, plant; B, cross section of thallus; C, section of thallus of *R. palmata*.

Thalli dark to brownish-green, 0.3-0.5 mm wide, up to 3 cm long, on soil, rocks, and logs in wet places such as bogs and swamps, widespread in North America.

2b Thallus irregularly or somewhat palmately branched, not distinctly pinnately branched, hyaline border absent or only single cell in width. Fig. 462C. ....
.... *Riccardia palmata* (Hedw.) Carruth.

Thalli dark green, 0.2-0.4 mm wide, up to 10 mm long, on rotting logs and peat in wet places, widespread in North America.

Order Sphaerocarpales

SPHAEROCARPACEAE
Sphaerocarpus

Sphaerocarpus texanus Aust. (Fig. 463)

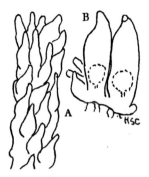

Figure 463 *Sphaerocarpus texanus*. A, male plant; B, female plant.

Female thalli caespitose, 3-5 mm wide, 4-8 mm long, suborbicular to cuneate with leaf-like lobes, bright green when moist to dingy green or olive-green when dry, antheridial thalli smaller, forking several times, spores in permanent tetrads, on damp soil, Washington to California, and Arkansas to Virginia south to the Gulf States. Five other species occur in the United States.

Riella

Riella americana M. A. Howe & Underw. (Fig. 464)

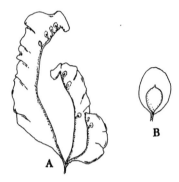

Figure 464 *Riella americana*. A, female thallus with sporophytes; B, sporophyte surrounded by involucre.

Thalli erect or ascending, 1-30 mm high, simple or dichotomously branched, with stem-like portion and a wing 2-5 mm wide, rounded-falciform at apex, slightly undulate, aquatic, submerged, South Dakota and Texas.

Order Marchantiales

TARGIONIACEAE
Targionia

Targionia hypophylla L. Fig. 465

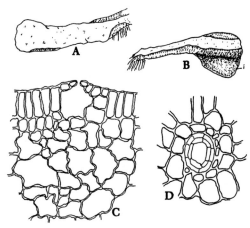

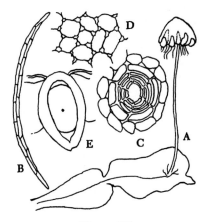

Figure 465

Figure 466

Figure 465 *Targionia hypophylla*. A, thallus, dorsal viw; B, female thallus showing involucre (i); C, vertical section through thallus; D, pore of thallus.

Thalli dark green, 2-5 mm wide, up to 2 cm long, underside purple, margins more or less incurved when dry, on soil, British Columbia south to California, Nevada, and Arizona.

AYTONIACEAE
Reboulia

Reboulia hemisphaerica (L.) Raddi (Fig. 466)

Figure 466 *Reboulia hemisphaerica*. A, plant; B, scale from upper end of stalk of receptacle; C, pore; D, cells of epidermis; E, involucre.

Thalli 6-7 mm wide, up to 3 cm long, green above, brownish-purple along sides and beneath, margins thin, scarcely or not at all incurved when dry, on soil and rocks, throughout North America.

Mannia

1a	Green tissue compact in structure; margin of thallus strongly incurved when dry. 2
1b	Green tissue loose in structure; margin of thallus not incurved when dry. 3
2a	Plants mainly of eastern North America; appendages of ventral scales white, forming a conspicuous apical cluster in fertile plants. Fig. 467. .... *Mannia fragrans* (Balbis) Frye & Clark

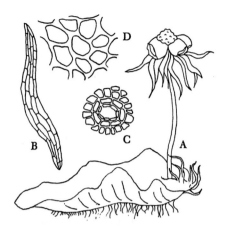

Figure 467

Figure 467 *Mannia fragrans*. A, plant; B, scale; C, pore; D, cells of epidermis.

Thalli green to glaucous green, purplish along margin, deep purple beneath, 2-3 mm wide, curling up along edges when dry, up to 2 cm long, on exposed, dry thin soil over rocks, Idaho to Quebec south to New Mexico, Texas, Missouri, and Alabama.

2b Plants of west coast of North America; appendages mostly purple. ....
...... *Mannia californica* (Gott.) Wheeler

Similar to the preceeding species, on wet rocks, California to Arizona.

3a Dorsal epidermis firm, persistent, leathery, not becoming ruptured and lacunose with age. ....
Mannia pilosa (Hornem.) Frye & Clark

Similar to the preceeding species, on rocks in arctic and alpine regions, Alaska to Quebec south to British Columbia, Minnesota, Wisconsin, and New England. *M. sibirica* (K. Müll.) Frye & Clark is a similar species found in arctic-alpine regions.

3b Dorsal epidermis delicate, obviously areolate, becoming ruptured and lacunose with age. .... *Mannia triandra* (Scop.) Grolle

Similar to the preceeding species, on calcareous rocks and soil, Minnesota to Ontario and Quebec south to Missouri, Illinois, and Ohio.

Asterella

Eight species of *Asterella* are reported from North America.

1a Female receptacle distinctly lobed, smooth or nearly so; spores 80-90 μ in diameter. Fig. 468. ....
.... *Asterella tenella* (L.) P. Beauv.

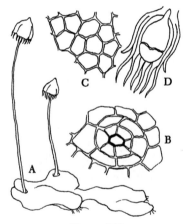

Figure 468

Figure 468 *Asterella tenella*. A, plant; B, pore; C, cells of epidermis; D, sporophyte and pseudoperianth.

Thalli 0.5-1.5 cm long, 1.5-3 mm wide, dichotomously branching, green, purplish-green along margin, on damp soil and rocks, along streams, in open fields, widespread in eastern North

America west of Nebraska, Oklahoma, and Texas.

1b Female receptacle hardly lobed, covered with low, coarse tubercles; spores 60-63 μ in diameter. .... *Asterella gracilis* (Web.) Underw.

Similar to the preceeding species, on soil and rocks, generally arctic-alpine, Alaska to Quebec south to California, Idaho, Utah, Minnesota, Nebraska, Michigan, and New York.

CONOCEPHALACEAE
Conocephalum

Conocephalum conicum (L.) Lindb. (Fig. 469)

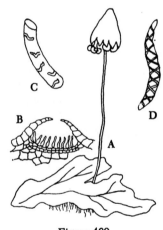

Figure 469

Figure 469 *Conocephalum Conicum*. A, plant; B, section of pore; C, rhizoid with pegs; D, elater.

Thalli pale to dark green above, purplish below, 1-2 cm wide, up to 20 cm long, dichotomously branching, upper surface with distinct polygonal areas, pores distinct, on moist rocks and soil, widespread in North America.

LUNULARIACEAE
Lunularia

Lunularia cruciata (L.) Dum. (Fig. 470)

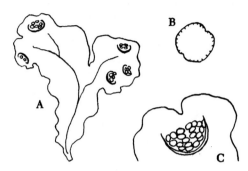

Figure 470

Figure 470 *Lunularia cruciata*. A, plant; B, gemma; C, gemma cup.

Thalli porose, green to yellowish-green, dichotomously branching, 1 cm wide, up to 3 cm long, margins somewhat crispate, gemmae cups cresent shaped, on soil, introduced from Europe, common in greenhouses in the United States.

CLEVEACEAE
Athalamia

Athalamia hyalina (Sommert.) Hatt. (Fig. 471)

Class II. Hepaticae 283

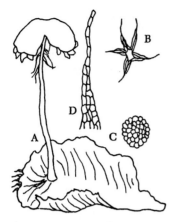

Figure 471

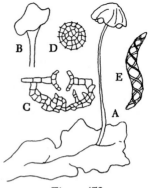

Figure 472

Figure 471 *Athalamia hyalina*. A, plant with archegonial umbrella; B, pore; C, spore; D, scale.

Thalli branching dichotomously, 2-6 mm wide, up to 1.5 cm long, dull green or glaucous green throughout, slightly purplish along margins and below, porose, polygonal areas visible on upper surface, on limestone rocks, British Columbia to Quebec south to California, Utah, Minnesota, and New England.

MARCHANTIACEAE
Pressia

Pressia quadrata (Scop.) Nees (Fig. 472)

Figure 472 *Preissia quadrata*. A, plant with capsules; B, antheridial receptacle; C, section of pore; D, spore; E, elater.

Thalli pale green above, sometimes purplish below, 0.5-1 cm wide, up to 3 cm long, porose, ventral scales present in two longitudinal rows, purplish-black, appendiculate, on calcareous soil and rocks, across North America south to Oregon, Colorado, Nebraska, Arkansas, Kentucky, Tennessee, and Virginia.

Marchantia

1a Thallus about 1 cm wide, without sclerenchymatous cells; gemmae cups with surface papillae; ventral scales in 6 or more rows. Fig. 473. .... *Marchantia polymorpha* L.

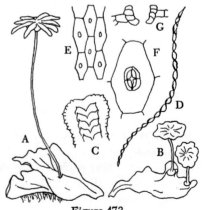

Figure 473

284   Class II. Hepaticae

Figure 473 *Marchantia polymorpha*. A, female plant; B, male plant; C, scales on under side of thallus; D, elater; E, surface of thallus; F, pore; G, section of pore.

Thalli 4-6 cm long, pale to dark green, often purplish below, pores barrell-shaped, gemmae cups round, male and female receptacles umbrella-like, on damp soil, rocks, along streams, in gardens and greenhouses and on burned over areas, throughout North America.

1b   **Thallus usually less than 0.8 cm wide, with sclerenchymatous cells; gemmae cups without surface papillae; ventral scales in 2-4 rows.** 2

2a   Inner openings of pores cross-shaped. .... *Marchantia paleacea* Bertol.

Similar to the preceeding species, on soil banks and rocks, Oklahoma, Texas, and Arizona.

2b   Inner openings of pores not cross-shaped, its sides nearly straight. .... *Marchantia domingensis* Lehm. & Lindenb.

Similar to the preceeding species, on soil banks, Oklahoma to Tennessee, south to Alabama and Texas.

Dumortiera

Dumortiera hirsuta (Sw.) Nees (Fig. 474)

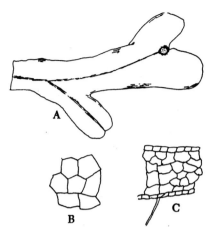

Figure 474

Figure 474 *Dumortiera hirsuta*. A, thallus; B, vestige of surface pore; C, cross-section of thallus.

Thalli dark green to yellowish-green, without or only with vestigial pores or polygonal areas, 1-2 cm wide, up to 20 cm long, hairs present on upper surface of female receptacles, on wet rocks and soil banks, Oklahoma and Missouri to Pennsylvania south to Louisiana, Alabama, and Florida.

RICCIACEAE
Ricciocarpus

Ricciocarpus natans (L.) Corda (Fig. 475)

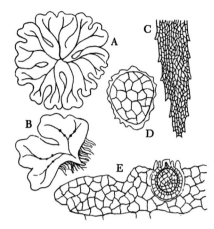

Figure 475

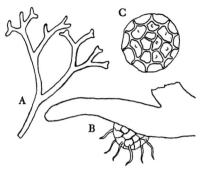

Figure 476

Figure 475 *Ricciocarpus natans*. A, land form; B, floating form; C, ventral scale; D, spore; E, section of thallus and sporophyte.

Thalli in floating forms obcordate to flabelliform, mostly 4-10 mm long and 4-9 mm wide, dark green above, orange or violet below, land thalli forming rosettes 20-35 mm in diameter or larger, British Columbia to Maine south to California, Texas, and the Gulf States.

Riccia

At least 25 species of *Riccia* have been reported from North America. Sporophytes are imbedded in thalli. Ripe spores are often necessary for identification and are usually present in the spring, but sometimes spores may be found in the decaying thalli at other times of the year.

1a   Plants in rosettes on earth, gardens, fields, or river banks; lobes 1-3 mm wide. 2

1b   Plants floating, dichotomously branching ribbons 1 mm wide, often in tangled masses. Fig. 476. .... *Riccia fluitans* L.

Figure 476 *Riccia fluitans*. A, plant; B, plant with capsule bulging out beneath; C, spore.

Thalli 1-5 cm long, green on both sides, throughout North America. *R. canaliculata* Hoffm., a terrestrial form with a thick, fleshy thallus has been reported from North America.

2a   Spores permanently in tetrads; thallus chambered. Fig. 478G. .... *Riccia curtsii* (Aust.) James

Thalli forming rosettes up to 3 cm in diameter, dichotomously branching 1-4 times, light green to yellow-green, on rich moist soil in ditches, North Carolina to Florida and Texas.

2b   Spores separate at maturity. 3

3a   Photosynthetic tissue composed of vertical filaments of compact tissue. 4

3b   Photosynthetic tissue loose, composed of irregular chambers separated by walls 1 cell thick. 5

4a   Thallus margins with distinct cilia and obtuse; spores irregularly reticulate and winged. Fig. 477. .... *Riccia beyrichiana* Hampe *ex* Lehm.

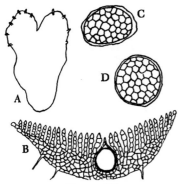

Figure 477

Figure 477 *Riccia beyrichiana*. A, thallus with cilia; B, section of thallus; C, spore; D, spore of *R. dictyospora*.

Plants gregarious or forming rosettes, 1-4 dichotomously branched, light green above, on soil of cultivated ground or rocky soil, British Columbia to New York south to California, Arizona, Texas, and the Gulf States.

4b   Thallus margins not ciliate; spores with wide, irregular, sinuous ridges. Fig. 477D. .... *Riccia dictyospora* M. A. Howe

Plants 0-2 dichotomously branched, 4-10 mm long, light green with a narrow purplish border, ultimate segments have a sharp, acute sulcus, on moist soil, Minnesota to Connecticut south to Georgia and Texas.

5a   Spores ellipsoid or subglobose, not aerolate or winged, but spiny. .... *Riccia membranacea* Gott. & Lindenb.

Thalli small, 1-2 dichotomously branched, 3-7 mm long, dark green, membraneous margin 1-5 cells wide and 1 cell thick, on wet ground in open woods, Missouri and Oklahoma to Connecticut south to New Mexico, Louisiana, Kentucky, and North Carolina.

5b   Spores angular, aerolate or ridged, not spiny. 6

6a   Lobes of thallus short, 1-2 mm broad, touching or overlapping; spores winged, 40-65 μ in diameter, areolate poorly formed, ridges rarely anastomose. Fig. 478A-D. .... *Riccia frostii* Aust.

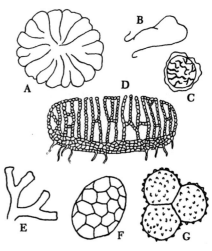

Figure 478

Figure 478 *Riccia frostii*. A, plant; B, lobe enlarged, with capsule; C, spore; D, section of thallus; E, portion of thallus of *R. sullivantii*; F, spore; G, spore of *R. curtsii*

Thalli small, in compact rosettes 4-15 mm in diameter, dark green, grayish-green or reddish-purple above, on silty deposits along rivers and streams, British Columbia to New England south to California, Colorado, Kansas, Arkansas, Illinois, Indiana, and Virginia.

6b   Lobes of thallus slender, 1 mm wide, usually wide apart; spores winged, 50-78 μ in diameter, with ridges anastomosing to form 5-7 areoles across the spore. Fig. 478E. .... *Riccia sullivantii* Aust.

Class II. Hepaticae

Thalli in rosttes 10-20 mm in diameter, 3-5 dichotomously branched, light to dark green above, on damp ground, cultivated fields, and margins of ponds, Ontario to Quebec south to Arkansas and Florida.

List of Synonyms

The scientific names of many bryophytes have been changed. The following list gives the names used in earlier editions of this book as well as those appearing in some recent manuals and indicates by the equals sign their equivalent name in this book.

Anthocerotae

Anthoceros crispulus = *A. punctatus*
Anthoceros laevis = *Phaeoceros laevis*

Hepaticae

Aneura pinguis = *Riccardia pinguis*
Apometzgeria pubescens = *Metzgeria pubescens*
Asterella ludwigii = *A. gracilis*
Calypogeia = *Calypogeja*
Calypogeia arguta = *Calypogeja sullivanti*
Cephalozia lammersiana = *C. bicuspidata* subsp. *lammersiana*
Cephalozia media = *C. lunulifolia*
Cephaloziella byssacea = *C. divaricata*
Chiloscyphus rivularis = *C. polyanthos* var. *rivularis*
Ceratolejeunea integrifolia = *C. latefusca*
Clevea hyalina = *Athalamia hyalina*
Drepanolejeunea bidens = *D. appalachiana*
Frullania asagrayana = *F. tamarisci* var. *asagrayana*
Frullania nisquallensis = *F. tamarisci* var. *nisquallensis*
Herberta adunca = *Herbertus adnuncus*
Jungermannia lanceolata = *J. leiantha*
Lejeunea patens = *L. lamacerina* subsp. *gemminata*
Lepidozia setacea = *Kurzia setacea*
Lepidozia sylvatica = *Kurzia sylvatica*
Leptocolea cardiocarpa = *Cololejeunea cardiocarpa*
Lophocolea bicuspidata = *L. cuspidata*
Lophocolea macounii = *L. heterophylla*
Lopholejeunea sargreana = *L. subfusca*
Lophozia porphyroleuca = *L. guttulata*
Mannia rupestris = *M. triandra*
Metzgeria hamata = *M. leptoneura*
Microlejeunea laetevirens = *Lejeunea laetevirens*
Microlejeunea ulicina = *Lejeunea ulicina*
Microlepidozia setacea = *Kurzia setacea*
Microlepidozia sylvatica = *Kurzia sylvatica*
Moerckia flotoviana = *M. hibernica*
Orthocaulis floerkei = *Barbilophozia floerkei*
Orthocaulis kunzeanus = *Barbilophozia kunzeana*
Orthocaulis quadrilobus = *Barbilophozia quadriloba*
Petalophyllum lamellatum = *P. ralfsii*
Plectocolea crenulata = *Junpermannia gracillima*
Plectocolea cruneliformis = *Junergmannia crenuliformis*
Plectocolea hyalina = *Jungermannia hyalina*
Plectocolea rubra = *Jungermannia rubra*
Radula andicola = *R. mollis*
Tricholea tomentella = *Tricholea tomentella*

Musci

Acaulon rufescens = *A. muticum* var. *rufescens*
Amblystegiella confervoides = *Platydictya confervoides*
Amblystegiella subtilis = *Platydictya subtile*
Amblystegium brevipes = *Leptodictyum brevipes*
Amblystegium compactum = *Rhynchostegiella campacta*
Amblystegium fluviatile var. *fluviatile* = *Hygroamblystegium tenax*

Amblystegium fluviatile var. *orthocladon* = *Hygroamblystegium tenax*
Amblystegium kochii = *Leptodictyum trichopodium*
Amblystegium laxirete = *Leptodictyum laxirete*
Amblystegium riparium = *Leptodictyum riparium*
Amblystegium tenax = *Hygroamblystegium tenax*
Amblystegium tenax var. *spinifolium* = *Hygroamblystegium tenax* var. *spinifolium*
Amblystegium trichopodium = *Leptodictum trichopodium*
Amblystegium vacillans = *Leptodictyum vacilllans*
Anomodon tristis = *Haplohymenium triste*
Aphanorhegma patens = *Physcomitriella patens*
Arctoa starkei = *Kiaeria starkei*
Atrichum macmillani = *A. angustatum*
Barbula cruegeri = *C. cancellata*
Bestia breweriana = *Isothecium cristatum*
Brachythecium flagellare = *B. plumosum*
Brachythecium lamprochryseum = *B. frigidum*
Brachythecium serrulatum = *Rhynchostegium serrulatum*
Braunia californica = *Pseudobraunia californica*
Bryum bimum = *B. pseudotriquetrum* var. *bimum*
Bryum cernuum = *B. uliginosum*
Bryum cuspidatum = *B. creberrimum*
Bryum pendulum = *B. algovicum*
Calliergonella schreberi = *Pleurozium schreberi*
Camptothecium nitens = *Tomenthypnum nitens*
Campylopus introflexus (Hedw.) Brid. = *C. pilitera*
Campylopus tallulensis = *C. flexuosus*
Chamberlainia acuminata = *Brachythecium acuminatum*
Chamberlainia cyrtophylla = *Brachythecium acuminatum* var. *cyrtophyllum*
Cirriphyllum boscii = *Bryoandersonia illecebra*
Climacium kindbergii = *C. americanum*

Crumia latifolia = *Scopelophila latifolia*
Desmatodon coloradensis = *D. obtusifolius*
Dicranella herminieri = *D. hiliariana*
Dicranella schreberi = *D. schreberiana*
Dicranella squarrosa = *D. palustris*
Dicranum bergei = *D. undulatum*
Dicranum sabuletorum = *D. condensatum*
Dicranum rugosum = *P. polysetum*
Dicranum strictum = *D. tauricum*
Didymodon recurvirostris = *Bryoerythrophyllum recuvirostrum*
Didymodon trifarius—Excluded
Ditrichum currituckii = *D. pallidum*
Ditrichum cylindricum = *Trichodon cylindricus*
Ditrichum giganteum = *D. flexicaule*
Ditrichum glaucescens = *Saelania glaucescens*
Ditrichum henryi = *D. rhynchostegium*
Drepanocladus intermedius = *D. revolvens*
Encalypta rhabdocarpa = *E. rhaptocarpa*
Encalypta streptocarpa = *E. procera*
Entodon compressus = *E. challengeri*
Entodon drummondii = *E. macropodus*
Entodon macropus = *E. macropodus*
Ephemerum sessile—Excluded
Eurhynchium brittoniae = *Stokesiella brittoniae*
Eurhynchium diversifolium = *E. pulchellum*
Eurhynchium oreganum = *Stokesiella oregana*
Eurhynchium praelonga var. *praelonga* = *Stokesiella praelonga* var. *praelonga*
Eurhynchium pulchellum var. *praecox* = *E. pulchellum* (form)
Eurhynchium rusciforme = *E. riparioides*
Eurhynchium serrulatum = *Rhynchostegium serrulatum*
Eurhynchium stokesii = *Stokesiella praelonga* var. *stokesii*

Eurhynchium strigosum = *E. pulchellum*
Fabronia imperfecta = *F. ciliaris*
Fissidens debilis = *F. fontanus*
Fissidens julianus = *F. fontanus*
Fissidens minutulus = *F. viridulus*
Fissidens pusillus = *F. viridulus*
Fissidens rufulus = *F. ventricosus*
Fontinalis duriaei = *F. hypnoides* var. *duriaei*
Fontinalis lescurii = *F. novae-angliae*
Funaria calvescens = *F. hygrometrica* var. *calvescens*
Grimmia apocarpa var. *alpicola* = *G. alpicola*
Heterocladium heteropteroides = *H. macounii*
Heterophyllium haldanianum = *Callicladium haldanianum*
Heterophyllium nemorosum = *H. affine*
Homalia jamesii = *H. trichomanoides*
Hygroamblystegium fluviatile = *H. tenax*
Hygroamblystegium irriguum = *H. tenax*
Hygroambmlystegium orthocladon = *H. tenax*
Hygrophypnum dilatatum = *H. molle*
Hygrophypnum montanum = *H. eumontanum*
Hygrohypnum palustre = *H. luridum*
Hyophila tortula = *H. involuta*
Hypnum crista-caestrensis = *Ptilium crista-castrensis*
Hypnum molluscum = *Ctnedium molluscum*
Hypnum patientiae = *H. lindbergii*
Hypnum reptile = *H. pallescens*
Isopterygium deplanatum = *Taxiphyllum deplanatum*
Isopterygium geophilum = *Taxiphyllum taxirameum*
Isopterygium micans = *I. tenerum*
Isothecium stoloniferum = *I. spiculiterum*
Leptodictyum riparium var. *laxirete* = *L. laxirete*
Leptodictyum trichopodium var. *kochii* = *L. trichopodium*
Leptodon nitilus = *Forsstroemia nitilus*
Leptodon ohioensis = *Forsstroemia ohioensis*
Leptodon trichomitrion = *Forsstroemia trichomitria*

Lescuraea incurvata = *Pseudoleskea incurvata*
Lescuraea patens = *Pseudoleskea patens*
Lescuraea radicosa = *Pseudoleskea radicosa*
Lescuraea stenophylla = *Pseudoleskea stenophylla*
Leskea arenicola = *L. polycarpa*
Leskea nervosa = *Leskeella nervosa*
Leskea tectorum = *Pseudoleskeella tectorum*
Leucodon sciuroides = *L. brachypus* var. *andrewsianus*
Leucodontopsis floridanus = *L. geniculata*
Macromitrium mucronifolium = *Groutiella mucronifolia*
Merceya latifolia = *Scopelophila latifolia*
Merceya ligulata = *Scopelophila ligulata*
Mnium affine = *M. affine* var. *ciliare*
Mnium menziesii = *Leucolepis menziesii*
Mnium orthorhynchum = *M. thomsonii*
Mnium serratum = *M. marginatum*
Myurella careyana = *M. sibirica*
Nanomitrium austini = *Micromitrium austini*
Neckera disticha = *Neckeropsis disticha*
Neckera menziesii = *Metaneckera menziesii*
Neckera undulata = *Neckeropsis undulata*
Oncophorus polycarpus var. *strumiferus* = *Cynodontium strumiferum*
Oreoweisia serrulata—Excluded
Orthotrichum elegans = *speciosum* var. *elegans*
Orthotrichum macounii = *O. laevigatum*
Oxyrhynchium hians subsp. *rappii* = *Eurhynchium hians*
Phascum cuspidatum var. *americum* = *P. cuspidatum*

Philonotis americana = *P. fontana* var. *americana*
Philonotis capillaris = *P. fontana*
Physcomitrium turbinatum = *P. pyriforme*
Pireella ludoviciae = *P. pohlii*
Plagiomnium drummondii = *Mnium drummondii*
Plagiomnium rostratus = *Mnium rostratum*
Plagiomnium ciliare = *Mnium affine* var. *ciliare*
Plagiomnium cuspidatum = *Mnium cuspidatum*
Plagiopus oederi = *P. oderiana*
Plagiothecium deplanatum = *Taxiphyllum deplanatum*
Plagiothecium elegans = *Isopterygium elegans*
Plagiothecium geophilum = *Taxiphyllum taxirameum*
Plagiothecium micans = *Isopterygium tenerum*
Plagiothecium muellerianum = *Isopterygiopsis muelleriana*
Plagiothecium roseanum = *P. cavifolium*
Plagiothecium striatellum = *Herzogiella striatella*
Pleuridium acuminatum = *P. subulatum*
Pogonatum capillae = *P. dentatum*
Polytrichadelphus lyallii = *Polytrichum lyallii*
Polytrichum gracile = *P. formosum* var. *aurantiacum*
Polytrichum longisetum = *P. formosum* var. *aurantiacum*
Polytrichum norvegicum = *P. sexangulare*
Porotrichum alleghaniense = *Thamnobryum alleghaniense*
Pottia heimii = *Desmatodon heimii*
Pseudisothecium stoloniferum = *Isothecium spiculiferum*
Pylaisia intricata = *Pylaisiella intricata*
Pylaisia polyantha = *Pylaisiella polyantha*

Pylaisia selwynii = *Pylaisiella selwynii*
Rhabdoweisia denticulata = *R. crispata*
Rhabdoweisia denticulata var. *americana* = *R. crispata*
Rhynchostegium hians = *Eurhynchium hians*
Rhynchostegium pulchellum = *Eurhynchium pulchellum*
Rhynchostegium riparioides = *Eurhynchium riparioides*
Roellii sandbergii = *Bryum sandbergii*
Schlotheimia sullivantii = *S. rugifolia*
Schwetschkeopsis denticulata = *S. fabronia*
Scleropodium caespitosum = *S. cespitans*
Sematophyllum carolinianum = *S. demissum*
Solmsiella biseriata = *Erpodium biseriatum*
Solmsiella kurzii = *Erpodium biseriatum*
Sphagnum warnstorfianum = *S. warnstorfii*
Syrrhopodon floridanus = *S. incompletus* var. *incompletus*
Taxiphyllum geophyllum = *Taxiphyllum taxirameum*
Tetraplodon pennsylvanicum = *Splachnum pennsylvanicum*
Thuidium microphyllum = *Haplocladium microphyllum*
Thuidium virginianum = *Haplodium virginianum*
Trichostomum cylindricum = *T. tenuirostre*
Ulota americana = *U. hutchinsiae*
Ulota ludwigii = *U. coarcta*
Weissia viridula = *Weissia controversa*
Weissia ludoviciana = *Astomum ludovicianum*
Weissia muhlenbergiana = *Astomum muhlenbergianum*

Index and Pictured Glossary

A

Acaulon, 98
  muticum, 98
    var. rufescens, 98
  rufescens, 289
Acrocarpi, 7
ACROCARPOUS: having the sporophyte at the end of a stem or ordinary leafy branch. Fig. 479.

Figure 479

ACUMEN: the tapering narrow point of an acuminate leaf.
ACUMINATE: tapering in the manner of Fig. 480. Note curvature of margin of leaf.

Figure 480

ACUTE: ending in a sharp angle, less than 90°. Fig. 481.

Figure 481

Adelanthaceae, 248
ALAR CELLS: the cells at the basal angle of the leaf. Fig. 482.

Figure 482

Alonia, 32, 101
  rigida, 101
Alsia, 155
  californica, 155
Amblyodon, 134
  dealbatus, 134
Amblystegiaceae, 177
Amblystegiella, 289
  confervoides, 289
  subtilus, 289
Amblystegium, 35, 55, 182
  brevipes, 289
  compactum, 289
  fluviatile, 289
    var. orthocladon, 290
  juratzkanum, 183
  kochii, 290
  laxirete, 290
  riparium, 290
  serpens, 183
  tenax, 290
    var. spinifolium, 290
  trichopodium, 290
  vacillans, 290
  varium, 182
Amphidium, 38, 141
  californicum, 141
  lapponicum, 141
  mougeotii, 141
AMPHIGASTRA: underleaves, on the ventral side of stem in liverworts. (See Fig. 439.)
Anacamptodon, 166
  splachnoides, 166
Anacolia, 136
  laevisphaera, 136
  menziessi, 136
ANASTOMOSE: union between lines or ridges in spores of liverworts.
Anastrophyllum, 257
  michauxii, 257
Andreaea, 27
  blytii, 28
  nivalis, 28
  rothii, 20, 28
  rupestris, 27
Andreaeidae, 7, 27
Aneura, 289
  pinguis, 289
Aneuraceae, 279
ANNULUS: a ring of thick walled cells between the mouth of the capsule and the lid, like the rubber gasket on a jar. Fig. 483.

Figure 483

Anoectangium, 89
  obtusifolium, 89
  sendtnerianum, 89
Anomodon, 171
  attenuatus, 172
  minor, 172
  rostratus, 171
  rugelii, 172
  tristis, 290
  viticulosus, 173
ANTHERIDIUM: the male reproductive organ containing the sperms. Fig. 484, 3, 12.

Figure 484

Anthoceros, 230
  crispulus, 289
  laevis, 289
  macounii, 230
  punctatus, 21, 230
Anthocerotaceae, 230
Anthocerotae, 3, 5, 6, 230
ANTICAL: upper surface of plant or upper margin of leaf.
Antitrichia, 157
  californica, 158
  curtipendula, 157
Aphanorhegma, 31, 114
  patens, 290
  serratum, 114
APICAL: belonging to the apex or tip.
APICULATE: ending in an abrupt, short, sharp point, but not stiff. Fig. 485.

Figure 485

Apometzgeria, 289
  pubescens, 289
APOPHYSIS: see hypophysis.
APPENDICULATE: of cilia with small transverse spurs at intervals along the margin. Fig. 486.

Figure 486

ARCHEGONIUM: the female reproductive organ containing the egg. Fig. 487. 3, 12

Figure 487

Archidales, 7, 67
Archidium, 67
  alternifolium, 68
  ohioense, 67
Arctoa, 290
  starkei, 290
AREOLE: one space or mesh of a network.
Arthodonteae, 7
Asterella, 282
  gracilis, 283
  ludwigii, 289
  tenella, 282
Astomum, 90
  ludovicianum, 90
  muhlengergianum, 90
  phascoides, 90
Athalamia, 283
  hyalina, 283
Atrichum, 223
  angustatum, 223
  crispum, 224
  macmillani, 290
  oerstedianum, 224
  selwynii, 224
  undulatum, 224
Aulacomniaceae, 133
Aulacomnium, 133
  androgynum, 134
  heterostichum, viii, 133
  palustre, 133
AURICLE: a lobe or bulge at the base of a leaf. Fig. 64, 488.

Figure 488

AUTOICOUS: having male and female organs on the same plant, the antheridia in a cluster just below the archegonia, or somewhere else along the shoot, or on a large or small branch. 12.
AWN: a bristle at the tip of a leaf. 11.
Aytoniaceae, 281

B

Barbella, 160
  pendula, 160
Barbilophozia, 253
  barbata, 253
  floerkei, 233, 255
  hatcheri, 254
  kunzeana, 255
  lycopodioides, 254
  quadriloba, 254
Barbula, 96
  cancellata, 97
  convoluta, 96
  cruegeri, 290
  cylindrica, 97
  ehrenbergii, 96
  fallax, 97
  unguiculata, 97
  vinealis, 98
Bartramia, 136
  ithyphylla, 136
  pomiformis, 136
  stricta, 137
Bartramiaceae, 135
Bartramiopsis, 225
  lescurii, 225
Bazzania, 243
  denudata, 244
  tricrenata, 244
  trilobata, 243
BEAK: a prolonged narrow tip of an operculum. Fig. 489.

Figure 489

Bestia, 290
  breweriana, 290
BIFURCATE: split into two branches.
BILOBED: with two divisions, especially rounded ones.
Blasia
  pusilla, 278
Blasiaceae, 278
Blepharostoma, 241
  arachnoideum, 241
  trichophyllum, 233, 241
Blindia, 73
  acuta, 73
BOG: a watery mass of decayed vegetation with acid reaction.
Bog mosses, 24
BORDERED: having the margin different from the rest of the leaf either in shape or color of cells. Fig. 490.

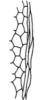

Figure 490

Brachyelyma, 151
  subulatum, 151
Brachymenium, 47, 123
  macrocarpum, 123
  systylium, 123
Brachytheciaceae, 190
Brachythecium, 55, 191
  acuminatum, 192
    var. *cyrtophyllum*, 192
  acutum, 196
  albicans, 193
  campestre, 194
  collinum, 197
  curtum, 196
  flagellare, 290
  frigidum, 193
  lamprochryseum, 290
  leibergii, 193
  oxycladon, 192
    var. *dentatum*, 192
  plumosum, 196
  populeum, 196
  reflexum, 195
  rivulare, 57, 194
  rutabulum, 195
  salebrosum, 194
  serrulatum, 290
  velutinum, 196
BRACT: a special leaflike structure at the base of a reproductive organ or cluster, 14.
BRACTEOLE: a small bract; a modified underleaf.
Braunia, 154
  californica, 290
  secunda, 154
BROOD-BODIES: detachable cells or organs which give rise vegetatively to new plants; gemmae. Figs. 120D, 135C, 147D.
Brothera, 79
  leana, 44, 79
Brotherella, 59, 207
  recurvans, 207
  roellii, 207
  tenuirostris, 207
Bruchia, 74
  brevifolia, 74
  flexuosa, 30, 74
  ravenelii, 74
  sullivantii, 75
  texana, 75
Bryaceae, 120
Bryhnia, 197
  graminicolor, 197
  novae-angliae, 197
Bryidae, 7, 28, 62
Bryoandersonia, 198
  illecebra, 198
Bryoerythrophyllum, 95
  recurvirostrum, 95
Bryoxiphiaceae, 72
Bryoxiphium, 72
  norvegicum, 72
Bryum, 12, 34, 47, 123
  algovicum, 126
  argenteum, viii, 113
  bicolor, 125
  bimum, 290
  caespiticium, viii, 125
  capillare, 126
  cernuum, 290
  creberrimum, 127
  cuspidatum, 290
  gemmiparum, 125
  mniatum, 124
  pallens, 126
  pallescens, 128
  pendulum, 290
  pseudotriquetrum, 127
    var. *bimum*, 127
    var. *crassirameum*, 127
  sandbergii, 124
  turbinatum, 127
  uliginosum, 126
  weigelii, 124
Buds, 2
Buxbaumia, 222
  aphylla, viii, 222
Buxbaumiaceae, 222
Buxbaumiales, 7, 222

C

CADUCOUS: falling off at a distinct line of dehiscence.
CAESPITOSE: plants arranged in thick mats or tufts.
Callicladium, 216
  haldanianum, 216
Callicostella, 163
  pallida, 163
Calliergon, 55, 188
  cordifolium, 188
Calliergonella, 189
  conradii, 189
  cuspidata, 189
  schreberi, 290
Calobryales, 6, 232
Calymperaceae, 87
Calymperes, 88
  richardii, 88
Calypogeia, 289
  arguta, 289
Calypogeja, 244
  fissa subsp. *neogaea*, 246
  muelleriana, 245
    subsp. *bloomquistii*, 246
  trichomanis, 21, 245
  . *sullivantii*, 245
Calypogejaceae, 244
CALYPTRA: the thin covering or hood fitted over the upper part of the capsule;

Index and Pictured Glossary

it is a part of the archegonium. Fig. 491C. 3, 5, 11.

Figure 491

CAMPANULATE: calyptras shaped like a bell.
Camptothecium, 290
  nitens, 290
Campylium, 177
  chrysophyllum, 178
  hispidulum, 177
  polygamum, 178
  radicale, 179
  stellatum, 178
Campylopus, 44, 77
  flexuosus, 78
  introflexus, 290
  tallulensis, 290
  pilifera, 77
CAPSULE: the spore-containing sac which, with the seta and foot compose the sporophyte. Fig. 492. (See Figs. 3, 10, 13) 3, 5, 11, 14.

Figure 492

Catoscopiaceae, 135
Catoscopium, 135
  nigritum, 135
Cephalozia, 289
  lammersiana, 289
Cephalozia, 246
  bicuspidata, 247
    subsp. lammersiana, 247
  catenulata, 248
  connivens, 247
  lunulifolia, 247
  media, 289
Cephaloziaceae, 246
Cephaloziella, 249
  byssacea, 289
  divaricata, 249
  hampeana, 249
Cephaloziellaceae, 249
Ceratodon, 71
  purpureus, viii, 71
Ceratolejeunea, 271
  integrifolia, 289
  laetefusca, 271
Chamberlainia, 290
  acuminata, 290
  cyrtophylla, 290
Chandonanthus, 253
  setiformis, 253
Chiloscyphus, 19, 251

pallescens, 251
  var. fragilis, 251
  polyanthus, 254
    var. rivularis, 251
  rivularis, 289
Chlorophyll, 2, 3
CHLOROPHYLLOSE: containing chloroplasts.
CHLOROPLAST: a green photo-synthetic particle in a cell.
CILIA: hair-like appendages, 2
Cinclidium, 132
  stygium, 132
CIRCINATE: bent around in more or less of a circle.
Cirriphyllum, 198
  boscii, 290
  cirrosum, 198
  piliferum, 198
Cladonia, 1
Cladopodiella, 246
  fluitans, 246
  francisci, 246
Claopodium, 173
  bolanderi, 173
  crispifolium, 173
  whippleanum, 173
Clasmatodon, 55, 167
  parvulus, 167
Clevea, 289
  hyalina, 289
Cleveaceae, 283
Climaciaceae, 153
Climacium, 153
  americanum, 153
  dendroides, 153
  kindbergii, 290
Codoniaceae, 275
Collecting locality, how to record, 9
COLLENCHYMATOUS: cell walls thickened at angles or corners (see Fig. 24).
Cololejeunea, 275
  biddlecomiae, 275
  cardiocarpa, 275
COLUMELLA: the central axis of the capsule, around which are the spores. Fig. 493.

Figure 493

COMPLANATE: flattened; more or less in one plane.
COMPLICATE: folded over one another as in leaves of liverworts.
COMPLICATE-BILOBED: with two lobes, the lobes folded together. (See Figs. 22, 23) 12, 13.
CONIC: shaped like a cone.
CONNIVENT: ends of leaf lobes approaching one another. (See Fig. 403 B, C).

Conocephalaceae, 283
Conocephalum, 283
  conicum, 283
Conostomum, 137
  tetragonum, 137
CONTIGUOUS: leaves or under leaves touching or nearly so.
CORDATE: heart-shaped, the broadest portion near the attachment. Fig. 494.

Figure 494

CORTICOLOUS: growing on bark of trees.
COSTA: the midrib of a moss leaf. 11.
COSTATE: having a costa. 11.
Cratoneuron, 177
  commutatum, 177
  filicinum, 177
CRISPATE: curled and wavy.
Crisped leaves, 45
Crossidium, 32, 99
  squamiferum, 99
Crossotolejeunea, 272
  bermudiana, 272
Crumia, 290
  latifolia, 290
Cryphaea, 154
  glomerata, 154
  nervosa, 154
Cryphaeaceae, 154
Ctenidium, 218
  molluscum, 218
CUCULLATE: forming a pocket opening on one side; of a calyptra usually cone-shaped and slit on one side only. Fig. 495.

Figure 495

CUSPIDATE: having a small, stiff, abrupt point.
CUTICLE: the outer covering of stems and leaves.
Cyclodictyon, 163
  varians, 163
Cynodontium, 79
  strumiferum, 79

D

DECURRENT: running down; the margin of a leaf extending below its point of attachment. Fig. 496.

Figure 496

Dendroalsia, 155
  abietina, 155
DENDROID: having an erect stem with branches, like a little tree.
DENTATE: toothed with the teeth pointing outward. Fig. 497.

Figure 497

DENTICULATE: dentate with little teeth. Fig. 498.

Figure 498

Desmatodon, 99
  coloradensis, 290
  heimii, 99
  latifolius, 100
  obtusifolius, 100
  plinthobius, 100
  porteri, 100
Dichelyma, 152
  capillaceum, 152
  falcatum, 152
  pallescens, 152
  uncinatum, 152
Dichodontium, 79
  olympicum, 80
  pellucidum, 79
Dicranaceae, 74
Dicranales, 7, 68
Dicranella, 75
  herminieri, 290
  heteromalla, viii, 77
  hiliariana, 76
  palustris, 75
Dicranodontium, 44, 78
  denudatum, 78

Index and Pictured Glossary   295

Dicranoweisia, 80
  cirrata, 80
  crispula, 80
Dicranum
  bergeri, 290
  bonjeanii, 83
  condensatum, 37, 86
  drummondii, 85
  flagellare, 37, 82
  fulvum, 84
  fuscescens, 85
  majus, 84
  montanum, 82
  muhlenbeckii, 85
  polysetum, 83
  rhabdocarpum, 83
  rugosum, 290
  sabuletorum, 290
  spurium, 85
  strictum, 290
  tauricum, 81
  undulatum, 83
  viride, 81
Didymodon, 95
  recurvirostris, 290
  rigidulus, 95
  trifarius, 290
DIMORPHIC: having two very different kinds of leaves on stems and branches.
DIOICOUS: having male and female organs on different plants. 12.
Diphyscium, 222
  cumberlandianum, 223
  foliosum, 222
Diplasiolejeunea, 274
  rudolphiana, 274
DIPLOID: having two homologous sets of chromosomes. 3
Diplolepideae, 7
Diplophyllum, 261
  albicans, 261
  apiculatum, 262
  taxifolium, 262
Disceliaceae, 112
Discelium, 112
  nudum, 112
DISTAL: farthest from the point of attachment.
Distichum, 72
  capillaceum, 72
  inclinatum, 72
Ditrichaceae, 68
Ditrichum, 69
  currituckii, 290
  cylindricum, 290
  flexicaule, 71
  giganteum, 290
  glaucescens, 290
  henryi, 290
  heteromallum, 71
  lineare, 69
  pallidum, 69
  pusillum, 70
  rhynchostegium, 71
DORSAL: the back or under side of a leaf; the upper side of a prostrate shoot.
DORSAL LAMINA: in Fissidentaceae the lamina on the side opposite the sheathing portion. (See Fig. 42.)
Drepanocladus, 4, 57, 184
  aduncus, viii, 185
    var. *kneiffii,* 185
    var. *polycarpus,* 185
  exannulatus, 186
  fluitans, 186
  intermedius, 290
  revolvens, 184
  uncinatus, 184
  vernicosus, 184
Drepanolejeunea, 272
  appalachiana, 272
  bidens, 289
Drummondia, 148
  prorepens, 148
Dumortiera, 285
  hirsuta, 285

E

ECOSTATE: without a midrib.
EGG: the female germ cell or gamete. 3
ELATER: elongate and usually spirally thickened cells mixed with the spores, in liverworts only. 13
ELONGATE: considerably longer than wide.
EMARGINATE: having a broad, shallow notch at the end.
EMBRYO: the many-celled product of the fertilized egg, still but little differentiated. 2
Encalypta, 88
  ciliata, 89
  procera, 88
  rhabdocarpa, 290
  rhaptocarpa, 89
  streptocarpa, 290
Encalyptaceae, 88
ENTIRE: with an even margin, not notched or toothed.
Entodon, 35, 203
  brevisetus, 203
  challengeri, 204
  cladorrhizans, 204
  compressus, 290
  drummondii, 290
  macropodus, 204
  macropus, 290
  seductrix, 203
Entodontaceae, 202
Entosthodon, 117
  drummondii, 117
Ephemeraceae, 113
Ephemerum, 113
  cohaerens, 113
  crassinervium, 113
  serratum, 31, 113
  sessile, 290
  spinulosum, 113
EPIPHRAGM: a membrane covering the mouth of the capsule under the operculum in Polytrichaceae. Fig. 499.

Figure 499

Epipterygium, 122
  tozeri, 122

EQUITANT: when the lower part of a leaf appears to be split so that it stands astride of the stem and the base of the next leaf above. *Fissidens.* Fig. 500.

Figure 500

Erpodiaceae, 140
Erpodium, 140
  acrifolium, 140
  biseriatum, 50, 140
Eubryales, 7, 120
Eucladium, 92
  verticillatum, 92
Eurhynchium, 200
  brittoniae, 290
  diversifolium, 290
  hians, 201
  oreganum, 290
  praelonga, 290
  pulchellum, 201
    var. *praecox,* 290
  riparioides, 201
  rusciforme, 290
  serrulatum, 290
  stokesii, 290
  strigosum, 290
EXANNULATE: without an annulus.
EXCURRENT: with the costa extending beyond the tip of the leaf. Fig. 501.

Figure 501

F

Fabronia, 165
  ciliaris, 166
  gymnostoma, 165
  imperfecta, 290
  pusilla, 165
  ravenelii, 165
  wrightii, 165
Fabroniaceae, 165
FALCATE: curved like a sickle. Fig. 502.
FALCATE-SECUND: each leaf falcate and all bent in the same direction.

Figure 502

FEN: a water-soaked area with lime in solution.
FILIFORM: thread-like. 11
Filing system, 9
Fissidens, 29, 62
  adianthoides, 66
  asplenioides, 65
  bryoides, 64
  bushii, 67
  cristatus, 66
  debilis, 290
  fontanus, 63
  garberi, 64
  grandifrons, 63
  julianus, 290
  kegelianus, 64
  limbatus, 64
  minutulus, 290
  obtusifolius, 64, 65
  osmundoides, 66
  polypodioides, 65
  pusillus, 290
  ravenelii, 64
  rufulus, 290
  taxifolius, 67
  ventricosus, 62
  viridulus, 65
Fissidentales, 7, 62
FLABELLIFORM: stems with reduced or vestigal leaves, thus forming flagella-like structures.
FLACCID: soft and flabby in texture.
FLAGELLA: slender whip-like branches.
FLEXUOSE: irregularly wavy.
Flowering moss, 1
Fontinalaceae, 149
Fontinalis, 149
  antipyretica, 149
  dalecarlica, 150
  duriaei, 290
  filiformis, 151
  flaccida, 151
  hypnoides, 151
    var. *duriaei,* 151
  lescurii, 290
  neomexicana, 149
  novae-angliae, 151
  sullivantii, 151
FOOT: the basal and absorbing portion of the sporophyte. 3, 6
Forsstroemia, 55, 155
  ohioense, 155
  trichomitria, 155
Fossombronia, 275
  cristula, 275
  foveolata, 276
  wondraczekii, 22, 275
FROND: a much-divided leaf, as of a fern.
FRONDOSE: resembling a frond.

FRUIT: a term often applied to the capsule—not strictly accurate.
Frullania, 266
  asagrayana, 289
  bolanderi, 266
  brittoniae, 269
  californica, 267
  eboracensis, 269
  franciscana, 267
  inflata, 268
  kunzei, 268
  nisquallensis, 289
  oakesiana, 268
  obcordata, 268
  plana, 268
  riparia, 269
  selwyniana, 268
  squarrosa, 269
  tamarisci, 267
    var. *asaprayana*, 267
    var. *nisquallensis*, 267
Funaria, 3, 34, 116
  americana, 117
  calvescens, 290
  flavicans, 116
  hygrometrica, viii, 116
    var. *calvescens*, 116
  serrata, 117
Funariaceae, 114
Funariales, 7, 112

G

GAMETOPHYTE: the plant bearing the gametes; the sexual generation. (See Fig. 13) 5.
Gasket, 5
GEMMA: a cell or cluster of cells, often bud-like, borne on the gametophyte, capable of reproducing the plant vegetatively; broad body. 2
GENE: a determiner of hereditary characters. 3
General references, 13
GENETICS: the study of heredity. 3
Geocalyaceae, 249
Geocalyx, 249
  graveolans, 249
GONIAUTOICOUS: monoicous plants with the male inflorescence small, gemmiform, and axillary.
GREGARIOUS: plants growing close together.
Grimmia, 45, 105
  affinis, 109
  alpicola, 105
  anodon, 110
  apocarpa, 106
    var. *alpicola*, 290
  calyptrata, 109
  donniana, 110
  laevigata, 108
  maritima, 105
  montana, 107
  olneyi, 107
  ovalis, 107
  pilifera, 109
  plagiopoda, 108
  pulvinata, 108
  rauii, 108

  tenerrima, 107
  torquata, 105
  trichophylla, 107
  wrightii, 108
Grimmiaceae, 104
Grimmiales, 7, 104
Groutiella, 148
  mucronifolia, 148
  tomentosa, 148
GUIDE CELL: a large empty cell in the middle of a midrib. (See Figs. 57, 58) 38, 43.
Gymnocolea, 257
  acutiloba, 257
  inflata, 257
Gymnomitriaceae, 261
Gymnostomum, 91
  aeruginosum, 91
  calcareum, 92
  recurvirostrum, 91

H

Habitat data, recording, 9
Haplocladium, 171
  microphyllum, 174
  virginianum, 174
Haplohymenium, 171
  triste, 171
HAPLOID: with one set of chromosomes, all different.
Haplomitriaceae, 232
Haplomitrium, 19, 23, 232
  hookeri, 232
Haplopepideae, 7
Harpanthus, 249
  scutatus, 249
Hedwigia, 37, 153
  ciliata, 153
Hedwigiaceae, 153
Helodium, 176
  blandowii, 176
  paludosum, 176
Hepaticae, 3, 5, 6, 232
Herberta, 241
  adunca, 289
  aduncus, 233, 241
Herpetineuron, 173
  toccoae, 173
Herzogiella, 218
  striatella, 218
Heterocladium, 171
  dimorphum, 171
  heteropteroides, 290
  macounii, 171
  procurrens, 171
Heterophyllium, 207
  affine, 207
  haldanianum, 290
  nemorosum, 290
Homalia, 162
  jamesii, 290
  trichomanoides, 162
Homalotheciella, 191
  subcapillata, 191
Homalothecium, 190
  megaptilum, 190
  nevadense, 191
  nuttallii, 190
Homomallium, 211
  adnatum, 211
  mexicanum, 211
HOMOMALLOUS: leaves or branches all pointed in the same direction.

Hookeria, 163
  acutifolia, 163
  lucens, 163
Hookeriaceae, 163
Hookeriales, 7, 163
Hornworts, 6, 230
Hoyer's solution, 8
Husnotiella, 92
  revoluta, 92
  torquescens, 92
HYALINE: colorless and clear.
HYDRIC: of very wet habitat, or in water.
Hygroamblystegium, 181
  fluviatile, 290
  irriguum, 290
  noterophilum, 181
  orthocladon, 290
  tenax, 181
    var. *spinifolium*, 182
Hygrohypnum, 186
  bestii, 187
  dilatatum, 290
  eugyrium, 188
  eumontanum, 187
  luridum, 187
  molle, 188
  montanum, 290
  novae-caesareae, 187
  ochraceum, 186
  palsutre, 290
Hylocomniaceae, 221
Hylocomnium, 221
  brevirostre, 222
  pyrenaicum, 221
  splendens, 221
  umbratum, 221
Hypnaceae, 209
Hypnobryales, 7, 164
Hypnum, 211
  callichorum, 212
  circinale, 214
  crista-castrensis, 290
  cupressiforme, 214
    var. *filiforme*, 215
  curvifolium, 213
  fertile, 212
  imponens, 215
  lindbergii, 212
  molluscum, 290
  pallescens, 215
  patientiae, 290
  pratense, 212
  reptile, 290
  revolutum, 213
  subimponens, 213
  vaucheri, 214
HYPOPHYSIS: a swelling of the seta immediately under the capsule. Fig. 503.

Figure 503

Hyophila, 95
  involuta, 95
  tortula, 290

I

IMBRICATE: leaves or under leaves overlapping. (See Fig. 61)
IMMERSED: of the capsule when the perichaetial leaves project beyond it. Fig. 504.

Figure 504

INCRASSATE: with thickened walls.
INCUBOUS: leaves overlapping like shingles on a roof if base of plant is at ridge and apex at the eaves. Fig. 505. 12, 13

Figure 505

INFLATED: of alar cells which are enlarged much beyond the size of neighboring cells. Fig. 506.

Figure 506

INVOLUCRE: a protective covering around the calyptra or perianth formed of bracts or a short tube.
INVOLUTE: having the margins rolled inward (upward). Fig. 507.

Figure 507

Isobryales, 7, 140
ISODIAMETRIC: with the same diameter in every direction. 13, Fig. 508.

Index and Pictured Glossary

Figure 508

Isopterygiopsis, 62, 216
  muelleriana, 216
Isopterygium, 62, 216
  deplanatum, 290
  elegans, 217
  geophilum, 290
  micans, 290
  tenerum, 216
Isothecium, 55, 191
  cristatum, 191
  spiculiferum, 191
  stoloniferum, 290

J

Jaegerina, 158
  scariosa, 158
Jamesoniella, 258
  autumnalis, 258
Jubula, 270
  pennsylvanica, 270
Jubulaceae, 266
JULACEOUS: cylindrical, and smooth or downy. Fig. 509.

Figure 509

Jungermannia, 259
  crenuliformis, 260
  exsertifolia, 260
  hyalina, 259
  lanceolata, 289
  leiantha, 22, 260
  pumila, 260
Jungermanniaceae, 253
Jungermanniales, 6, 232

K

KEEL: a sharp ridge as when a leaf is folded along the midrib or main division. Fig. 510.

Figure 510

Kiaeria, 81
  starkei, 81

Kurzia, 242
  setacea, 243
  sylvatica, 242

L

Label, 10
Labeling, 9
LACINIATE: slashed; cut into narrow lobes. Fig. 511.

Figure 511

LACUNOSE: irregular pits or depressions on the surface of liverworts.
LAMELLA: (pl. ae) thin sheets of cells usually standing perpendicular to the surface of a leaf. Fig. 512.

Figure 512

LAMINA: the flat, green part of the leaf; blade.
LANCEOLATE: lance-shaped; 4-6 times longer than wide, broadest at base and tapering to a point. Fig. 513.

Figure 513

Leaf, 11
Lejeunea, 273
  cavifolia, 273
  cladogyna, 274
  flava, 274
  laetivirens, 273
  lamacerina, 274
  patens, 289
  ulicina, 273
Lejeuneaceae, 270
Lepidozia, 243
  reptans, 243
  setacea, 289
  sylvatica, 289
Lepidoziaceae, 242
Leptobryum, 123
  pyriforme, viii, 123
Leptocolea, 289

cardiocarpa, 289
Leptodictyum, 179
  brevipes, 180
  laxirete, 180
  riparium, 179
    fo. elongatum, 180
    fo. fluitans, 179
    fo. longifolium, 180
    fo. obtusum, 179
    var. laxirete, 290
  sipho, 180
  trichopodium, 180
    var. kochii, 290
  vacillans, 180
Leptodon, 290
  nitilus, 290
  ohioensis, 290
  trichomitrion, 290
Lescuraea, 169, 291
  incurvata, 291
  patens, 291
  radicosa, 291
  stenophylla, 291
Leskea, 167
  arenicola, 291
  australis, 167
  gracilescens, 168
  nervosa, 291
  obscura, 168
  polycarpa, 168
  tectorum, 291
Leskeaceae, 167
Leskeella, 168
  nervosa, 168
Leucobryaceae, 86
Leucobryum, 86
  albidum, 86
  glacum, 29, 86
Leucodon, 156
  brachypus, 156
    var. andrewsianus, 156
  julaceus, 156
  scuiroides, 291
Leucodontaceae, 156
Leucodontopsis, 157
  floridanus, 291
  geniculata, 157
Leucolejeunea, 271
  clypeata, 271
  unciloba, 271
Leucolepis, 128
  menziesii, 128
Lichens, 1
LID: see operculum.
Life Cycle, 2
Lindbergia, 167
  brachyptera, 167
LINEAR: very narrow, with parallel sides.
LINGULATE: tongue-shaped.
Liverworts, 4, 6, 232
LOBE: a division (especially a rounded one), as of a leaf.
Lophocolea, 250
  bicuspidata, 289
  bidentata, 250
  cuspidata, 251
  heterophylla, 5, 6, 233, 250
  macounii, 289
  minor, 250
Lophocoleaceae, 250
Lopholejeunea, 270
  muelleriana, 270
  sargreana, 289
  subfusca, 271
Lophozia, 255

  excisa, 255
  guttulata, 255
  incisa, 255
  porphyroleuca, 289
  ventricosa, 256
Luisierella, 101
  barbula, 101
LUMEN: the cavity of a cell.
Lunularia, 283
  cruciata, 283
Lunulariaceae, 283

M

Macromitrium, 291
  mucronifolium, 291
MAMILLA: a single large prominence, covering the cell and including an extension or bulge of the cell cavity.
MAMILLOSE: having mamillae.
Mannia, 281
  californica, 282
  fragrans, 281
  pilosa, 282
  rupestris, 289
  sibirica, 282
  triandra, 282
Marchantia, 4, 19, 284
  domingensis, 285
  paleacea, 285
  polymorpha, 285
Marchantiaceae, 284
Marchantiales, 6, 239, 281
MARGINED: see bordered.
Marsupella, 261
  emarginata, 261
  sphacelata, 261
Mastigolejeunea, 270
  auriculata, 270
MEDIAN LEAF CELL: a cell from the middle of the lamina, as distinguished from alar or apical cells.
Meesia, 134
  triquetra, 135
  uliginosa, 134
Meesiaceae, 134
Merceya, 291
  latifolia, 291
  ligulata, 291
MESIC: of moist habitat, neither very wet or very dry.
Metaneckera, 161
  menziesii, 161
Meteoriaceae, 159
Metzgeria, 277
  conjugata, 278
  furcata, 278
  hamata, 289
  leptoneura, 278
  pubescens, 277
Metzgeriaceae, 277
Metzgeriales, 6, 238, 275
Microlejeunea, 289
  laetevirens, 289
  ulicina, 289
Microlepidozia, 289
  setacea, 289
  sylvatica, 289
Micromitrium, 114
  austinii, 114
MIDRIB: the middle vein of the leaf.

Mielichhoferia, 120
  macrocarpa, 120
  mielichoferi, 120
MITRATE: in the form of a peaked cap with undivided margin, or with margin equally and several times cleft. (See Fig. 491.)
Mittenothamnium, 219
  diminutivum, 219
Mniaceae, 128
Mnium, 19, 129
  affine, 291
    var. ciliare, ix, 132
  cuspidatum, viii, x, 132
  drummondii, 131
  glabrescens, 129
  hornum, 130
  insigne, 132
  marginatum, 131
  medium, x, 132
  menziesii, 291
  orthorhynchum, 291
  punctatum, 129
    var. elatum, 130
  serratum, 291
  spinulosum, 131
  stellare, 130
  thomsonii, 131
  venustum, 131
Moerckia, 276
  flotoviana, 289
  hibernica, 276
MONOICOUS: having male and female organs on the same plant. See autoicous, paroicous and synoicous.
Moss
  pink, 1
  rose, 1
MUCRO: a short, stout, abrupt, point.
Musci, 3, 5, 6, 24
Mylia, 257
  anomala, 257
  taylori, 258
Myurella, 164
  careyana, 291
  julacea, 164
  sibirica, 165
  tenerrima, 165

N

Nanomitrium, 291
  austini, 291
Nardia, 258
  lescurii, 258
NECK: of the capsule, the sterile portion, if any, between seta and urn. Fig. 514n. (See Fig. 13). 5

Figure 514

Neckera, 160
  complanata, 160
  disticha, 291
  douglasii, 160
  pennata, 160
  menziesii, 291
  undulata, 291
Neckeraceae, 160
Neckeropsis, 55, 161
  disticha, 161
  undulata, 162
Nematodonteae, 7
NODOSE: with rounded thickenings at intervals.
Notothylas, 231
  orbicularis, 231
Nowellia, 248
  curvifolia, 248

O

OBOVATE: similar to ovate, but broadest at the distal end.
OBTUSE: blunt or rounded at the end. Fig. 515.

Figure 515

Octoblepharum, 87
  albidum, 87
Odontoschisma, 248
  denudatum, 248
  prostratum, 233, 248
Oedipodiaceae, 117
Oedipodium, 33, 117
  giffithianum, 117
Oligotrichum, 225
  aligerum, 225
  hercynicum, 225
  parallelum, 225
Oncophorus, 80
  polycarpus,
    var. strumiferus, 291
  wahlenbergii, 80
OPERCULUM: the lid or cover of the capsule. Fig. 516. 5

Figure 516

Oreoweisia, 291
  serrulata, 291
Orthothecium, 202
  chryseum, 202
Orthocaulis, 289
  floerkei, 289
  kunzeanus, 289
  quadrilobus, 289
Orthotrichaceae, 140
Orthotrichum, 3, 34, 141
  affine, 146
  anomalum, 144
  consimile, 142
  cupulatum, 145
  diaphanum, 141
  elegans, 291
  hallii, 145
  laevigatum, 143
  lyellii, 143
  macounii, 291
  obtusifolium, 142
  ohioense, 146
  pulchellum, 143
  pumilum, 145
  pusillum, 145
  rivulare, 144
  rupestre, 146
  soridum, 291
  speciosum, 143
    var. elegans, 144
OVAL: broadly elliptical. Fig. 517.

Figure 517

OVATE: egg-shaped, with the broader end downward. Fig. 518.

Figure 518

Oxyrhynchium, 291
  hians, 291
    subsp. rappii, 291

P

Packets, how to fold, 9
Pallavicinia, 276
  lyellii, 22, 276
Pallavicinaceae, 276
PAPILLA: a tiny lump or knob on a cell wall. (See Fig. 17) 10
Papillaria, 159
  nigrescens, 51, 159
PAPILLOSE: rough with papillae. Fig. 519.

Figure 519

PARACYST: an enlarged or brightly colored cell, very different from the surrounding cells.
Paraleucobryum, 44, 78
  longifolium, 78

PARAPHYLLIA: thread-like or tiny leaf-like growths among the leaves. Fig. 520.

Figure 520

PARENCHYMATOUS: cells joined with broad ends, walls thin, intercellular spaces often conspicuous.
PAROICOUS: with antheridia in axils of perichaetial leaves just below the archegonia. 12
Peat moss, 4, 24
Pellia, 277
  endiviifolia, 277
  epiphylla, 5, 6, 277
  neesiana, 21, 277
Pelliaceae, 277
PELLUCID: translucent.
PERCURRENT: reaching to the apex but not beyond; percurrent costa. Fig. 521.

Figure 521

PERIANTH: a sheath surrounding the archegonia or young sporophyte. Fig. 522. 5, 14

Figure 522

PERICHAETIUM: the special leaves or bracts surrounding the archegonium or base of the seta. Fig. 523.

Figure 523

PERISTOME: the fringe of teeth around the mouth of the capsule. Fig. 524. 5, 12

Figure 524

Petalophyllum, 276
  ralfsii, 276
  lamellatum, 289
Phaeoceros, 231
  laevis, 5, 6, 231
Phascum, 98
  cuspidatum, 98
    var. americanum, 291
Philonotis, 138
  americana, 291
  capillaris, 291
  fontana, 138
    var. americana, 139
  glaucescens, 138
  longiseta, 138
  marchica, 138
Photosynthesis, 2
Physcomitriella, 31, 114
  californica, 114
  patens, 114
Physcomitrium, 31, 115
  hookeri, 115
  immersum, 115
  pyriforme, viii, 115
  turbinatum, 291
Picture-Key to the Bryophytes of North America, 19
PINNATE: having numerous branches on each side of an axis. Fig. 525.

Figure 525

Plagiochila, 251
  asplenioides, 252
  floridana, 253
  sullivantii, 233, 252
  virginica, 253
  undata, 252
Plagiochilaceae, 251
Plagiomnium, 129
  ciliare, 291
  cuspidatum, 291
  drummondii, 291
  rostratus, 291
Plagiopus, 135
  oederi, 291
  oederiana, 135
Plagiotheciaceae, 205
Plagiothecium, 205
  cavifolium, 206
  deplanatum, 291
  denticulatum, 206
  elegans, 291
  geophilum, 291
  micans, 291
  muellerianum, 291
  piliferum, 206

roseanum, 291
striatellum, 291
undulatum, 205
PLANE: flat, not rolled.
Platydictya, 183
  confervoides, 183
  subtile, 183
Platygyrium, 209
  repens, 209
Plectocolea, 289
  crenulata, 289
  crenuliformis, 289
  hyalina, 289
  rubra, 289
Pleuricarpi, 7
Pleuridium, 68
  acuminatum, 291
  subulatum, 68
PLEUROCARPOUS: having the seta rising from a short, lateral special branch. Fig. 526. 7

Figure 526

Pleurochaete, 94
  squarrosa, 94
Pleurozium, 204
  schreberi, 204
PLICATE: folded in longitudinal pleats. Fig. 527.

Figure 527

Pogonatum, 226
  alpinum, 227
  brachyphyllum, 226
  capillare, 291
  contortum, 226
  dentatum, 227
  pensilvanicum, 226
  urnigerum, 227
Pohlia, 120
  annotina, 121
    var. decipiens, 121
  cruda, 121
  elongata, 122
  nutans, 122
  proligera, 121
  rothii, 120
  wahlenbergii, 122
POLYGONAL: many sided.
POLYMORPHIC: many shaped.
Polytrichaceae, 223

Polytrichadelphus lyallii, 291
Polytrichales, 7, 223
  Polytrichum, 227
  commune, 3, 229
  formosum, 228
    var. aurantiacum, 229
  gracile, 291
  juniperinum, xi, 228
  longisetum, 291
  lyallii, 227
  norvegicum, 291
  ohioense, 229
  piliferum, 228
  sexangulare, 228
PORE: the opening through the epidermis into the air chamber of a liverwort. Fig. 528.

Figure 528

Porella, 265
  cordaeana, 265
  navicularis, 265
  pinnata, 265
  platyphylla, 266
  platyphylloidea, 266
  roellii, 265
Porellaceae, 265
POROSE: of thick walls with thin spots (pores). Fig. 529.

Figure 529

Porotrichum
  alleghaniense, 291
Portulaca, 1
POSTICAL: lower surface of the plant or lower margin of leaf.
Pottia, 98
  heimii, 291
  truncata, 98
Pottiaceae, 89
Pottiales, 7, 87
Pressia, 284
  quadrata, 284
Propagula 2 (see also gemmae).
PROSENCHYMATOUS: cells narrow with ends dove-tailed into each other. (See Fig. 66 F-I).
PROSTRATE: lying flat on the substrate.
PROTONEMA: the green, branched alga-like threads

growing from a spore. (See Fig. 13). 2, 5
Pseudisothecium
  stoloniferum, 291
Pseudobraunia, 154
  californica, 154
Pseudolepicoleaceae, 241
Pseudoleskea, 169
  incurvata, 169
  patens, 169
  radicosa, 170
  stenophylla, 170
Pseudoleskeella, 169
  tectorum, 169
PSEUDOPARAPHYLLIA: small, unistratose, filiform or lanceolate structures resembling paraphyllia. (See Fig. 69). 62
PSEUDOPODIUM: a leafless branch resembling a seta, bearing the capsule in Sphagnum and Andreaea. Fig. 530.

Figure 530

Pterigynandrum, 170
  filiforme, 170
  sharpii, 170
Pterobryaceae, 158
Pterogoneuron, 32, 99
  ovatum, 99
  subsessile, 99
Pterogonium, 158
  gracile, 158
Ptilidiaceae, 241
Ptilidium, 241
  californicum, 241
  ciliare, 242
  pulcherrimun, 233, 242
Ptilium, 219
  crista-castrensis, 219
Ptychocoleus, 270
  heterophyllus, 270
Ptychomitriaceae, 140
Ptychomitrium, 5, 45, 140
  drummondii, 140
  gardneri, 140
  incurvum, 140
  leibergii, 140
  serratum, 140
Pylaisia, 291
  intricata, 291
  polyantha, 291
  selwynii, 291
Pylasiella, 210
  intricata, 210
  polyantha, 210
  selwynii, 210
Pyrimidula, 115
  tetragona, 115

Q

QUADRATE: square (cubical) or nearly so.

R

RADICLES: filaments on stems, mostly brown, and running into the ground; rhizoids.
Radula, 264
  andicola, 289
  bolanderi, 264
  complanata, 264
  mollis, 264
  obconica, 264
Radulaceae, 264
Reboulia, 281
  hemisphaerica, 21, 281
Rectolejeunea, 272
  maxonii, 272
REFLEXED: bent slightly backward. Fig. 531. 11

Figure 531

RENIFORM: shaped like a kidney.
REVOLUTE: rolled backward and under, as the margins of leaves. Fig. 532. 11

Figure 532

Rhabdoweisia, 79
  crenulata, 79
  crispata, 79
  denticulata, 291
    var. americana, 291
Rhacomitrium, 37, 110
  aciculare, 111
  canescens, 111
  fasciculare, 111
  heterostichum, 111
  lanuginosum, 111
Rhizogoniaceae, 132
Rhizogonium, 132
  spiniforme, 132
RHIZOIDS: thread-like growths, simple or branched, which serve for absorption and anchorage. Fig. 533. (See Fig. 13). 2, 5

Figure 533

RHIZOME: a root-like, horizontal stem. Fig. 534r.

Figure 534

Rhodobryum, 128
  roseum, viii, 128
Rhynchostegiella, 200
  compacta, 200
Rhynchostegium, 55, 200
  hians, 291
  pulchellum, 291
  riparioides, 291
  serrulatum, 200
Rhytidiaceae, 219
Rhytidiadelphus, 220
  loreus, 220
  squarrosus, 221
  triquetrus, viii, 220
Rhytidiopsis, 220
  robusta, 220
Rhytidium, 219
  rugosum, 219
Riccardia, 279
  multifida, 279
  palmata, 279
  pinguis, 279
Riccia, 286
  beyrichiana, 286
  curtsii, 286
  dictyospora, 287
  fluitans, 286
  frostii, 287
  membranacea, 287
  sullivantii, 287
Ricciaceae, 285
Ricciocarpus, 285
  natans, 285
Riella, 280
  americana, 280
Roellii, 291
  sandbergii, 291
ROUGH: same as papillose.
RUGOSE: roughened with transverse wrinkles.

S

Saelania, 71
  glaucescens, 71
Scapania, 262
  bolanderi, 262
  curta, 263
  irrigua, 263
  nemorosa, 262
  undulata, 263
Scapaniaceae, 261
Schistostega, 119
  pennata, 119
Schistostegaceae, 119
Schistostegales, 7, 119
Schlothemia, 148
  lancifolia, 149
  rugifolia, 148
  sullivantii, 291
Schwetschkeopsis, 166
  denticulata, 291
  fabronia, 166
Sciaromium, 182
  lescurii, 182
  tricostatum, 182
Schleropodium, 198
  caepitans, 199
  caespitosum, 199
  obtusifolium, 199
  tourettei, 199
    var. colpophyllum, 199
Scopelophila, 90
  cataractae, 90
  latifolia, 90
  ligulata, 90
Scorpidium, 189
  scorpioides, 189
  turgescens, 189
Scouleria, 104
  aquatica, 104
  marginata, 105
Seaweed, 2
SECTION: a thin slice.
SECUND: turned to one side.
SEGMENTS: the divisions of the inner peristome.
Seligeria, 72
  calcarea, 73
  campylopoda, 73
  donniana, 72
Seligeriaceae, 72
Sematophyllaceae, 207
Sematophyllum, 69, 208
  adnatum, 208
  caespitosum, 209
  carolinianum, 291
  demissum, 208
  marylandicum, 208
SERRATE: the margin cut into teeth pointing forward. Fig. 535.

Figure 535

SERRULATE: very finely serrate.
SESSILE: sitting close, without a stalk.
SETA: the stalk of the capsule or sporophyte. (See Fig. 13.) 3, 11, 13
SETACEOUS: tip of leaf bristle-like.
Solmsiella, 291
  biseriata, 291
  kurzii, 291
Spanish moss, 1
SPERM: the active, coiled, male reproductive cell or gamete. 3
Sphaerocarpaceae, 280
Sphaerocarpales, 6, 239, 280
Sphaerocarpus, 22, 280
  texanus, 280
Sphagnidae, 7, 24
Sphagnum, 4, 24
  capillaceum, 26
  compactum, 25
  cuspidatum, 25
  fimbriatum, 26
  fuscum, 26
  girgensohnii, 26
  imbricatum, 20, 24
  macrophyllum, 25
  magellanicum, 20, 24
  palustre, 20, 24
  papillosum, 24
  recurvum, 25
  subsecundum, 26
  warnstorfianum, 291
  warnstorfii, 26
SPINDLE-SHAPED: tapering to each end. Fig. 536.

Figure 536

SPINOSE: having spines.
Splachnaceae, 118
Splachnum, 33, 118
  ampullaceum, 119
  luteum, 119
  pennsylvanicum, 119
SPORANGIUM: the capsule.
SPORE: a microscopic reproductive body, in mosses 1-celled and borne in the capsule. (See Fig. 13.) 2, 3, 12
SPOROPHYTE: the spore-bearing part or phase, composed of foot, seta and capsule. (See Fig. 13.) 3, 5, 12
SQUARROSE: of leaves with midrib bent back at right angles to the stem. Fig. 537.

Figure 537

STEREID: a slender fiber cell with walls thicker than the lumen, in the midrib or margin of a leaf. (See Fig. 58.) 43
Stereophyllum, 55, 205
  leucostegum, 205
  radiculosum, 205
STERILE: without sporophyte or spores.
Stokesiella
  brittoniae, 201
  oregana, 202
  praelonga, 202
    var. stokesii, 202

STOMA: (pl. stomata) an opening through the epidermis bordered by two special cells. Fig. 538. 3

Figure 538

STRIATE: marked with fine longitudinal lines.
STRUMA: a swelling on one side of the base of a capsule. Fig. 539.

Figure 539

STRUMOSE: having a struma.
SUBSTRATUM: the material on which the plant grows.
Succubous, 12, 13 (See Fig. 21.)
SUCCUBOUS: with leaves overlapping like shingles on a roof if base of plant is at eaves and apex at the ridge. (See Fig. 21.) 12
SULCUS: groove or furrow.
SWAMP: low ground saturated with water, but usually not covered with it, producing more or less of shrubs and trees.
SYNOICOUS: with antheridia and archegonia mingled. 12
Syrrhopodon, 87
  floridanus, 291
  incompletus, 87
  texanus, 35, 88

T

Takakia, 23, 232
  lepidozioides, 232
Takakiaceae, 232
Takakiales, 6, 232
Targionia, 281
  hypophylla, 281
Targioniaceae, 281
Taxiphyllum, 62, 217
  deplanatum, 217
  geophilum, 291
  taxirameum, 217
Taxithelium, 209
  planum, 209
Tayloria, 33, 118
  serrata, 118
Telaranea, 243
  nematoides, 243
TENIOLAE: intramarginal border of linear cells.
TERETE: cylindrical; round in cross section.
TESSELATE: marked in checkered squares. Fig. 540.

Figure 540

TETRAD: a group of four spores forming a three sided pyramid (tetrahedron).
Tetraphidaceae, 223
Tetraphidales, 7, 223
Tetraphis, 29, 223
  geniculata, 223
  pellucida, 223
Tetraplodon, 33, 118
  mnioides, 118
THALLUS: a plant body not differentiated into stem and leaf. Fig. 541. 5

Figure 541

Thamnobryum, 162
  alleghaniense, 162
Thelia, 164
  asprella, 51, 164
  hirtella, 51, 164
  lescurii, 164
Theliaceae, 164
Thuidiaceae, 171
Thuidium, 35, 174
  abietinum, 175
  delicatulum, 176
    var. radicans, 176
  microphyllum, 291
  minutulum, 174
  pygmaeum, 174
  recognitum
  scitum, 174
  virginianum, 291
Timmia, 139
  austriaca, 139
  megapolitana
    var. barvarica, 139
Timmiaceae, 139
Timmiella, 93
  anomala, 93
  crassinervis, 93
Tomenhypnum, 190
  nitens, 190
TOOTH: (teeth) the processes composing the peristome, or the outer row of such processes when the peristome is double. Fig. 542.

Figure 542

Tortella, 34, 93
  fragilis, 94
  humilis, 93
  inclinata, 94
  tortuosa, 94
Tortula, 101
  bistratosa, 102
  bolanderi, 103
  fragilis, 104
  latifolia, 102
  mucronifolia, 103
  muralis, 103
  norvegica, 104
  obtusissima, 104
  pagorum, 101
  papillosa, 102
  princeps, 104
  ruralis, 104
  subulata, 103
TRACHEA: a thick-walled water tube made of many cells in a row.
TRACHEIDS: single celled water tubes with thickened walls.
Trematodon, 75
  ambiguus, 75
  boasii, 75
  longicollis, 75
Trichocolea, 242
  tomentella, 233, 289, 242
Trichocoleaceae, 242
Trichodon, 69
  cylindricus, 69
Trichostomum, 92
  cylindricum, 291
  tenuirostris, 92
TRIGONE: a thickening of cell walls where three or four cells come together. Fig. 543. 13

Figure 543

Tritomaria, 256
  exsecta, 256
  exsectiformis, 256
  quinquedentata, 256
True mosses, 6
TRUNCATE: ending abruptly as if cut off.
TUBERCULATE: small, knob-like projections, often in spores of liverworts.

U

Ulota, 38, 147
  americana, 291
  coarcta, 147
  crispa, 148
  hutchinsiae, 147
  ludwigii, 291
  obtusiuscula, 148
  phyllantha, 147
UNDERLEAF: a small leaf on the under side of the stem in liverworts. Fig. 544. 12

Figure 544

UNDERLOBE: in a folded complicate-bilobed leaf the part lying nearest the substratum. Fig. 545a.

Figure 545

UNDULATE: with wavy margin or surface.
URN: the spore-bearing part of the capsule. (See Fig. 13.) 5
Uses, 4
Usnea, 1

V

VAGINANT LAMINA: in Fissidentaceae the part of the leaf that embraces the next higher leaf.
VENTRAL: the upper (inner, adaxial) side of a leaf; the lower (under) side of a prostrate shoot.
Vesicularia, 218
  vesicularis, 218
Vessels, 6
VITTA: strips formed by elongated cells in liverworts.
Voitia, 118
  nivalis, 118

W

Weissia, 91
  controversa, 30, 91
  ludoviciana, 291
  muhlenbergianum, 291
  viridula, 291

X

XERIC: of a dry habitat.

Z

Zygodon, 38, 140
  viridissimus, 140
    var. rupestris, 140
ZYGOMORPHIC: bilaterally symmetrical.
ZYGOTE: the germ cell resulting from the fusion of egg and sperm. 2

NOTES